T0337862

QoS and QoE Management in UMTS Cellular Systems

QoS and QoE Management in UMTS Cellular Systems

Edited by

David Soldani
Nokia Networks, Nokia Group, Finland

Man Li
Previously *Nokia Research Center, Nokia Group, Boston, USA*
Currently *JumpTap, Inc., Cambridge, USA*

Renaud Cuny
Nokia Networks, Nokia Group, Finland

JOHN WILEY & SONS, LTD

Other Wiley Editorial Offices

John Wiley & Sons, Inc., 111 River Street, Hoboken, NJ 07030, USA

Jossey-Bass, 989 Market Street, San Francisco, CA 94103-1741, USA

Wiley-VCH Verlag GmbH, Boschstr. 12, D-69469 Weinheim, Germany

John Wiley & Sons Australia Ltd, 42 McDougall Street, Milton, Queensland 4064, Australia

John Wiley & Sons (Asia) Pte Ltd, 2 Clementi Loop #02-01, Jin Xing Distripark, Singapore 129809

John Wiley & Sons Canada Ltd, 6045 Freemont Blvd, Mississauga, Ontario, Canada L5R 4J3

Wiley also publishes its books in a variety of electronic formats. Some content that appears
in print may not be available in electronic books.

British Library Cataloguing in Publication Data

A catalogue record for this book is available from the British Library

ISBN-13 978-0-470-01639-8 (HB)
ISBN-10 0-470-01639-6 (HB)

Project management by Originator, Gt Yarmouth, Norfolk (typeset in 10/12pt Times).
Printed and bound in Great Britain by Antony Rowe Ltd, Chippenham, Wiltshire.
This book is printed on acid-free paper responsibly manufactured from sustainable forestry
in which at least two trees are planted for each one used for paper production.

Contents

5 QoS Functions in Access Networks   141
David Soldani, Paolo Zanier, Uwe Schwarz, Jaroslav Uher,
Svetlana Chemiakina, Sandro Grech, Massimo Barazzetta and
Mariagrazia Squeo

6   QoS Functions in Core and Backbone Networks   209

Renaud Cuny, Heikki Almay, Luis Alberto Peña Sierra and Jani Lakkakorpi

10 Optimisation   385
David Soldani, Giovanni Giambiasi, Kimmo Valkealahti, Mikko Kylväjä,
Massimo Barazzetta, Mariagrazia Squeo, Jaroslav Uher, Luca Allegri and
Jaana Laiho

Preface

Wireless mobile networks have come a long way from providing voice-only services to offering a proliferation of multimedia data services. Mobile data services are rapidly becoming an essential component of mobile operators' business strategies and are growing very quickly alongside traditional voice services. For example, in the second quarter of 2005, total average data revenue per month from leading US operators reached $575 million – more than double the amount for the same period in 2004. With the spread of WCDMA and EGPRS, availability of new services, functionality-rich handsets and convergence of various technologies, this trend is poised to gain pace in the future. According to a Yankee Group study, US mobile data revenue streams in 2009 are expected to reach $15.9 billion.

Mobile office applications, browsing and multimedia messaging services (MMS) are expected to be the major contributors together with many small contributions from other current and future applications. The following list covers some of the revenue generating data services available today and some to come in the near future:

- Short messages (SMS).
- Multimedia messages.
- Community chats, forums.
- Web browing.
- Email access, email to SMS access.
- Ring tones and graphics download.
- SMS votes, alerts.
- Interactive gaming.
- Video and game download.
- Streaming.
- Video sharing (VS).
- Mobile office (email, browsing, etc.).
- M-payments, m-banking, m-booking, m-brokering, m-ticketing.
- Proximity services.
- Push to talk over cellular (PoC).
- Presence.
- Conferencing.
- Instant messaging.

It is becoming increasingly evident that data are equivalent to 'gold' on balance sheets.

As the requirements for different applications vary, this growth of non-voice services has posed a new challenge to managing their performance in more effective ways. This is essential in order to provide best-of-class services to the end-user without overdimensioning precious network resources.

'Quality of experience' (QoE) is the term used to describe user perceptions of the performance of a service. Quality of service (QoS), on the other hand, is the ability of the network to provide a service at an assured service level. In order to provide the best QoE to users in a cost-effective, competitive and efficient manner, network and service providers must manage network QoS and service provisioning efficiently and effectively.

Enterprises and network providers that provide superior QoE enjoy a significant competitive advantage, while companies that ignore the importance of QoE may suffer unnecessary costs, lost revenue and diminished market perception. A survey by a famous consulting firm suggests that around 82% of customer defections ('churning' to the competition) are due to frustration over the product or service and the inability of the provider/operator to deal with this effectively. Moreover, this leads to a chain reaction, because, on average, 1 frustrated customer will tell 13 other people about their bad experience. An operator cannot afford to wait for customer complaints to assess the level of its service quality. Surveys have shown that for every person who calls with a problem, there are another 29 who will never call. About 90% of customers will not complain before defecting – they will simply leave (churn) once they become unsatisfied. This churn directly affects the profitability and image of the operator, especially if it happens in the early stage of their induction. So, the only way to prevail in this situation is to devise a strategy to constantly manage and improve QoE and QoS.

QoE and QoS management can be classified in four interdependent categories: network planning, service and QoS provisioning, QoE and QoS monitoring and optimisation. There has been rich research and development in this field, and the purpose of this book is to introduce the principles, practices and research undertaken in these four areas. The book is intended for both academic and professional audiences.

Acknowledgements

We would like to acknowledge the contributions and time invested by our colleagues working at Nokia. Apart from the editors, the contributors were Anna Sillanpää, Paolo Zanier, Giovanni Giambiasi, Carolina Rodriguez, Jaana Laiho, Kimmo Valkealahti, Davide Chiavelli, Jaroslav Uher, Heikki Almay, Noman Muhammad, Uwe Schwarz, Massimo Barazzetta, Martin Kristensson, Luis Alberto Peña Sierra, Mariagrazia Squeo, Mikko Kylväjä, Sandro Grech, Svetlana Chemiakina, Jani Lakkakorpi and Luca Allegri.

We would like to express our gratitude to our employer, Nokia, for general permission, support and encouragement, and for providing some illustrations. In particular, Lauri Oksanen is acknowledged for giving us the opportunity to spend several years in dealing with QoE and QoS management issues in UMTS cellular systems, and for letting us collect a part of the attained results in this manuscript.

The publishing team at John Wiley & Sons, Ltd, led by Mark Hammond, has done an outstanding job in the production of this book. We are especially grateful to Sarah Hinton and Jennifer Beal for their patience, support, guidance and assistance.

Ultimately, we would like to dedicate this book to our families for their love, patience and assistance during this endeavour.

The editors and authors welcome any comments and suggestions for improvement or changes that could be implemented in possible future editions.

David Soldani, Man Li and Renaud Cuny
Espoo, Finland and Boston, Massachusetts, USA

Abbreviations

16QAM	16 Quadrature Amplitude Modulation
1G	1st Generation
2G	2nd Generation
3G	3rd Generation
3GPP	3rd Generation Partnership Project
3GPP2	3rd Generation Partnership Project 2
3GSM	3G GSM or UMTS
8-PSK	Octagonal Phase Shift Keying
AAA	Authentication, Authorisation and Accounting
AAL	ATM Adaptation Layer
AB	Access Burst
ABR	Available Bit Rate
AC	Admission Control
ACK	Positive Acknowledgment
AD	Access Delay
AF	Activity Factor, Assured Forwarding, Application Function
AF-AI	AF-Application Identifier
AG	Absolute Grant
AGCH	Access Grant Channel
AH	Authentication Header
AICH	Acquisition Indication Channel
ALCAP	Access Link Control Application Part protocol
AM	Acknowledged Mode
AMC	Adaptive Modulation and Coding
AMR	Adaptive Multi Rate speech codec
AMR-WB	Wideband AMR
ANSI	American National Standard Institute
AP	Access Point
API	Application Programming Interface
APN	Access Point Name
ARED	Adaptive RED
ARFCN	Absolute Radio Frequency Number
ARIB	Association of Radio Industries and Businesses
ARP	Allocation Retention Priority
ARQ	Automatic Repeat reQuest
AS	Access Stratum
ASC	Access Service Class
AST	Active Session Throughput
ATIS	Alliance for Telecommunications Industry Solutions
ATM	Asynchronous Transfer Mode
AuC	Authentication Centre (Register)
AVP	Attribute Value Pairs
AWND	Advertised Window
BBS	Broad Band System
BCCH	Broadcast Control Channel
BCH	Broadcast Channel

BCS	Block Check Sequence
BE	Best Effort PHB, Back End
BEC	Backward Error Correction
BER	Bit Error Rate (Ratio)
BG	Border Gateway
BH	Busy Hour
BLER	Block Error Ratio
BMC	Broadcast/Multicast Control
BMS	Business Management Systems
BN	Backbone Network, Bit Number
BP	Blocking Probability
BR	Bit Rate
BRR	Bucket Round Robin
BS	Bearer Service, Base Station (or Node B)
BSC	Base Station Controller
BSIC	Base Station Identity Code
BSS	Base Station Subsystem
BSSGP	Base Station Subsystem GPRS Protocol
BTFD	Blind Transport Format Detection
BTS	Base Transceiver Station (2G Cell)
BVC	BSSGP Virtual Connection
BVCI	BSS Virtual Connection Identifier
C-ID	Cell IDentification
C/T	Channel Type
C1	Criterion for cell reselection
C31, C32	Packet criteria for cell reselection
CAMEL	Customised Applications for Mobile network Enhanced Logic
CAP	Camel Application Part
CAPEX	CAPital EXpenditure
CBM	Cell Broadcast Message
CBR	Call Block Ratio, Constant Bit Rate
CBS	Cell Broadcast Service
CC	Call Control, Convolutional Coding
CCCH	Common Control Channel
CCH	Common Channel (s)
CCR	Call Completion Rate
CCSA	China Communications Standards Association
CCTrCH	Coded Composite Transport Channel
CCU	Channel Control Unit
CDF	Cumulative Distribution Function
CDMA	Code Division Multiple Access
CDR	Call Drop Ratio
CDR	Charging Data Record
CE	Congestion Experienced
CFN	Connection Frame Number
CID	Cell Identifier
CIR	Carrier Interference Ratio, Committed Information Rate
CLI	Command Line Interface
CM	Configuration or Connection Management
CmCH-PI	Common Channel Priority Indicator
CN	Core Network
COPS	Common Open Policy Service
COPS-PR	COPS policy Provisioning
CoS	Class of Service
CP	Control Plane
CPCH	Common Packet Channel
CPICH	Common Pilot Channel (Perch Channel)
CQI	Channel Quality Indicator
CQM	Customer QoS Management
CR	Capacity or Change Request, CRedit
CRC	Cyclic Redundancy Check

CRF	Charging Rules Function
CRNC	Controlling RNC
CRRR	Capacity Request Rejection Ratio
CS	Circuit Switched, Coding Scheme
CSD	Circuit Switched Data
CSV	Comma Separated Value
CT	Core network and Terminal
CTCH	Common Traffic Channel
CWND	Congestion Window
DCCA	Diameter Credit Control Application
DCCH	Dedicated Control Channel
DCH	Dedicated Channel
DCP	Delay Control Parameter
DCS	Digital Communication System
DDI	Data Description Indicator
DHO	Diversity Handover
DiffServ	Differentiated Services
DL	Down Link
DLCI	Data Link Connection Identifier
DLL	Data Link Layer
DM	Device Management
DMZ	DeMilitarised Zone
DNS	Domain Name Server
DPCCH	Dedicated Physical Control Channel
DPCH	Dedicated Physical Channel
DPDCH	Dedicated Physical Data Channel
DQ	Data Quality
DRNC	Drifting RNC
DRR	Deficit Round Robin
DRX	Discontinuous Reception
DSCH	Downlink Shared Channel
DSCP	Differentiated Services Code Point
DSL	Digital Subscriber Line
DT	Discard Timer
DTCH	Dedicated Traffic Channel
DTM	Dual Transfer Mode
DTX	Discontinuous Transmission
E-AGCH	E-DCH Absolute Grant CHannel
E-DCH	Enhanced Dedicated CHannel
E-DPCCH	E-DCH Dedicated Physical Control CHannel
E-DPCH	Enhanced Dedicated Physical CHannel
E-DPDCH	E-DCH Dedicated Physical Data CHannel
E-HICH	E-DCH HARQ Acknowledgement Indicator CHannel
E-RGCH	E-DCH Relative Grant CHannel
E-RNTI	E-DCH Radio Network Temporary Identifier
E-TCH/F	Enhanced Full rate Traffic CHannel
E-TFC	E-DCH Transport Format Combination
E-TFCI	E-DCH Transport Format Combination Indicator
E2E	End to End
ECN	Explicit Congestion Notification
EDF	Earliest Detect First
EDGE	Enhanced Data Rates for GSM Evolution
EF	Expedited Forwarding
EFL	Effective Frequency Load
EFR	Enhanced Full Rate
EGPRS	Enhanced GPRS
EGSM	Extended GSM
EIR	Equipment Identity Register
EM	Element Management
EMS	Element Management System
EPS	Encapsulation Security Payload

EQoS	Enhanced Quality of Service
ES	Enterprise System
ESP	Encapsulating Security Payload
ETFC	Enhanced Transport Format Combination
eTOM	enhanced Telecom Operations Map
ETSI	European Telecommunications Standards Institute
EUPA	Enhanced Uplink Packet Access
EVDO	Evolution Data Only or Evolution Data Optimised
EWMA	Exponential Weighted Moving Average
EXP	Experimental [field in the MPLS shim header]
F-DPCH	Fractional Dedicated Physical Channel
FACCH	Fast Associated Control CHannel
FACCH/F	Fast Associated Control CHannel/Full rate
FACCH/H	Fast Associated Control CHannel/Half rate
FACH	Forward Access Channel
FB	Frequency Correction Burst
FBC	Flow Based Charging
FC	Flow Control
FCCH	Frequency Correction CHannel
FDD	Frequency Division Duplex
FDMA	Frequency Division Multiple Access
FEC	Forward Error Correction
FER	Frame Erasure Ratio
FH	Frequency Hopping
FHS	Frequency Hopping Sequence
FIFO	First In First Out
FMT	Field Measurement Tool
FN	Frame Number
FP	Frame Protocol
FQ	Fair Queuing
FR	Fair Resources, Frame Relay, Full Rate
FRED	Flow RED
FT	Fair Throughput
FTP	File Transfer Protocol
FW	Fire Wall
GB, GBR	Guaranteed Bit Rate
GCID	GPRS Charging Identifier
GERA	GSM/GPRS Edge Radio Access
GERAN	GSM/Edge Radio Access Network
GGSN	Gateway GPRS Support Node
GMM	GPRS Mobility Management
GMM/SM	GPRS Mobility Management and Session Management
GMSC	Gateway MSC
GMSK	Gaussian Minimum Shift Keying
GPRS	General Packet Radio Service
GRE	Generic Routing Encapsulation
GSM	Global System for Mobile Communications
GSMA	GSM Association
GSN	GPRS Support Node
GTP C	GPRS Tunnelling Protocol for the Control plane
GTP	GPRS Tunnelling Protocol
GTP-C	GTP Control plane
GTP-U	GTP User plane
GUI	Graphical User Interface
GW	Gateway
GW/TPF	Traffic Plane Function
HARQ	Hybrid Automatic Repeat reQuest
HC	Handover Control
HCS	Hierarchical Cell Structure
HLR	Home Location Register
HMA	Hybrid Multiple Access

HO	Handover
HR	Half Rate
HS	High Speed
HS-DPCCH	High Speed Dedicated Physical Control CHannel
HS-DSCH	High Speed DSCH
HS-PDSCH	High Speed Physical DSCH
HS-SCCH	High Speed Shared Control CHannel
HSCSD	High Speed Circuit Switched Data
HSDPA	High Speed Downlink Packet Access
HSN	Hopping Sequence Number
HSS	Home Subscriber Server
HSUPA	High Speed Uplink Packet Access
HTML	Hypertext Markup Language
HTTP	Hyper Text Transfer Protocol
HTTPS	Secure HTTP
HW	HardWare
IAB	Internet Architecture Board
IANA	Internet Assigned Numbers Authority
ICID	IMS Charging Identifier
ICMP	Internet Control Message Protocol
ID	Identity
IE	Information Element
IEEE	Institute of Electrical and Electronic Engineering
IESG	Internet Engineering Steering Group
IETF	Internet Engineering Task Force
IFHO	Intra or Inter Frequency Handover
IKE	Internet Key Exchange
IM	Instant Messaging
IMA	Inverse Multiplexing for ATM
IMS	IP Multimedia Subsystem
IMSI	International Mobile Subscriber Identity
IMT	International Mobile Telephony
IntServ	Integrated Services
IP	Internet Protocol
IPsec	IP security
IRSG	Internet Research Steering Group
IRTF	Internet Research Task Force
IS	Information System
IS-2000	IS-95 evolution standard (cdma2000)
IS-95	cdmaOne, 2G system, mainly in Americas and Korea
ISAKMP	Internet Security Association and Key Management Protocol
ISCP	Interference Signal Code Power
ISDN	Integrated Service Digital Network
ISHO	Inter System Handover
ISOC	Internet Society
ISP	Internet Service Provider
ISUP	ISDN User Part
ITC	Information Transfer Capability
ITU	International Telecommunications Union
kbps	kilobits per second
KFI	Key Financial Indicator
KPI	Key Performance Indicator
KQI	Key Quality Indicator
L1, L2, L3, L4, L5, L6, L7	Layers 1–7
L2TP	Layer 2 Tunnelling Protocol
LA	Link Adaptation
LAC	Link Access Control
LAN	Local Area Network
LAU	Location Area Update
LC	Load Control, Congestion Control
LDAP	Lightweight Directory Access Protocol

LLC	Logical Link Control protocol
LLE	Logical Link Entity
LMDR	Lightweight Mobility Detection and Response
LoCH	Logical CHannel
LS	Liaison Statement
LSA	Localised Service Area
M	Mandatory presence of an information element
MA	Mobile Allocation
MAC	Medium Access Control
MAC-c	Common Medium Access Control protocol
MAC-d	Dedicated Medium Access Control protocol
MAC-e/es	Enhanced Medium Access Control protocol
MAC-hs	High Speed Medium Access Control protocol
MAIO	Mobile Allocation Index Offset
MAP	Mobile Application Part
MbA	Mobile based Agent
MBMS	Multimedia Broadcast/Multicast Service
MBR	Maximum Bit Rate
MC	Management Class
MCC	Mobile Connection Control unit
MCCH	MBMS point-to-multipoint Control Channel
MCS	Modulation and Coding Scheme
MCU	Main Control Unit
MDC	Macro Diversity Combiner
MEHO	Mobile Evaluated Handover
MICH	MBMS Indicator Channel
MM	Mobility Management, Multi Media
MMS	Multimedia Messaging Service
MMSC	Multimedia Messaging Service Centre
MNO	Mobile Network Operator
MOS	Mobile Objective Score, Mean Opinion Score
MoU	GSM Memorandum of Understanding
MPLS	Multi-protocol Label Switching
MQA	Mobile Quality Analyser, Mobile QoS Agent
ms	millisecond
MS	Mobile Station
MSC	Mobile Switching Centre
MSCH	MBMS point-to-multipoint Scheduling CHannel
MSG	Mobile Specification Group
MSISDN	Mobile Subscriber Integrated Services Digital Network Number
MSS	Mobile Satellite Spectrum, Maximum Segment Size
MT	Mobile Termination
MTCH	MBMS point-to-multipoint Traffic Channel
MTMS	Mobile Terminal Management Server
MTU	Maximum Transfer Unit
MVNO	Mobile Virtual Network Operation
NACC	Network Assisted Cell Change
NACK	Negative Acknowledgment
NAS	Non Access Stratum
NASREQ	Network Access Server Requirements
NB	Normal Burst, Node B
NBAP	Node B Application Part protocol
NBR	Nominal Bit Rate
NC	Network Control order, Non Controllable
NC0, NC1, NC2	Network Control Orders 1–3
NCCR	Network Controlled Cell Reselection
NCH	Notification CHannel
NE	Network Element
NEHO	Network Evaluated HandOver
NGB, NGBR	Non Guaranteed Bit Rate
NGN	Next Generation Network

NM	Network Manager
NMS	Network Management System
NMT	Nordic Mobile Telephone
NNA	Network Non Accessibility
NR	Network Resources
NRT	Non Real Time
NS	Network Service
NS-VC	Network Service Virtual Channel
NSAPI	Network layer Service Access Point Identifier
NSE	Network Service Entity
NSEI	Network Service Entity Identifier
NT	Non Transparent data
NW	Network
O	Optional presence of an information element
O&M	Operation and Maintenance
OH	Overhead
OLPC	Outer Loop Power Control
OLPCE	Outer Loop Power Control Entity
OMA	Open Mobile Alliance
OPEX	OPEration and Management EXpenditure
OS	Operating System or Operation System
OSS	Operation Support System
OTA	Over-The-Air
P-CCPCH	Primary Common Control Physical CHannel
P-CSCF	Proxy-Call Session Control Function
PACCH	Packet Associated Control CHannel
PAGCH	Packet Access Grant CHannel
PBCCH	Packet Broadcast Control CHannel
PC	Power Control, Personal Computer, Policy Consumer
PCC	Policy and Charging Control
PCCCH	Packet Common Control CHannel
PCCH	Paging Control CHannel
PCEF	Policy and Charging Enforcement Function
PCF	Policy Control Function
PCH	Paging CHannel
PCPCH	Physical Common Packet CHannel
PCRF	Policy and Charging Rules Function
PCS	Personal Communication Systems (2G system in Americas)
PCU	Packet Control Unit
PD	Packet or Protocol Data
PDC	Personal Digital Cellular (2G system in Japan)
PDCH	Physical Data CHannel
PDCP	Packet Data Convergence Protocol
PDF	Policy Decision Function, Probability Density Function
PDG	Packet Data Gateway
PDN	Packet Data Network
PDP	Packet Data Protocol, Policy Decision Point
PDR	Peak Data Rate
PDSCH	Physical Downlink Shared CHannel
PDTCH	Packet Data Traffic CHannel
PDU	Protocol Data Unit
PEP	Performance Enhancing Proxy, Policy Enforcement Point
PFC	Packet Flow Context
PFI	Packet Flow Identifier
PHB	Per Hop Behaviour
PHS	Personal Handy phone System
PI	Performance Indicator
PICH	Paging Indicator CHannel
PIMRC	Personal Indoor and Mobile Radio Communications
PLMN	Public Land Mobile Network
PLR	Packet Loss Ratio

PM	Performance Monitoring or Management
PNCH	Packet Notification CHannel
PoC	Push to talk over Cellular
PPCH	Packet Paging CHannel
PPP	Point-to-Point Protocol
PR	Power Ratio
PRACH	Physical Random Access CHannel, Packet Random Access CHannel
PRACK	PRovisional ACKnowledgement
PrC	Process Call function
PS	Packet Switched, Packet Scheduler
PSD	Packet Switched Data
PSI5	Packet System Information 5
PSK	Phase Shift Keying
PSTN	Public Switched Telephone Network
PTCCH/D	Packet Timing advance Control Channel/Downlink
PTCCH/U	Packet Timing advance Control Channel/Uplink
PTM	Point To Multipoint
PTM-M	Point To Multipoint–Multicast
PTMSI	Packet Temporary Mobile Subscriber Identity
PTP	Point To Point
PTT	Push To Talk (or PoC)
PU	Payload Unit
PVC	Permanent Virtual Circuit
QoE	Quality of (end-user) Experience
QoS	Quality of Service
QPM	QoS Policy Manager or Policy based Management
QPSK	Quadrature Phase Shift Keying
R5	Release 5
R6	Release 6
R7	Release 7
R97	Release 1997
R98	Release 1998
RA	Routing Area
RAB	Radio Access Bearer
RACH	Uplink Random Access CHannel
RADIUS	Remote Authentication Dial In User Service
RAI	Routing Area Identity
RAN	Radio Access Network
RANAP	RAN Application Part protocol
RAT	Radio Access Technology, Radio Access Type
RAU	Routing Area Update
RB	Radio Bearer, Radio Block
RBR	Re Buffering Ratio
RDI	Restricted Digital Information
RED	Random Early Detection
RF	Radio Frequency, Reduction Factor
RFC	Request For Comments
RFCH	Radio Frequency CHannel
RG	Relative Grant
RGSM	Railway GSM
RL	Radio Link
RLC	Radio Link Control
RLS	Radio Link Set
RM	Rate Marching
RMA	Rate Matching Attribute
RNAS	RAN Access Server
RNC	Radio Network Controller
RNF	Radio Network Feedback
RNL	Radio Network Layer
RNP	Radio Network Planning
RNS	Radio Network Subsystem

RNSAP	Radio Network Subsystem Application Part protocol
RNTI	Radio Network Temporary Identity
ROHC	Robust Header Compression
RR	Round Robin, Resource Request
RRC	Radio Resource Control
RRI	Radio Resource Indication
RRM	Radio Resource Manager
RRP	Radio Resource Priority
RSCP	Received Signal Code Power
RSN	Retransmission Sequence Number
RSVP	ReSerVation Protocol
RT	Real Time
RTCP	Real Time Control Protocol
RTO	Retransmission Time Out
RTP	Real Time Protocol
RTSP	Real Time Streaming Protocol
RTT	Round Trip Time
RTVS	Real Time Video Sharing
RV	Redundancy Version
S-CCPCH	Secondary Common Control Physical CHannel
S-CSCF	Serving-Call Session Control Function
SA	Service Accessibility
SACCH	Slow Associated Control CHannel
SACCH/C	Slow Associated Control CHannel/Combined
SACCH/TF	Slow Associated Control CHannel/TCH Full rate
SACCH/TH	Slow Associated Control CHannel/TCH Half rate
SACK	Selective ACKnowledge
SAP	Service Access Point
SAPI	Service Access Point Identifier
SAW	Stop And Wait
SB	Synchronisation Burst
SBLP	Service Based Local Policy
SC	Service Configurator
SCH	Synchronisation CHannel
SCTP	Stream Control Transmission Protocol
SDCCH	Standalone Dedicated Control CHannel
SDP	Session Description Protocol
SDU	Service Data Unit
SE	Spectral Efficiency
SeCR	Session Completion Rate
SF	Spreading Factor
SFN	System Frame Number
SG	Serving Grant
SGSN	Serving GPRS Support Node
SHO	Soft Handover
SI	System Information
SI3	System Information 3
SIM	Subscriber Identity Module
SINR	Signal to Noise plus Interference Ratio
SIP	Session Initiation Protocol
SIR	Signal to Interference Ratio
SLA	Service Level Agreement
SLS	Service Level Specification
SM	Session Management, Service Management
SMA	Service Management Application
SMG	Special Mobile Group
SMS	Short Message Service, Service Management System
SNDCP	Sub Network Dependent Convergence Protocol
SNMP	Simple Network Management Protocol
SNR	Signal Noise Ratio
SP	Strict Priority

SPA	Self Provided Applications
SPI	Scheduling Priority Indicator
SpQ	Speech Quality
SPR	Subscriber Profile Repository
SQM	Service Quality Management, Service Quality Manager
SRB	Signalling Radio Bearer
SRNC	Serving RNC
SRNS	Serving RNS
SS	Supplementary Service
SSL	Secure Socket Layer
SSS	Scheduling Step Size
ST	Setup Time
SW	SoftWare
SWIS	See What I See
T	Transparent data
TB	Transport Block
TBC	Token Bucket Counter
TBF	Temporary Block Flow
TBS	Transport Block Size
TBSS	Transport Block Set Size
TC	Traffic Class, QoS Class, Time slot Capacity
TCH/F	Traffic Channel/Full rate
TCH/H	Traffic Channel/Half rate
TCP	Transmission Control Protocol
TD-CDMA	Time Division-Code Division Multiple Access
TD-SCDMA	Time Division-Synchronous Code Division Multiple Access
TDD	Time Division Duplex
TDM	Time Division Multiplexing
TDMA	Time Division Multiple Access
TE	Terminal Equipment, Traffic Engineering
TEID	Tunnel Endpoint Identifier
TF	Transport Format, Translation Factor
TFC	Transport Format Combination
TFCI	Transport Format Combination Indicator
TFCS	Transport Format Combination Set
TFI	Transport Format Indicator, Temporary Flow Identifier
TFRC	Transport Format Resource Combination, TCP Friendly Rate Control
TFRI	Transport Format Resource Indication
TFS	Transport Format Set
TFT	Traffic Flow Template
THP	Traffic Handling Priority
TI	Transaction Identifier
TID	Tunnel IDentifier
TLLI	Temporary Logical Link Identity
TLS	Transport Layer Security
TLV	Type Length Value coding
TM	Transparent Mode
TMF	Tele Management Forum
TMN	Telecommunications Management Network
TMSI	Temporary Mobile Subscriber Identity
TN	Timeslot Number
TNL	Transport Network Layer
TOM	Telecom Operations Map
ToS	Type of Service
TPC	Transmit Power Control
TPF	Traffic Plane Function
TR	Technical Report
TrCH	Transport CHannel
TRX	Transmitter Receiver or Transceiver
TS	Time Slot, Technical Specification
TSC	Training Sequence

TSG	Technical Specification Group
TSG-GERAN	TSG responsible for GSM and EDGE Radio Access Networks
TSG-RAN	TSG responsible for new Radio Access Networks
TSG-SA	TSG responsible for Services and system Aspects
TSG-T	TSG responsible for Terminals
TSG-TN	TSG responsible for Core network and Terminals
TSL	Time SLot
TSN	Transmission Sequence Number
TTA	Telecommunications Technology Association
TTC	Telecommunication Technology Committee
TTI	Transmission Time Interval
TTL	Time To Live
TTV	Transmission Turn Value
UARFCN	Absolute Radio Frequency Channel Number
UBR	Unspecified Bit Rate
UDI	Unrestricted Digital Information
UDP	User Datagram Protocol
UE	User Equipment
UEP	Unequal Error Protection
UL	Up Link
UM	Unacknowledged Mode
UMA	Unlicensed Mobile Access
UMTS	Universal Mobile Telecommunication System
UNC	UMA Network Controller
UP	User Plane
UPH	UE Power Headroom
URA	UTRAN Registration Area
URI	Uniform Resource Identifier
URL	Uniform Resource Locator
USIM	UMTS Subscriber Identity Module
UTRA	UMTS Terrestrial Radio Access
UTRAN	UMTS Terrestrial Radio Access Network
VAS	Value Added Service
VBR	Variable Bit Rate
VC	Virtual Channel
VLAN	Virtual Local Area Network
VLR	Visitor Location Register
VoIP	Voice Over IP
VPN	Virtual Private Network
VS	Video Sharing
VTC	Vehicular Technology Conference
WAG	WLAN Access Gateway
WAN	Wide Area Network
WAP	Wireless Application Protocol
WARC	World Administrative Radio Conference
WBXML	Wireless Binary XML
WCDMA	Wideband Code Division Multiple Access
WFQ	Weighted Fair Queuing
WG	Working Group
WLAN	Wireless Local Area Network
WRC	World Radio communication Conference
WRED	Weighted RED
WRR	Weighted Round Robin
WSP	Wireless Session Protocol
WWW	World Wide Web
XML	eXtensible Markup Language

1

Introduction

Noman Muhammad, Davide Chiavelli, David Soldani and Man Li

Browsing through the literature, one may find many different definitions for quality of end-user experience (QoE) and quality of service (QoS). Some try to define the terms from a business perspective whereas others do so from a technical perspective. In the context of this book, QoE is the term used to describe the perception of end-users on how usable the services are. QoS, on the other hand, describes the ability of the network to provide a service with an assured service level. In order to provide the best QoE to users in a cost-effective, competitive and efficient manner, network and service providers must manage QoS and services in proper and appropriate ways.

QoS and QoE are so interdependent that they have to be studied and managed with a common understanding, from planning to implementation and engineering (optimisation). In short, the aim of the network and services should be to achieve the maximum user rating (QoE), while network quality (QoS) is the main building block for reaching that goal effectively. QoE, however, is not just limited to the technical performance of the network, there are also non-technical aspects, which influence the overall user perception to a great deal. Figure 1.1 shows an example of technical and non-technical factors affecting the QoE. This book will only deal with the technical aspects of QoE in detail.

In subsequent chapters of this book, the relationship between QoS and QoE and their management is discussed in length. Although the specific focus of the discussion is on Universal Mobile Telecommunication System (UMTS) cellular networks, many of the covered concepts are generic in nature and, as such, they are effectively applicable to other types of fixed and cellular networks.

1.1 QoE value chain

The overall QoE depends on how well the operator orchestrates the entire value chain as seen by the user. This value chain comprises the following:

- Mobile content providers, the content originators, websites, WAP sites, games, video, audio, portals, etc.

QoS and QoE Management in UMTS Cellular Systems
Edited by David Soldani, Man Li and Renaud Cuny © 2006 John Wiley & Sons, Ltd

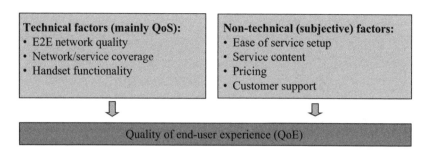

Figure 1.1   QoE is affected by the technical (QoS) and non-technical aspects of service.

- Service and network providers – that is, mobile network operators (MNOs), mobile network virtual operators (MVNOs) and mobile Internet service providers (ISPs), which are often owned by the operators themselves and transmit the content to the user.
- User device and application software that enables the user to experience the content.
- The network infrastructure providers and system integrators who, although not seen by the user, enable the above three components in the value chain.

The QoE value chain is depicted in Figure 1.2.

Although everyone in the mobile data service value chain should focus on optimising the experience, it is the MNO who has the main stake. Ignoring QoE, when designing a system/service, and waiting for users to vote with their loyalty is expensive and can have even worse consequences.

Mobile operators sit in the middle of this chain and orchestrate all four of these components in order to provide an overall 'experience' to the user. At the same time they have the highest stakes in ensuring an excellent QoE.

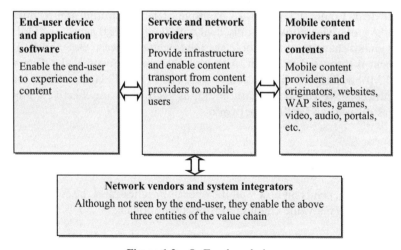

Figure 1.2   QoE value chain.

As already pointed out, QoE and QoS are integral parts of each other and no discussion on QoE would be complete without referring to QoS. These two terms have been described in different ways in different forums, leading to confusion for some readers in many cases. It is very important to understand their interaction – as discussed in the next section. An effort has been made to present various definitions and view points to come to a common understanding. Throughout this book, the terms 'QoS' and 'QoE' will be used in the context described in the following section.

1.2 QoS and QoE

This section defines and explains the differences between QoS and QoE. This will help us to understand the requirements of the operator and of the end-user:

- *QoS* is defined as the ability of the network to provide a service at an assured service level. QoS encompasses all functions, mechanisms and procedures in the cellular network and terminal that ensure the provision of the negotiated service quality between the user equipment (UE) and the core network (CN).
- *QoE* is how a user perceives the usability of a service when in use – how satisfied he or she is with a service in terms of, for example, usability, accessibility, retainability and integrity of the service. Service integrity concerns throughput, delay, delay variation (or jitter) and data loss during user data transmission; service accessibility relates to unavailability, security (authentication, authorisation and accounting), activation, access, coverage, blocking, and setup time of the related bearer service; service retainability, in general, characterises connection losses.

The term 'QoE' refers to the perception of the user about the quality of a particular service or network. It is expressed in human feelings like 'good', 'excellent', 'poor', etc. On the other hand, QoS is intrinsically a technical concept. It is measured, expressed and understood in terms of networks and network elements, which usually has little meaning to a user. QoE and QoS concepts are shown in Figure 1.3.

Although a better network QoS in many cases will result in better QoE, fulfilling all the traffic QoS parameters will not guarantee a satisfied user. An excellent throughput in one part of a network might not help if there is no coverage a short distance away.

As far as measures are concerned, these statistics tell an operator very little about the level of customer satisfaction. Flawless transmission of garbled packets does not make for happy users. So, the inference that QoE is improved because QoS mechanisms are used to reduce jitter or average packet delivery delay may not be accurate in all circumstances. What is important is good user experience or QoE, and the goal of QoS should be to deliver a high QoE.

Delivering high QoE depends on gaining an understanding of the factors contributing to the user's perception of the target services, and applying that knowledge to define the operating requirements. This top-down approach reduces development costs and the risks of user rejection and complaint, by ensuring that the device or system will meet user requirements.

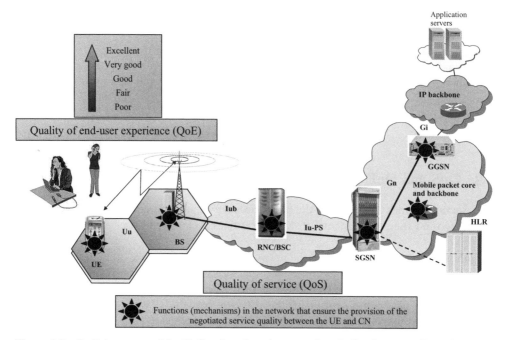

Figure 1.3 QoE is expressed in 'feelings' rather than metrics. QoS relates to all mechanisms, functions and procedures in the network and terminal that implement the quality attributes (bearer service) negotiated between the UE and CN.

QoS is often treated as a bottom-up process, consisting of a concatenation of point-to-point performance differentiation methodologies with little consideration for what happens on an end-to-end basis. The top-down approach is based on the premise that it is the end-user who is the ultimate beneficiary of QoS. In order to meet end-user expectations, the implementation of QoS in actual networks must be focused on the end-user perspective and provide the service performance levels necessary for a high QoE for the user. In practice, this means focusing on the customer – that is, the person who pays the bill – understand end-user expectations for QoS performance (QoE), and use these to drive requirements for specific QoS mechanisms (functions) for individual network domains such as UE, access, core, backbone and external packet data networks, and corresponding interfaces. The conceptual models of end-to-end QoE and QoS adopted in this book are illustrated in Figure 1.4.

There are many different end-to-end scenarios for delivering end-to-end QoS that may occur from a UE connected to a UMTS cellular network. The network architecture for the most significant contexts presented in this book is as depicted in Figure 1.4. Although the backbone IP network is shown as a single domain, it may consist of a number of separate domains. The structure of the local UE includes cases from a simple host to a gateway to a network such as a local area network (LAN). If the UE is acting as a gateway, it is responsible for providing the IP bearer service management towards the extended network.

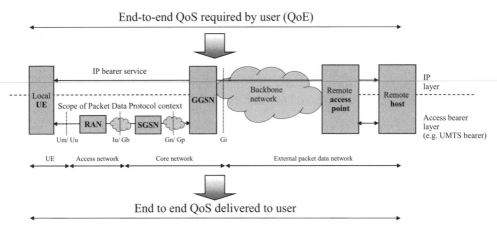

Figure 1.4 Top-down approach and E2E QoS definition.

In Figure 1.4 the remote end is shown as a simple host. Other more complex cases at the remote end, such as private LAN with overprovisioning, or possibly LAN priority marking, and DiffServ and/or Reservation Protocol (RSVP) capable routing elements are not depicted. The reference point shown at the UE is at the interface to the UE. Within the UE, QoS control could be derived from any of the mechanisms that occur across that reference point, or it could use a different mechanism internally. Although the scenarios currently identified are mainly using DiffServ in the backbone IP network, it is not mandated that DiffServ must be used for this. Other mechanisms, for example, over-provisioning and aggregated RSVP may be used.

The scenario presented in Chapters 3–5 will give examples of concatenating QoS mechanisms in different parts of the network that together can deliver an end-to-end QoS (consistent treatment and interworking between QoS mechanisms implemented in different network domains). *End to end* is by definition intended from a reference point – say, a service access point (SAP) – between two immediately above and below (adjacent) protocol layers, such as, for example, IP and TCP/UDP, through which the layers can exchange data in the UE to the corresponding SAP between the peers (same protocol entity layers) located in the remote host. For an example see SAP2–SAP3 in Figure 4.6.

1.3 QoE and QoS management

QoE and QoS management can be classified in four interdependent categories: network planning, QoS provisioning, QoE and QoS monitoring, and optimisation.

1.3.1 Network planning

The planning process includes network dimensioning and detailed network planning. Network dimensioning (or initial planning) provides an estimate of the required number of radio, transmission and core network elements and the capacity of related interfaces.

The calculations thereof are based on the operator's requirements for coverage, capacity and QoS. In detailed network planning, capacity and coverage are analysed for each relevant part of the network and interfaces between the entities in communication. This requires real traffic estimates and a network topology for each analysed area, the utilisation of accurate models for signal and user data transmission, and the actual network element's characteristics, functionalities and parameters.

1.3.2   QoS provisioning

QoS provisioning is a process that deploys QoS in networks and mobile terminals. The process translates planning results into mechanisms and parameters understandable by network elements and mobile terminals and it further configures them on equipment or devices.

QoS provisioning can be classified in three categories: radio, core and transport QoS provisioning that configures the QoS mechanisms inside the network; service QoS provisioning that maps services into QoS profiles (set of bearer service attributes); and terminal QoS provisioning that provides service application specific QoS information to terminals.

1.3.3   QoE and QoS monitoring

With the growth of mobile services, it has become very important for an operator to measure the QoS and QoE of its network accurately and improve it further in the most effective and cost-efficient way to achieve customer loyalty and maintain competitive edge. A poor QoE will result in unsatisfied customers, leading to a poor market perception and, ultimately, brand dilution.

Although QoE is very subjective in nature, it is very important that a strategy is devised to measure it as realistically as possible. The ability to measure QoE will give the operator some sense of the contribution of the network's performance to the overall level of customer satisfaction in terms of reliability, availability, scalability, speed, accuracy and efficiency.

Together, these elements define QoE and competitive advantage across today's packet-based communications networks. The experience is expressed in human terminology rather than metrics. An experience can be excellent, very good, good, fair or poor.

Two practical approaches to measuring QoE are the following:

1. Service level approach using statistical samples of a population of terminals.
2. Network management system (NMS) approach using QoS parameters.

The first relies on a statistical sample of overall network users to measure the QoE for all the users in the network. This process involves:

- Determining the weighting of key applications.
- Identifying and weighting QoE key performance indicators (KPIs).
- Devising a proper statistical sample (geographic areas, traffic mix, time of day, etc.) and taking KPI measurements accordingly.

- Utilising mobile agents in the handsets to make the results more accurate.
- Giving an overall QoE score (index) from KPI values for each separate service and a service mix.

The second is a methodology whereby hard QoS performance metrics from various parts of the network are mapped onto user-perceptible QoE performance targets. These QoS measurements are made using an NMS, collecting KPI figures from the network elements and comparing them with the target levels. The process involves:

- Identifying the relationship between QoS KPIs and their effect on QoE.
- Measuring QoS KPIs in the network.
- Rating users' QoE through measured QoS KPIs using some mapping rules.

1.3.4 Optimisation

Cellular network optimisation can be seen as a process to improve the overall network quality as experienced by the mobile subscribers and to ensure that the network resources are efficiently utilised. This includes performance measurements, analysis of measurement results and updates of the network configuration and parameters. The optimisation process can be initiated because of several reasons, the most typical are:

- New technologies, elements or features of particular network elements are taken into use.
- External edge conditions have changed.
- Detection of decreased QoS performance in a particular network area.
- As a part of daily network operation process.

1.4 Organisation of the book

The purpose of this book is to introduce the principles, practices and researches in mobile service planning, provisioning, performance monitoring and optimisation in QoE and QoS management. This book is intended for both academic and professional audiences (industry).

The first part of the book (Chapters 2–6) lays the foundations for further understanding of QoS and QoE management.

In Chapter 2 we describe some of the existing and upcoming multimedia services and their performance in detail to provide a general understanding of mobile service applications and performances.

Chapter 3 presents the latest mobile network QoS standardisation in the 3rd Generation Partnership Project (3GPP).

Chapter 4 describes the packet data transfer across second-generation (2G) and third-generation (3G) cellular systems, adopter protocols to implement the QoS and radio interface channels.

Chapters 5 and 6 discuss QoS mechanisms in the radio, core and backbone networks, respectively.

In the second part of the book (Chapters 7–10), we discuss the four elements of QoE and QoS management in detail.

Chapter 7 'Service and QoS aspects in radio network dimensioning and planning' addresses radio dimensioning and detailed planning issues that may arise when deploying multimedia services in wideband code division multiple access (WCDMA) and enhanced General Packet Radio Service (EGPRS) cellular systems.

Chapter 8 'QoS provisioning' examines service and QoS provisioning in detail. QoE and QoS measurements including more insights into these methods are discussed in Chapter 9 'QoE and QoS monitoring'.

The objectives, concepts, processes, algorithms and means for QoS and QoE optimisation across UMTS cellular networks are discussed in Chapter 10 'Optimisation'.

2

Mobile Service Applications and Performance in UMTS

Renaud Cuny, Man Li and Martin Kristensson

The purpose of this chapter is to introduce the functional aspects of circuit-switched (CS) and packet-switched (PS) based services and their performance over Universal Mobile Telecommunication System (UMTS).

The convergence phenomenon that started some years ago and that is accelerating will gradually remove the traditional boundaries between fixed and mobile communication systems. In the future, personal communication devices will be multiradio-capable while broadband wireless coverage will extend considerably. On the other hand, Internet Protocol (IP) based services are expanding all the time and now overlap with those provided by CS systems, either fixed or mobile, Voice over IP (VoIP) being one of the most remarkable examples. In short, the global trend is that communications will be PS and that more and more devices and networks will support wireless mobility. Getting there, however, will require effort and time; so, if the trend is clear, we shall not yet ignore established CS systems and services as they will continue to have a key role for many years.

The functional aspects and performance requirements of mobile services are important pieces in the overall end-user quality-of-experience management. For instance, the necessary procedures to set up a service will determine how long session establishment will take depending on the actual underlying network technology (e.g., cellular or fixed access network). Also, once the session has been established, there will typically be specific throughput and latency requirements to fulfil, which in turn may impose particular services to be used only in certain types of networks. As an example, some action games may be well-supported in wideband code division multiple access (WCDMA) but not in the General Packet Radio Service (GPRS) because of too long latencies.

Evaluating the overall service performance in a given low- to medium-loaded network may be one starting point when considering the launch of a new service. If the required performance is not fulfilled, even in the best conditions, it is probably not worth moving on to the next phase of service deployment. At the same time, it is important to note that

QoS and QoE Management in UMTS Cellular Systems
Edited by David Soldani, Man Li and Renaud Cuny © 2006 John Wiley & Sons, Ltd

low service performance may not always be due to network limitations, but may also, for instance, be due to less than optimal terminal or server implementations. In certain cases, optimising the application or the transport layer or applying, for example, efficient header or data compression mechanisms may be enough to achieve satisfactory quality-of-experience levels.

This chapter concentrates mostly on PS services. The main reason being that the CS service's functional aspects, requirements and performance in UMTS have already been widely documented elsewhere (e.g., [1]). Thus, Section 2.1 summarises some key points in the area and largely refers to other sources. The remaining sections – Sections 2.2 and 2.3 – focus solely on the PS service's functional aspects, requirements and performance in UMTS.

2.1 CS service applications

The UMTS network architecture is logically divided between [2] the radio access network (RAN) and the core network (CN). The CN is itself separated into the PS CN domain, the CS CN domain and the IP multimedia CN subsystem (IMS) (see Chapter 3). The CS domain is an evolution of the Global System for Mobile Communications (GSM) technology that was developed under the European Telecommunication Standard Institute (ETSI) until 2000, and pursued later in the 3rd Generation Partnership Project (3GPP) standardisation body. In CS systems, resources are allocated at service session setup and reserved during the entire session duration. This approach is well-suited for applications, such as (multimedia) telephony, that have stringent and stable requirements in terms of delay and bandwidth and when traffic sources are sending or receiving data during the majority of the session. PS systems, on the other hand, can efficiently support any type of service applications, including those that may be very bursty in nature, and possibly idle a great part of the session (e.g., web browsing), by dynamically allocating network resources on a need basis.

2.1.1 CS telephony

CS telephony is specified in [3] together with other services – such as emergency calls, Short Message Service (SMS).

Speech codecs for mobile communications have improved over the years bringing significant gains both in terms of speech quality and capacity. The adaptive multirate (AMR) consists of a family of codecs with different bit rates operating in GSM full-rate (FR) and half-rate (HR).

AMR speech codec (adopted in 3GPP R99 as a mandatory codec) performance is characterised in [4]: unlike previous GSM speech codecs (full-rate, enhanced full-rate and half-rate) that operate at a fixed rate and at a constant error protection level, the AMR speech codec can adapt its error protection level to the radio channel and traffic conditions. AMR selects the optimal channel (half- or full-rate) and codec mode (speech and channel bit rates) to deliver the best combination of speech quality and system capacity. This flexibility provides a number of important benefits:

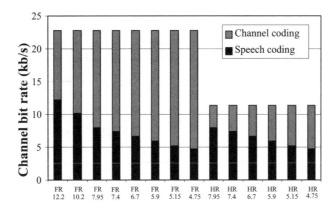

Figure 2.1   AMR codec modes.

- An improved speech quality because of codec mode adaptation, by varying the balance between speech and channel coding for the same gross bit rate.
- The ability to trade speech quality and capacity smoothly and flexibly by a combination of channel and codec mode adaptation; this can be controlled by the network operator on a cell-by-cell basis.
- Improved robustness to channel errors under marginal radio signal conditions in full-rate mode. This increased robustness to errors and hence to interference may be used to increase capacity by operating a tighter frequency reuse pattern in GSM.
- Ability to tailor AMR operation to meet the different needs of operators.
- Potential for improved handover and power control resulting from additional signalling transmitted rapidly in-band.

The AMR codec concept is adaptable not only in terms of its ability to respond to changing radio and traffic conditions but also to be customised to the specific needs of network operators. The AMR codec modes are illustrated in Figure 2.1.

Wideband AMR (AMR-WB) was introduced in 3GPP R5 and provides further speech quality enhancements that are essentially due to the larger speech-coding bandwidth (from 50 to 7000 Hz). These improvements make AMR-WB also suitable for applications having high-quality audio requirements.

2.1.2 CS multimedia telephony

The requirements for multimedia CS calls are defined in 3GPP [5]. Some of these are provided below:

- CS multimedia is based on an International Telecommunication Union (ITU) H.324 terminal.
- All call scenarios are supported: mobile-originating and mobile-terminating call, against a mobile Integrated Services Digital Network (ISDN) or Public Switched Telephone Network (PSTN) call party.
- Single and multiple numbering.

- If the setup of the multimedia call fails the call will be set up as a speech call.
- In-call modification: modification of call type from speech call to multimedia call (and *vice versa*) during the call. Service degradation and upgrading.
- End-to-end user rate negotiation.
- H.324 and H.323 (for PS multimedia) interworking.

H.324 consists of two mandatory components [6]: H.223 for multiplexing and H.245 for the control. Other optional components are H.263 video codec, G.723.1 speech codec and V.8 bis. MPEG-4 video and AMR were later added to the system as optional codecs.

CS video quality is more sensitive to the block error rate than voice due to the nature of video compression and, thus, the allowed residual bit error rate is typically very small (e.g., 10^{-5}) to ensure a good quality of experience [6].

2.2 Packet-switched service applications

Before going into the details of PS services, it is worth introducing very briefly the IMS which is described in more detail in Chapter 3. The IMS, added in 3GPP R5, contains all CN elements needed for providing multimedia (IP-based) services [7], [8]. The IMS enables operators to offer subscribers multimedia services based on and built upon Internet applications, services and protocols. The Session Initiation Protocol (SIP) is the application layer control protocol used for establishing, modifying and terminating peer-to-peer service sessions [9]. A peer-to-peer service session could be a simple two-way telephone call, a collaborative multimedia conference session or, for instance, a network game. SIP is an Internet Engineering Task Force (IETF) driven signalling protocol based on the request–response scheme similar to Hypertext Transfer Protocol (HTTP) signalling.

2.2.1 Browsing

The browsing service allows a mobile user to browse the Web by using a browser installed on a mobile phone. The user may either type in a URL or click a link in order to access a web page.

The Open Mobile Alliance (OMA) browsing enabler is based on Wireless Application Protocol (WAP) standards from the WAP Forum and is migrating towards Internet protocols. Hence, depending on the software implemented, a mobile phone may use:

- HTTP 1.1 to communicate directly with a web server [10];
- Wireless Profile HTTP to communicate with a WAP 2.0 gateway that in turn contacts a web server [11]; or
- Wireless Session Protocol (WSP) to communicate with a WAP 1.0 or 2.0 gateway that in turn contacts a web server [12].

All three protocols are based on the HTTP 1.1 request and response paradigm [10]. As shown in Figure 2.2, a web page is requested by sending a GET request message to a web server (or a WAP gateway that in turn sends the request to the server). If the operation is

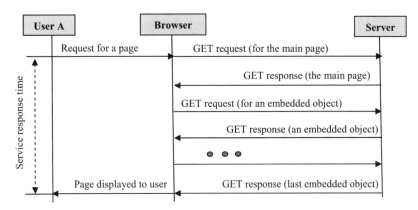

Figure 2.2 Browsing service message flows.

successful, the web server will reply with a GET response message that contains the requested web page. A web page may have multiple embedded objects such as images, in which case each object is fetched with a separate GET request message. A GET response message from a web server contains a status code to indicate the result of the operation.

2.2.2 Multimedia Messaging Service (MMS)

The MMS is a system application by which a client is able to conduct messaging operations with a rich set of media contents, such as image, video, etc. MMS is a non-real time delivery system, though it is expected that a multimedia message should be deliverable within a reasonable time frame. MMS can interoperate with other messaging systems – such as the traditional email on the Internet. In other words, when one endpoint of the service is an MMS client, the other endpoint for the multimedia message may be another MMS client, a client on a legacy wireless messaging system or an email client. MMS is specified in OMA in [13]–[16].

An MMS proxy-relay is a network element that interacts with MMS clients on mobile phones to provide MMS services. In addition, it provides access to an MMS server that stores messages. Furthermore, it also serves as a gateway when interacting with other messaging systems (e.g., Internet email). Some implementations may combine MMS proxy-relay and MMS server into one physical element.

There are two ways a message may be retrieved by a receiving MMS client depending on service profiles and/or device settings. Immediate retrieval of a new message means that a receiving MMS client immediately retrieves the message via the MMS proxy-relay upon receiving a new message notification. In contrast, deferred retrieval means that a receiving MMS client first acknowledges a new message notification and at some later point retrieves the message via the MMS proxy-relay.

Figure 2.3 shows the MMS transactions for immediate retrieval. In the figure, both the origination and termination points are MMS clients. The MMS protocol data unit (PDU) can be carried by either WSP or HTTP. The message notification (M-Notification.ind) is sent via WAP.

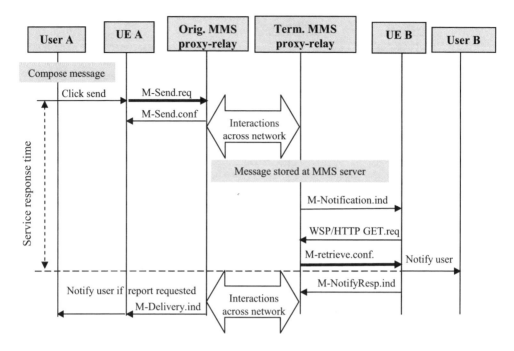

Figure 2.3   MMS with immediate retrieval.

After a user composes a message, he/she clicks the send button on the mobile phone. The MMS client transmits the message to the MMS proxy-relay. Upon receiving the message with no error, the MMS proxy-relay sends back a confirmation to the MMS client. The multimedia message is then forwarded across the mobile network to the terminating MMS proxy-relay and is stored in its associated MMS server. The terminating MMS proxy-relay then notifies the originating MMS client.

For immediate retrieval, the MMS client tries to fetch the message with a GET operation. After the message is retrieved, the MMS client sends a Notify response back to the MMS proxy-relay. This response is forwarded to the original MMS proxy-relay, which then sends a delivery indication (M-Delivery.ind) to the originating MMS client to acknowledge the status of the message delivery. If the user has requested a delivery report, the MMS client will inform the user upon receiving the M-Delivery.ind. Otherwise, the user is informed only when the delivery is not successful.

For deferred retrieval, a receiving MMS client responds to a message notification immediately. But the message is not retrieved until a later time – for example, when the user asks to read the message.

Depending on the service settings on the receiving mobile device, sending of a delivery report back to the original user may be denied. In other words, the original MMS client cannot assume that a delivery report (M-Delivery.ind) is guaranteed for every message sent.

2.2.3 *Content download*

OMA is defining a generic content download over-the-air specification [17]. As specified, a download user agent is an agent or software function in the device responsible for downloading a media object. A download descriptor contains information about a media object and instructions to the download agent about how to download it. It allows the client device to decide whether it has the capabilities to install and render/execute the media object.

The OMA specification supports two scenarios:

- Download with separate delivery of download descriptor and media object.
- Download with co-delivery of download descriptor and media object.

In both scenarios, the download agent may send a notification to the server in order to confirm the status of the transaction. This notification is likely to be mandatory in a pay-per-transaction model where the confirmation of a successful installation of a media object typically triggers server-end billing actions.

Figure 2.4 shows the message flows for a separate delivery of download descriptor and media object. A user is initially presented with a reference to the download descriptor. The reference may be on a web page, inside an email or an MMS message, or stored in memory.

If the user has the intention of pursuing the download, he/she will click the reference to obtain the descriptor. The download descriptor is then transferred to the mobile device. The transfer mechanism or protocol may be HTTP or secure HTTP (HTTPS) but can also be through MMS, email or some instant messaging protocol. The download agent

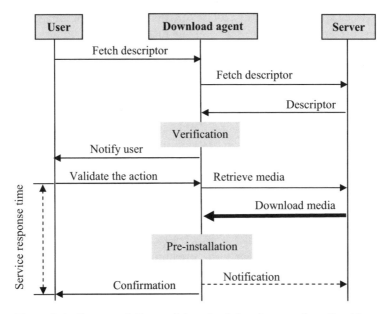

Figure 2.4   Separate delivery of download descriptor and media object.

uses the information in the descriptor (e.g., media type and size) to check whether the device is capable of using or rendering the media object. If the device is able to handle the media object, the user is then presented with more detailed information about the media (type, size, vendor, etc.) and is asked to confirm that she indeed wants to download and install the media object. On confirmation from the user, the media object is retrieved, typically using HTTP or HTTPS, according to the instructions from the download descriptor. The device then tries to pre-install the media. If an installation notification has been requested in the download descriptor, the device prepares the media object for rendering or execution to the largest extent possible without actually allowing it to be used. This is to ensure, as much as possible, that the media can be installed and consumed by the end-user without error. It then sends an installation notification to the server to report the outcome of the previous step. A set of status codes to be included in the notification is defined in [17]. If the installation notice is successfully sent, the media object will be made available for execution and the user is notified with a confirmation. Otherwise, the media object will be removed from the device.

Different servers may provide the download description and the media object. The mobile agent may send the notification to yet another server according to the download descriptor.

In co-delivery of the download descriptor and media object, when an end-user requests a download, the download descriptor and the media are delivered together to the mobile device. The remaining installation and notification processes are the same as described above.

2.2.4   Streaming

The streaming service description below follows the 3GPP transparent end-to-end PS streaming service specification [18]–[20]. A streaming service contains a set of one or more streams presented to a user as a complete media feed. The content is transported using Real time Transport Protocol (RTP) over User Datagram Protocol (UDP). Control for the session setup and for the playing of media (PLAY, PAUSE) is via the Real Time Streaming Protocol (RTSP) [21].

Figure 2.5 shows the streaming service message flows. When a user starts a streaming service by either clicking a link in a web page or entering a uniform resource identifier (URI) of a streaming server and content address, the streaming client on the mobile phone must first obtain a presentation description that contains information about one or more media streams within a presentation, such as encoding, network addresses and information about the content. This presentation description may be obtained in a number of ways – for example, via MMS or RTSP signalling. 3GPP mandates that the description be in the form of a Session Description Protocol (SDP) file [22].

Once the presentation description is obtained, the streaming client establishes a session for each media stream. Specifically, it may try to establish a secondary PDP context (depending on UE capabilities) for each streaming media and also sends a SETUP request message to the media server in order for the server to allocate resources for the stream. The SETUP reply message contains a session identifier, server port for displaying the media and other information required by a client for playback of the media stream.

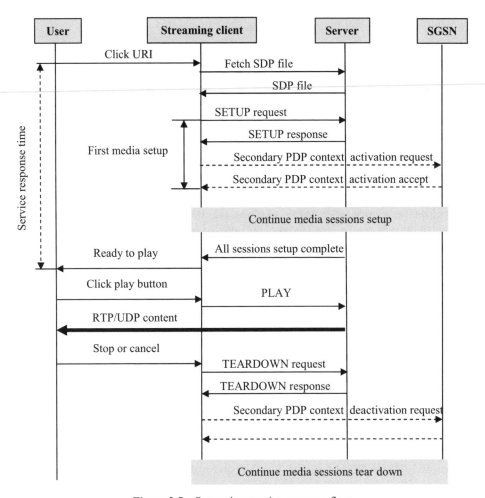

Figure 2.5 Streaming service message flow.

After all media stream sessions and their required PDP contexts are established, the user may click the play button to start playing the synchronised media. The user can also pause, resume or cancel the streaming service at any time. The RTSP PLAY, PAUSE and TEARDOWN messages are sent to the server for the corresponding action.

2.2.5 Gaming

Gaming services allow end-users to play games on mobile devices over mobile networks.

A *solo game* is a game where a single game player interacts with a game server. A *multiplayer game* is a game where multiple players are involved. A *person-to-person game* is a game where two or more players interact with each other without the intervention of a game server. A *server-based game* has a game server that is responsible for game synchronisation between players, updating the game status to all players, keeping high scores, etc.

There are already many mobile games out there and new games are being released every day. Each game is different with its own game logic, session management, etc. In addition, different types of games have different end-user requirements [23]. For example, a real time shooting game would have more stringent requirements than a turn-based game such as chess. Furthermore, game applications may run on top of different transport protocols such as HTTP, TCP, UDP, SMS, WAP push, etc.

We note that OMA is working on gaming service standardisation. The working group has been specifying a gaming architecture [24], a server framework [25] and a client/server protocol [26]. The client/server protocol is in its very early stage and is not completely consistent with the other OMA gaming specifications.

In multiplayer games, since not every player is suited to match with every other player, prospective players need a place to meet and stage a game. A lobby is such a place. There can be many lobbies for a specific game. A game room is where a game is actually played and there can be many game rooms in a lobby. Depending on the design of a game, a player may or may not be allowed to create a new lobby. A player can join an existing game room, or create a game room to start a new game. Both lobbies and game rooms are in fact multicasting groups where a message or a move is broadcast to all members in the group.

To play the game, the player must first connect to a game server, log in and be authenticated. Then, he/she can browse the list of lobbies, enter a specific lobby, join an existing game room or create a new game room, and play a game.

2.2.6 Business connectivity

Business connectivity is about enabling end-users to access corporate Intranet or Internet services from a wireless device, in a secure manner. The access network may be, for instance, EGPRS, WCDMA or WLAN. The end-user device may be a smartphone or a laptop.

Security can be ensured with a virtual private network (VPN). For large enterprises that seek end-to-end security, a client VPN is needed in the mobile device. Small companies may on the other hand decide that encryption between the mobile operator's domain and their enterprise's domain is sufficient.

The security can be offered through either IP security (IPsec) protocols [27]–[30] or the secure socket layer (SSL) [31] protocol. In this section we only consider IPsec-based VPN.

IPsec VPN protects IP packets by offering:

- *Packet confidentiality* – packets are encrypted before being sent over the network so that only authorised entities can read them.
- *Packet integrity* – packets are protected so that any alterations during transmission over the network can be detected.
- *Packet origin authentication* – packets are protected to ensure that they are indeed from the claimed sender whose IP address is contained in the source address of the IP header.

- *Protection against replay* – packets are protected from being captured and re-sent at some later time.

The above protections are achieved through the use of one or a combination of two security protocols: the encapsulating security payload (ESP) [29] and the authentication header (AH) [30]. Note that AH alone does not provide confidentiality protection.

There are network-based and client-based VPN services. In a network-based service, a secure IPsec tunnel is established between a security gateway at the Gi interface and another gateway at the corporate network premises. The IPsec tunnel thus protects data when they travel through the Internet. In a client-based service, a piece of VPN software, referred to as the 'VPN client' in the following, is installed on a mobile device together with the IPsec VPN policy. The client creates IPsec tunnels between the mobile device and the security gateway at the corporate network premises, according to the installed VPN policy. The tunnels offer security protection to packets that travel through it. In addition, dictated by the VPN policy, the client can also drop incoming and/or outgoing packets. For example, if a received packet does not pass data authentication, it will be dropped by the VPN client.

In order to use the mobile VPN service, VPN access points have to be configured on mobile devices. Each access point must also have a reference to the VPN policy. VPN access points can be used in the same way as any other access points.

As shown in Figure 2.6, to access enterprise services from a mobile device, an end-user simply activates the service in the same way as when using other services. When being asked to select an access point, the user chooses a VPN access point. The mobile device first establishes the PDP context for Internet access. Then, it negotiates, according to the VPN policy associated with the VPN access point, with a security gateway at the enterprise in order to establish a secure tunnel. The process includes two phases: Phase 1 is called 'Internet key exchange (IKE)' and Phase 2 involves negotiations [28]. In the first phase, the two ends agree on encryption and authentication algorithms for a secure IKE tunnel, authenticate each other and derive security keys for encryption and authentication. Phase 1 negotiation may involve either three or six Internet Security Association and Key Management Protocol (ISAKMP) messages with different payloads depending on whether aggressive or main mode is used. When this phase begins, the user is also asked to provide authentication information to be used in Phase 1 negotiation. It can be a static password or a secure-id pass-code. The authentication information is exchanged in the last two messages of IKE Phase 1 negotiation. Depending on the configuration, the gateway may use a Radius server, an internal database, a Lightweight Directory Access Protocol (LDAP) directory or something else to handle the authentication requests received during IKE Phase 1 negotiations. At the end of Phase 1 negotiation, an IKE tunnel is set up.

IKE Phase 2 negotiation is conducted under the protection of the IKE tunnel. The second phase aims at establishing two unidirectional IPsec tunnels, based on either ESP, AH or a combination of both, to protect service or application traffic. And it is these IPsec tunnels that provide confidentiality, integrity, authentication and replay protections of application traffic to and from corporate networks. The second phase of negotiation typically consists of three ISAKMP messages. After the IPsec tunnels are

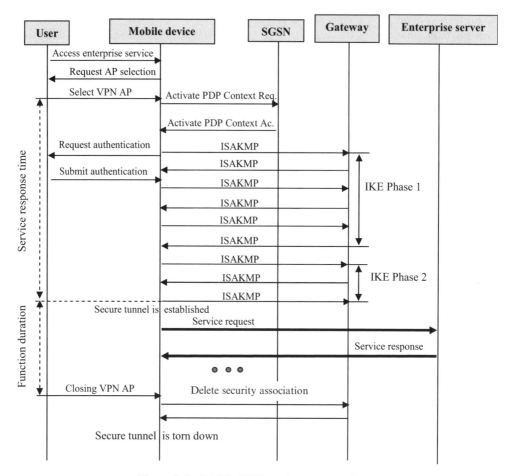

Figure 2.6 Mobile VPN service message flow.

established, legitimate data traffic can travel between the mobile device and the corporate network. The security tunnels can be closed when they are not needed.

During the lifetime of the IPsec tunnels, a VPN client can also drop incoming and outgoing packets according to the security policy. For example, outgoing packets that are destined to the corporate network but are not supposed to be carried by the IPsec tunnel, according to the policy, are dropped. Incoming packets that are damaged during the transmission cannot pass data authentication and hence are dropped.

2.2.7 Push to talk over Cellular (PoC)

PoC enables a real time one-to-one and one-to-many voice communication service in a cellular network. Users can select the person or talk group they wish to talk to, and then press the push to talk (PTT or PoC) key to start talking. The call is connected in real time. PoC calls are one-way communication: while one person speaks, the other(s) only listens.

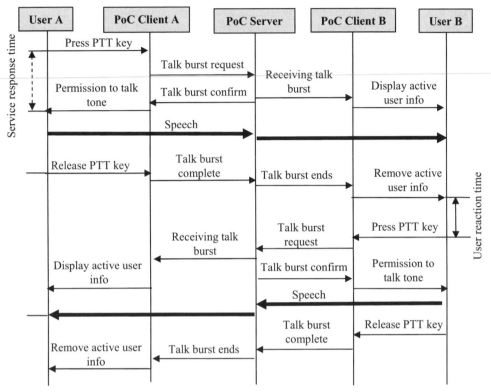

Figure 2.7 PoC service requests and indications.

The turns to speak are requested by pressing the PTT key and are granted on a first-come-first-serve basis. Hence, if User A has requested to speak prior to User B, the request from User B goes to a queue and a queuing tone is given to User B to indicate that the floor was granted to another user, if queuing is supported by the PoC client of User B. The user should release the PTT key and wait for the other person to finish. If he still wants to talk to the group, he must push the PTT key again. At the receiving end, the PTT speech is connected without the recipient(s) answering and is heard through the phone's built-in loudspeaker. The receiving user may respond to the call by pushing the PTT button.

While popular, PoC is an evolving service in terms of technology, standardisation and service features. The following description is based on the OMA PoC architecture [32].

Figure 2.7 shows the PoC requests/responses/indications message flows following the OMA PoC architecture [32] document. In practice, there may be multiple PoC servers involved that perform different functions – for example, handling signalling, distributing the media stream, etc. For simplicity, only one PoC server is shown in the figure.

When a user presses the PTT key, a talk burst request is sent by the PoC client to the PoC server in order to request permission to send a talk burst. The request contains a PoC session identifier and an indication on whether queuing is supported by the PoC client.

The PoC server grants the floor by sending a talk burst confirm response that contains the session identifier. If the PoC server cannot grant the floor, a talk burst reject response is sent to the PoC client. The rejection response contains the session identifier and a rejection reason. The reason may be:

- another PoC client has already been given permission to send a talk burst and no queuing of the request is allowed;
- another PoC client has already been given permission to send a talk burst and the queue is full.
- the PoC client is not allowed to request permission to send a talk burst at the moment; and
- only one participant in the PoC session – for example, if only one participant is left in a PoC session (hence no one is listening)

If the PoC server and the PoC client support queuing of the talk burst request and the particular request is queued, the PoC server sends back a talk burst request queued response to indicate that the request is queued. The indication includes the session ID and the position at which the request is queued.

At the end of a talk spurt, the user releases the PTT key and a talk burst completed indication is sent to the PoC server to indicate the end of the talk burst. The indication also includes a session ID.

At the terminating end, when the PoC server grants the floor to one client, it sends a receiving talk burst indication to all the other PoC clients in a PoC session in order to inform them that another PoC client has permission to send a talk burst and that the PoC clients must be prepared for receiving a talk burst. The receiving talk burst indication includes a PoC session identifier and the identity of the PoC participant at the PoC client sending the talk burst – that is:

- the PoC address when the sender does not want to be anonymous; and
- the display name of the PoC participant at the PoC client sending a talk burst.

The PoC client displays the name of the PoC participant to the PoC client sending a talk burst.

The PoC server can also revoke the permission to talk by sending a stop talk burst indication to the PoC client. The possible reason can be, for example, there is only one user in the PoC session (hence no one is listening) or the talk burst is too long and has exceeded the maximum duration allowed. The stop talk burst indication includes a PoC session identifier and may include a reason code and a retry-after time value indicating how long the PoC client has to wait before a request to send a talk burst will be confirmed. This parameter is only present when permission to send a talk burst is revoked due to the talk burst being too long.

2.2.8 Video sharing (VS)

VS is defined as a peer-to-peer, unidirectional, PS multimedia streaming service where at least one of the actors is using a mobile device. In VS, video from a live camera or a

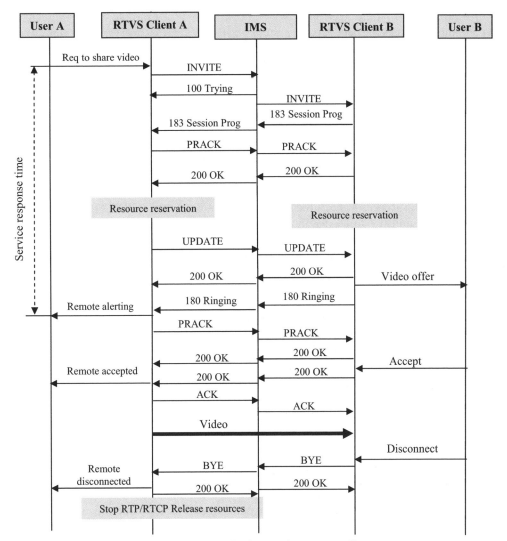

Figure 2.8 Video sharing service message flows.

multimedia file stored in the originating device is sent to a destination device. The multimedia data are streamed from one device to the other and are consumed in real time, creating the experience of 'sharing the moment'. One use case for VS is to enrich a CS voice call by sharing live video or pre-recorded video clips during the voice call.

Though the VS service is not being standardised, implementation may follow the standard IMS specification [7] that employs SIP [9] as its session management protocol. Video media are carried by RTP, and RTCP is used to provide video performance feedbacks in order to adjust media delivery according to network conditions.

Figure 2.8 shows a possible message flow for VS. The VS client in the figure represents all the software functions on a mobile device that handle VS applications. During a CS voice call, User A would like to share video content by, for example, selecting a 'share video' button. The VS Client A in turn sends a SIP INVITE message that contains all the codecs that are supported by the device. The VS Client B answers with a 183 Session Progress message that contains all the codecs its device supports. Then, Client A sends a provisional acknowledgement (PRACK) message that contains the final codec selected and Client B answers with 200 OK confirming the codec selection. If 183 Session Progress contained only one codec, then PRACK and 200 OK do not include any codec information. At this point, both sides start to reserve needed resources for the session – for example, activating the PDP context. The IMS may add an authorisation token to the INVITE and 183 Session Progress messages. In such a case, Clients A and B must use the tokens in PDP context activations (see Chapter 3).

Client A indicates a successful reservation to Client B with an UPDATE message. When Client B succeeds in resource reservation, it answers the UPDATE with 200 OK. Client B then indicates to User B that a video is offered by User A, and also sends a 180 Ringing message to indicate to Client A that user B is being alerted. Client A acknowledges the message with a PRACK. Finally, when User B accepts the offer, Client B sends a 200 OK message in reply to the first INVITE message. Client A answers with an ACK message and starts sending video. One may notice that there are many 200 OK reply messages in the figure. Every 200 OK message contains a sequence number that matches that of the message it replies. Hence, there is no potential confusion.

The UPDATE and PRACK messages are specified in [33] and [34], respectively. During video sharing, any of the users can pause the video stream for a period of time. The pause and resume commands are all carried inside an INVITE message that is in turn acknowledged by a 200 OK message.

2.2.9   Voice over IP (VoIP)

VoIP is a technology used to transfer voice over an IP network. VoIP does not refer to any particular application, but can be used in different scenarios. The main drivers for VoIP are cost savings, wireless coverage expansion and richer communication services enabled by better programmability. VoIP technology can be used over different networks such as fixed broadband (DSL/cable), WLAN (IEEE 802.11) and cellular 3G.

VoIP technology is fragmented, and there are several VoIP implementations that do not interoperate properly together. One example of differences between VoIP implementations is the signalling protocol used. IETF, 3GPP/3GPP2 standard systems use SIP, while other systems use different, non-interoperable protocols.

VoIP technology also has a place in the cellular environment. The first VoIP over cellular services are already in use. The popular PoC service has been launched by several operators around the world. With 3G networks and handsets, conversational full-duplex VoIP services become feasible. This opens up new possibilities for rich communication services. 3GPP networks will be able to offer an adequate level of quality for VoIP services [35]. 3G enhancements, such as high-speed uplink packet access (HSUPA) and high-speed downlink packet access (HSDPA), will further increase cellular networks' capabilities to provide rich communication services.

VoIP is not mandatory for conversational-rich communication – that is, Rich Call – services in a cellular network environment. One alternative is to first combine CS voice with an IP data session. Devices will be responsible for launching and managing Rich Call sessions, such as instant messaging, document sharing or real time VS. Even though Rich Call services use a CS bearer in the cellular domain, the services are not tied to it, but run over any bearer supporting IP such as WLAN.

VoIP service setup in cellular may be similar to the one presented for VS in Section 2.2.8, in case the session setup uses SIP.

2.2.10 *Presence*

The ability and willingness to be reached for communication is defined by items of information known as 'presence information' [36]. Presence information may be related to the mobile network connection status; however, it represents much more than just end-user network coverage. With presence it is possible to define a set of access rules to control access to presence information. A user's profile may be either publicly visible or visible for a restricted group of users. Some examples of the information that could be stored in the profile are listed below:

- Personal status or phone profile (available, busy, on holiday, in a meeting).
- Terminal status and capabilities (status: switched off, out of coverage, engaged in a videoconference; capabilities: supports chat and instant messaging).
- Location, which does not necessarily refer to the geographical location (in the office, at home, on-the-move).
- Other information (name, address, telephone number, email address, logo).
- Mood (happy, frustrated, angry, sad).
- List of games the user can play (he/she has downloaded and/or has general interest in playing it).

By supporting a presence service in the network, the operator has the capability to offer an exciting range of advanced presence-based services and applications. For instance, presence information may be used by such applications as online gaming. Instead of blindly logging on to a games master server, it would be possible to see who is playing and invite other users to join a game. A user simply has to subscribe to the presence information of friends. This presence information can be displayed in a list format showing who is playing, who is online and available, and who is not. Third-party call control could even automatically initiate a gaming session when everybody is ready.

SIP presence is defined as 'subscription to and notification of changes in the communications state of a user'. Presence holds the means of communication in which the user can be reached – that is, instant messaging (IM), call, email – or it can provide other information of the user: home page, picture, announcement text, mood, etc. In SIP presence, User A is able to subscribe (using SUBSCRIBE method) to the status of User B and be notified (using NOTIFY method) when there is a change in User B's status.

2.2.11 Instant messaging (IM)

IM is defined as the exchange of content between a set of participants in real time. IM typically differs from, for example, email in that instant messages are usually grouped together into brief live conversations, consisting of numerous small messages sent back and forth. There are several different messaging schemes:

- First, there is the distinction between one-shot messaging and conversational messaging. One-shot messaging means that there are only a few (usually 1–2) messages, which are exchanged between clients. IM and multimedia messaging usually belong to this category.
- Another messaging scheme is session-based messaging in a separate SIP session, which is established using SIP INVITE in a standard way. Actual messages may be carried inside the session using, for example, the RTP text, TCP-pipe or SIP MESSAGE method.

So, the MMS described in Section 2.2.2 may also be provided using IMS and SIP. 3GPP R6 defines even tighter integration of the MMS with the IMS especially for addressing and using SIP as a way to notify the UE of the MMS received [8].

2.3 PS service performance in UMTS

End-user requirements on wireless packet data connections are simple – it should be possible to set up connections quickly, provide high and stable bit rates for instant service delivery on the move, and the price should be affordable. Finding the right cost/performance trade-off requires vendor and operator understanding of the complete end-to-end chain. This includes aspects such as application design, application requirements, network features affecting the application performance and how to deliver a sufficient end-user experience in a nationwide network with multiple services. Figure 2.9 shows different aspects in end-user experience delivery.

QoE is a function of multiple protocol layers and network elements. The radio interface is typically the bottleneck, with its restrictions in bandwidth and coverage. Looking at WCDMA and enhanced data rates for GSM evolution (EDGE) networks with applications in mind provides a number of insights that reveal possibilities and challenges. One key finding is that both WCDMA and EDGE already enable many services today. The strong technology evolution in WCDMA and EDGE over the past couple of years decreases the difference between fixed and wireless networks and hence further simplifies service delivery in the wireless domain.

On the one hand, there are many services and it is not possible to handle each service independently in terms of planning, monitoring, etc. On the other hand, services have different delay and bit rate requirements. By exploiting the fact that requirements are different, network capacity may be increased. A reasonable solution is therefore to provide a number of 'bit-pipes' where each bit-pipe hosts multiple services with similar QoS requirements and where the network prioritises these bit-pipes differently.

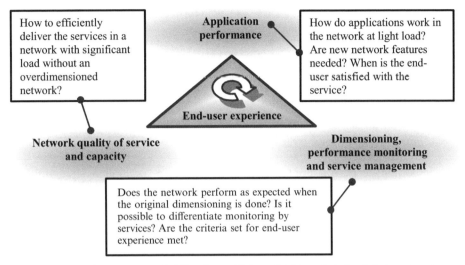

How to efficiently deliver the services in a network with significant load without an overdimensioned network?

Application performance

How do applications work in the network at light load? Are new network features needed? When is the end-user satisfied with the service?

End-user experience

Network quality of service and capacity

Dimensioning, performance monitoring and service management

Does the network perform as expected when the original dimensioning is done? Is it possible to differentiate monitoring by services? Are the criteria set for end-user experience met?

Figure 2.9 Three aspects of delivering sufficient QoE in a (wireless) data network.

In this section we look at the end-to-end chain by first considering the general application performance aspects and key performance indicators (KPIs) of content-to-person and person-to-person applications. Then, we describe how applications behave in WCDMA, EDGE and in multiradio environments. In the WCDMA and EDGE context both commercial networks and improvements tied to the technology evolution are covered. More information on access network performance is provided in Chapter 5.

2.3.1 General application performance

This section briefly introduces some QoE aspects for various service applications and describes how they relate to both service functional characteristics and mobile network performance.

2.3.1.1 Content-to-person applications: bit rates and round trip time

For many content-to-person applications – for instance, web browsing and file down-loading – the KPI is the time from the moment the user clicks a key until the service (webpage) is delivered. Download times for a typical Internet webpage and MP3 file are given in Table 2.1. The numbers assume that the stated application level bit rates are reached directly when the user clicks the key.

Because web browsing is very much an interactive service, the total download time of a webpage should be lower than 4–10 s. The acceptable level depends on if the click-to-content time refers to the time when the first parts of the webpage are shown (text), or the total page (text and pictures). The 4-s requirement is fulfilled for EDGE and WCDMA in Table 2.1. Downloading an MP3 song is less interactive and it is expected that the user is satisfied with the download time if it is a couple of minutes. This is also fulfilled with EDGE and WCDMA.

Table 2.1 The higher bit rates in EDGE and WCDMA yield low click-to-content times for applications with large data amounts as well.

Application	40 kb/s (GPRS)	200 kb/s (EDGE)	384 kb/s (WCDMA)	2 Mb/s (WCDMA/HSDPA)
Laptop browsing (200 kB page)	40 s	8 s	4 s	1 s
MP3 song download (4MB)	13′	2.6′	1.3′	20 s

Other content-to-person applications such as audio and video streaming are in addition to the click-to-content times characterised by the bit rate during the service. Video streaming to a mobile station with a limited screen size requires constant/stable bit rates higher or equal to 64 kb/s for good quality with current codecs. With codecs based on the current available 3GPP specifications, the streaming quality is further improved for mobile station based streaming up to 128 to 384 kb/s.

A common function in streaming applications is adaptation of the content bit rate. It is needed because end-user experience for streaming depends a lot on the bit rate, which varies in and between networks.

What the end-users experience is the average application level bit rate, which is often lower than the maximum bit rate provided by the radio technology. One explanation for this is that in browsing, HTTP and TCP reside in between the application and the radio protocols.

The so-called packet round trip time (RTT) is one fundamental property that affects the efficiency of HTTP and TCP. It is the time it takes to send a small packet from a computer to a server and back again. If the packet RTT is large, it takes a long time before there is a response back from the server. The packet RTT determines how fast TCP can establish a connection and, in some cases, also the maximum sustainable bit rate. The end-user can hence experience the network as slow even if the network offers high radio bit rates.

A sender and a receiver must continuously control the transmission rate when exchanging data. This is in order not to congest the network when multiple users share the resources, but also in order not to send data faster than the bandwidth of the narrowest link in the transmission path. To cope with this, TCP goes through the adaptation states illustrated in Figure 2.10 and described below:

- *Synchronisation.* In this phase the sender and receiver negotiate, for example, the sizes of the sender and receiver buffers.
- *Slow-start.* The sender gradually increases the transmission rate when it receives acknowledgements from the receiver.
- *Steady state.* The sender transmits new packets as soon as it has received acknowledgements from the receiver that the receiver is ready to accept more packets.

A short RTT speeds up the synchronisation and the slow-start because of the faster information exchange. During the steady state the RTT does not affect the performance

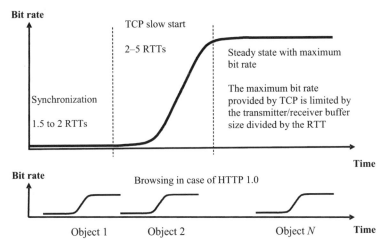

Figure 2.10   Three different states of a TCP/IP connection and the corresponding bit rate.

as long as, for example, the transmitter buffer size is large enough to keep all the unacknowledged packets 'traveling in the network' in its memory.

As noted in Figure 2.10 it takes about two RTTs before the steady state bit rate is reached. With RTTs typically less than 100 ms in fixed networks, the end-user does in general not notice any effects of the slow-start. However, in cellular radio networks where RTTs today are around 200–700 ms and even longer, the end-user experience may be affected. This is particularly true when browsing a webpage consisting of multiple objects, each requiring a separate slow-start in HTTP1.0. To cope with this it is possible to use HTTP1.1 instead, where multiple objects may be transferred within the same TCP session.

2.3.1.2   Person-to-person communication over IP

In PoC (see also Section 2.2.7), a communication session consists of a number of relatively short one-directional 'talk bursts' between the users. Because PTT is not a full duplex communication service, the requirement on the end-to-end delay is slightly less stringent than for a full-duplex communication. This means that the end-to-end delay may be longer than 400 ms, which is a common requirement for full-duplex voice. However, it is important that it is quick to push a button on the mobile and to get the corresponding confirmation back from the application that it is now possible to start talking. This requires that the network is fast to establish an initial communication link between the end-users.

For full-duplex VoIP (see Section 2.2.9), mouth-to-ear delay must be low enough to avoid users talking over each other. It is also crucial to minimise call setup times. End-user research indicates that the vast majority of users satisfied in a full-duplex voice communication requires a mouth-to-ear delay lower than 200 ms. Although many users are happy with delays up to 300 ms, communication is possible with up to 400-ms mouth-to-ear delays. When voice is carried over the PS domain the mouth-to-ear delay is approximately equal to the packet RTT plus the processing times in the mobile stations. A benchmark number for the call setup time is that it today takes around 7 s to establish a CS voice call between two mobile stations. A common requirement for person-to-person services such as VoIP and PTT is that the mobile station must be 'always-on' – that is, always connected to the network – and with a PDP context established with the corresponding IP address. This makes it possible to reach users when making a call. With always-on applications it is practical for terminals and networks to support multiple parallel PS connections. This sets functionality and capacity requirements on CN, radio network and mobile stations.

Another use case scenario in person-to-person communication that sets requirements on the networks is VS (see Section 2.2.8). In VS it is assumed that two users initially establish a CS voice call and during the voice call one of the users would then like to show something to the other user. This is done by setting up a one-way streaming connection between the users. For this to be feasible the network and mobile station must support concurrent CS and PS connections.

2.3.1.3 Stable and variable application bit rate

Some services, like audio/video streaming and PTT, need a constant minimum bit rate without interruptions throughout the service delivery. If the constant minimum bit rate is not delivered, the user experiences irritating breaks or distortions. The stable bit rate is achieved by keeping the radio connection active, by providing minimal mobility interruption times and by controlling network resources with QoS procedures.

2.3.1.4 Key performance indicators and network requirements

Table 2.2 provides some end-user KPIs and network requirements. Application performance requirements depend on user expectations and vary from market to market.

2.3.2 WCDMA and service application performance

This section provides performance figures for some selected service applications in low-loaded WCDMA networks.

2.3.2.1 Applications in commercial networks

Commercial WCDMA networks and phones provide stable bit rates of up to 384 kb/s in the downlink and in the range of 64 to 128 kb/s in the uplink direction. With seamless mobility and stable packet RTTs of below 200 ms the WCDMA bit-pipe already provides good quality for services such as browsing, streaming, laptop connectivity, PTT and VoIP.

With phones and networks supporting concurrent CS and PS connections it is also possible to provide services like end-to-end VS during voice calls, which require a CS connection for the voice and a PS connection for the end-to-end video stream between the users.

Table 2.2 Summary of services, end-user KPIs and network requirements/aspects.

Application	KPIs	Requirements (user/network)
Mobile station browsing	Click-to-content	Click-to-content time preferably 4 s and maximally 10 s. High bit rate, short initial connection setup time and packet RTT.
Audio and video streaming	Click-to-content, number of breaks during the service delivery, picture/audio quality	Bit rates in the range of 64 to 128 kb/s enable video streaming to a mobile phone with today's displays and 3GPP codecs. Proprietary codecs may provide better quality at lower speeds. Less than 3 to 5-s breaks in the connection during the service delivery. Network requirements on stable bit rate and small bit rate variations.
PTT (PoC)	Start-to-talk time, voice through delay, speech RTT, voice quality	Stable minimum bit rate of around 8 kb/s, start-to-talk times <1 to 2 s, speech RTT delay <4 s. Network requirements on initial and subsequent bearer setup times, mobility procedures and minimum bit rates, always-on PDP contexts.
VoIP	Mouth-to-ear delay, mean opinion score for the voice quality, call setup time	Mouth-to-ear delays of less than 200 to 300 ms set requirements on packet RTTs to be around 150 to 250 ms. Bit rates on the order of 16 to 64 kb/s depending on compression and codecs. Call setup time comparable with CS domain of around 7 s, always on PDP contexts.
Web browsing	Click-to-content	High bit rates (uplink and downlink), indoor coverage, and short packet RTTs. Downlink bit rates around 200 to 400 kb/s and packet RTTs <200 to 300 ms.
Gaming	Response times and bit rates	Strategy games require packet RTTs around 500 ms, while action-based games require on the order of 70 to 200 ms depending on user experience.

2.3.2.2 Click-to-content times in today's networks

To provide low click-to-content times for the end-user it is necessary to provide both high bit rates and low setup times. Depending on network settings and use cases, different setup procedures are needed before the maximum bit rate is delivered to the application/user. Figure 2.11 illustrates the approximate time it takes to download a webpage depending on the state at which the mobile station starts. The following different

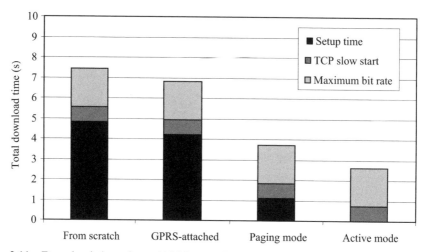

Figure 2.11 Download times for a 100-kB page in WCDMA networks starting from different always-on alternatives.

cases are considered in Figure 2.11:

- *Starting from scratch.* It is assumed that the mobile station starts from scratch in the PS domain and it hence performs both GPRS-attach and PDP context activation before the download starts.
- *GPRS-attached.* In this case the mobile station has performed GPRS-attach at, for example, power-on.
- *Paging mode.* In technical terms this means that the mobile station is kept in the radio resource control state denoted Cell_PCH or URA_PCH (see Chapter 4). The mobile station has performed GPRS-attach and PDP context activation beforehand. In addition, the radio network is aware of the mobile station location, but the mobile station is kept in 'paging' mode to save radio resources and battery power.
- *Active mode.* In technical terms this means that the mobile station is kept in the radio resource control state denoted Cell_DCH. The mobile station has performed GPRS-attach and PDP context activation beforehand. In addition, the radio network keeps a dedicated channel active for the mobile station. This is typically the state that the mobile station is in when the user browses multiple pages in a row.

As can be seen in Figure 2.11 the time to download a page varies significantly depending on the state at which the mobile station starts. The figure clearly illustrates that, in order to minimise the click-to-content time, it is beneficial to have an end-to-end view on the download process and not only to consider the maximum bit rates offered by the radio interface.

In the first two options, denoted 'from scratch' and 'GPRS-attached', the mobile station does not have any radio connection when the user presses the key to download the page. In the third option, denoted 'paging mode', the mobile station has a radio connection that is kept in paging mode when the user presses the key. All these first three options are very resource-efficient, because radio resources are only allocated once the user presses the key.

In the last option, denoted 'active mode', it is assumed that the mobile station has a dedicated radio connection allocated when the download starts. This makes the download time very short. Though, in case the mobile station is kept in 'active mode' too long, radio resources are then consumed and the battery power is drained.

2.3.2.3 Improved bit rates with HSDPA/HSUPA

Most services benefit from the higher uplink and downlink bit rates that are provided by HSDPA and HSUPA technologies. With HSDPA, packet RTTs will improve to below 100 ms while with HSUPA they may be as low as 50 ms.

Improved packet RTTs do not as such enable new services, except maybe for action-based gaming. However, because of the properties of TCP, it makes sense to improve the bit rates and reduce the RTTs in order for the end-user to perceive the whole end-to-end system as faster. For example, synchronising email clients may involve exchange of multiple small messages, and for such applications the lower RTTs make a difference.

The higher uplink speed provided by HSUPA is beneficial, particularly for laptop use cases where emailing with attachments requires high uplink bit rates. From an operator perspective HSDPA and HSUPA provide not only higher maximum bit rates, but also higher average bit rates for all data users in the cell because of the increased cell capacity that these techniques bring. With the introduction of fast over-the-air retransmissions in HSDPA and HSUPA, these techniques provide both low delays for the application and high spectral efficiency, thus lowering the cost per bit (see Chapters 4 and 5 for details).

2.3.2.4 End-to-end performance analysis of PTT in WCDMA

This section provides an example of end-to-end service performance analysis in WCDMA for PTT (PoC) (see [37] for further details). We consider two possible alternatives for the start-to-talk procedure: the 'RTP' solution that uses RTP messages to initiate the call; and the 'SIP REFER' that uses SIP and RTCP messages to initiate the call. We present performance estimates for both solutions.

From the performance point of view the PoC service may be divided into two phases: the initial service log-on and the performance when using the service. At PoC service log-on (SIP registration) the terminal first establishes an interactive PDP context for SIP signalling. During PDP context establishment the terminal is allocated a terminal IP address [38]. This PoC service log-on delay is not such a relevant indicator for the end-user because this procedure can be done long before the PoC service is actually initiated. The service log-on delay is therefore not covered in the performance analysis in the sequel. When using the PoC service the user is logged on and establishes connections to one or many users by pushing the PTT button. The three most important performance measures while using the PoC service are:

- *Start to talk time*: the delay from the time the user presses the PTT key to the time the handset indicates permission to talk (handset beep and graphic indication).

- *Voice through delay*: the delay measured from the time the originating user starts to speak to the time the receiving user hears the speech from the loudspeaker.
- *Speech RTT*: the delay from the time the originating user stops speaking and releases the button to the time the originating user starts hearing speech from User B. If the speech RTT is too high, end-users will hesitate and error cases will occur. If the start-to-talk delay is too long, users may release the key too early and therefore lose the talk permission.

A significant portion of PoC delays in WCDMA comes from changing from one radio resource control (RRC) state to another (e.g., from Cell_PCH to Cell_DCH). These transitions are managed by the radio network controller (RNC). In this study we have assumed the following delays:

- Transition from Cell_PCH state to Cell_FACH state is $\sim 350\,$ms.
- Transition from Cell_FACH state to Cell_DCH state is $\sim 900\,$ms.

Note that these figures are just estimates and may vary significantly depending on the product release or vendor.

In order to limit the number of scenarios, it is assumed that when starting a conversation (User A presses the PTT key), the UE will be in Cell_PCH state. Then, depending on the amount of data to be sent, the UE may be directed to Cell_FACH or Cell_DCH. The transition from Cell_FACH to Cell_DCH occurs if the amount of data to transmit exceeds 128 bytes (typical default threshold value). Note also that this threshold may be operator-configurable. When evaluating PoC performance we consider current DCH setup delays, as stated earlier. The additional assumptions are as follows: paging delay is 640 ms (average value in WCDMA). The SIP signalling message (used in SIP REFER) triggers the dedicated channel (DCH), but an RTP message (used in RTP) does not. This assumption is based on typical default values in current products, as explained above. Such values can be changed and, therefore, performances would also vary accordingly. The performance estimate summary is presented in Table 2.3.

As can be seen, the first speech RTT is larger in the RTP solution, but the first start to talk is shorter. This is due to the fact that the DCH for the originating mobile in the RTP

Table 2.3 PoC delays in WCDMA (one interactive PDP context).

Performance metrics	SIP REFER (ms)	RTP (ms)
Start to Talk I	1711	694
Start to Talk II	280	70
Voice Through Delay I	1755	2487
Voice Through Delay II	652	652
Speech RTT I	3987	4509
Speech RTT II	2644	2484

case is set up with the first voice packet, whereas in the SIP REFER case the DCH is set up during the start-to-talk procedure. A detailed delay analysis is provided in [37].

PoC terminals that are logged on to a PoC server should avoid going to idle state as otherwise start-to-talk delays may be unacceptable (up to 10 s). In current WCDMA networks, the UE may never switch to idle state (i.e., inactivity time when going from Cell_PCH to idle state can be set to infinity). The system will then keep the UE in Cell_PCH as long as there are available resources. If the capacity limit is reached, then the UE that has been longest in Cell_PCH state will be dropped. One simple way to ensure that terminals logged on to a PoC server never switch to idle state is to send them periodically a dummy packet. This dummy packet could be sent by the PoC server or eventually by the mobile network if it is aware that a certain PDP context is used for PoC only. The usefulness of this solution based on the UE probability of switching to idle state versus inactivity time should however be checked.

Finally, it can be mentioned that PoC performance figures in EGPRS are quite close to those presented above, although WCDMA typically outperforms EGPRS with respect to active (or consecutive) voice through delay and speech RTT. This is because of the shorter network latency in WCDMA.

2.3.3 EDGE and service application performance

This section provides service performance estimates in low-loaded EGPRS networks and illustrates the performance effects of various network enhancements.

2.3.3.1 Commercial EDGE networks

Today's EDGE networks and mobile stations provide bit rates up to and above 200 kb/s in the downlink direction and lower in the uplink direction. The exact speed depends on mobile station and network capabilities.

The shared and flexible channel approach in the EDGE radio interface provides slightly faster setup times than those usually obtained when using DCHs in WCDMA. These fast setup times result in low click-to-content times for browsing. The low click-to-content times and the bit rates provided by EDGE enable services like browsing, streaming, PTT and laptop connectivity.

2.3.3.2 Enhanced end-user experience with low RTTs

EDGE networks and mobiles are now being complemented with the 3GPP R4 feature 'extended uplink TBF mode' (see Section 3.2.2.8). This feature:

- halves the average packet RTTs from around 600 to 700 ms before the feature, down to and even below 300–500 ms;
- decreases the variation in packet RTT;
- provides stable uplink streaming connections.

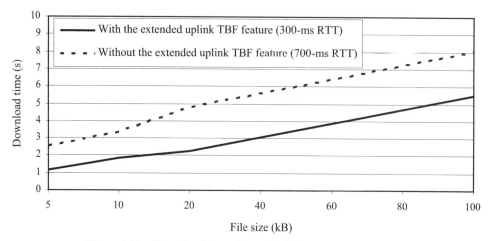

Figure 2.12   Download time for a 200-kb/s EDGE connection.

Halving and minimising the variance in the packet RTT for EDGE provide significant gains in end-user perceived throughput for TCP-based applications. As an example, Figure 2.12 shows the download time as a function of the file size for two different RTTs.

The stable uplink streaming connection that the extended uplink TBF mode feature provides is particularly beneficial for services like PTT in order to always deliver the bit rate required for the speech frames without interruptions. It is worth noting that the extended uplink TBF mode requires both network and mobile station support. The effect of this feature is hence gradually introduced when users upgrade their mobile stations.

2.3.3.3   Towards seamless mobility

Most of today's commercial EDGE networks and mobile stations result in non-negligible breaks in the radio connection during mobility procedures. The duration of these breaks depends on whether the mobile changes routing and location areas during the cell change, or whether the mobile is only performing an ordinary cell reselection.

Without routing/location area updates, in today's commercial networks and legacy mobiles the cell reselection break is around 2–3 s. In case the mobile station also changes routing area when changing the cell, the cell reselection time may be a couple or several seconds longer.

To hide the cell reselection times for end-users, most applications that require seamless mobility in EDGE, like audio/video streaming, have large enough buffers to cover for the break during the mobility procedures. In addition, advanced features in the CN, like 'large routing area support', enable flexible design to minimise the number of cell reselections that include routing area updates. For applications running on top of TCP, the mobility breaks obviously degrade the end-user experience, particularly because TCP may extend the breaks due to retransmission procedures. But, because applications using TCP are usually of non-real time nature, the mobility breaks are anyway less problematic for TCP-based services than for streaming services.

With network-assisted cell change (NACC), which is introduced in 3GPP R4, the mobility breaks decrease from today's 2–3 s down to around half a second for ordinary cell reselections. The NACC feature hence delivers seamless or close to seamless mobility. This improves end-user perceived quality in audio and video streaming because it reduces the number of breaks in service delivery. It also significantly improves the speech quality of PTT for users that are moving at medium to high speeds in the network. The NACC feature requires both mobile station and network support. The effects of this feature will hence be seen gradually when both networks and mobiles fully support 3GPP R4.

2.3.3.4   Enabling concurrent PS and CS services

From the start EDGE did not support services requiring concurrent CS and PS connections. However, with the so-called 'dual-transfer mode' (DTM) feature, concurrent CS and PS connections are possible in EDGE as well. When both networks and mobiles support this feature in the future, use cases like browsing during an ordinary phone call are enabled. The DTM feature simplifies matters for operators and end-users, particularly in environments with both WCDMA and EDGE coverage. This is because both WCDMA and EDGE can then host services that require parallel CS and PS connections.

2.3.4   Multiradio environments and application performance

This section provides some insight on service performance in the multiradio environment. More information on the topic is given in Chapter 5.

2.3.4.1   Intersystem PS handovers

With currently available and forthcoming performance improvements in EDGE, the problem of some packet data applications working only in WCDMA decreases. However, there are and will still be differences between WCDMA and EDGE in terms of service capabilities, network capacity and coverage. It is therefore still important to be able to direct services to different networks.

While the CS domain provides seamless handover for voice connections between WCDMA and GSM, the corresponding PS domain does not. Features available today to decrease the intersystem handover times for PS connections are the support of the Gs interface between the mobile switching centre (MSC) and the serving GPRS support node (SGSN) in the CN, and the packet system information (SI) status that allows the mobile to acquire SI faster.

In commercial networks, a connection break around 6–7 s is achievable when moving from EDGE to WCDMA and *vice versa*. For TCP-based services a connection break is not critical as long as it does not happen too often (ping-pong effects). On the other hand, for real time services like audio and video streaming it is crucial to minimise the delay when changing networks. Otherwise, there will be service breaks when the mobile station buffer is shorter than the intersystem connection break.

See Chapter 5 for further details on intersystem PS handover performance.

2.3.4.2   Bit rate adaptation

In multiradio scenarios, bit rate adaptation of the streamed content is necessary to avoid always streaming audio/video at constant bit rates constrained by the 'worst case' scenario/system. Bit rate adaptation is also beneficial in being able to deliver maximum quality in lightly loaded networks and sufficient quality at high network load.

At first glance, bit rate adaptation seems easy. Though, note that bit rate adaptation is dependent on many network elements and that network vendors have different radio resource management solutions. It then becomes clear that it is not trivial to design a stable adaptation algorithm that always streams at the maximum possible bit rate.

In most bit rate adaptation solutions it is the streaming server that adapts the content rates based on feedback information from the mobile station. To be robust, the bit rate adaptation algorithm must have reliable feedback information and the network must provide a reasonably stable connection. To aid the bit rate adaptation algorithms there is in 3GPP R5 an optional buffer model defined for mobile clients. This buffer model enhances the performance of bit rate adaptation because the server then knows how the mobile station reacts to bit rate changes. In 3GPP R6 there is further improved support with explicit feedback from the mobile station to the server about the exact buffer level in the mobile station. With this addition, bit rate adaptation should also be efficient in the challenging wireless environment.

The radio access networks and CNs can enhance the stability of the bit rate adaptation solution by providing a low-delay connection with minimal buffering and stable bit rates.

2.3.5   *Transport Protocol performance in wireless*

The IETF provides informational and best current practice requests for comments (RFCs) that describe the effects of various link characteristics on TCP performance [39], [40] and their implications for optimal TCP parameter settings and options usage [41]. Moreover, the IETF continues to develop TCP, and a number of standard extensions that aim at improving performance and robustness are added on a regular basis [42]–[45], while (at the time of writing) other features are still under development [46], [47]. Nowadays, many popular operating systems (e.g., Linux, FreeBSD, Microsoft) support the most advanced TCP versions (e.g., TCP NewReno) that are optimised for various environments including wireless. Some are also able to dynamically tune their parameters (e.g., sender and receiver buffer size) depending on the detected network type and properties, in order to achieve best end-user performance as well as optimal memory usage. Likewise, TCP implementation and configuration in mobile device operating systems (e.g., Symbian) are usually largely optimised for wireless environment. These are encouraging facts and it appears that TCP performance over wireless is, generally, not an issue anymore, provided that both TCP clients and servers implement state-of-the-art recommendations. There are of course few remaining challenges – such as radio link outage time during' for example, cell reselection in EGPRS or intersystem change (say, from EGPRS to WCDMA) – but luckily these are not so common. Moreover, as presented in Section 2.3.3.3, outage times during mobility events will gradually decrease as related network procedure performance improves (e.g., NACC or PS handover). These improvements are of course very beneficial from the TCP viewpoint.

The IETF has more recently standardised the TCP-friendly rate control (TFRC) scheme [48]. The standard explains that:

TFRC is a congestion control mechanism for unicast flows operating in a best-effort Internet environment. It is reasonably fair when competing for bandwidth with TCP flows, but has a much lower variation of throughput over time compared with TCP, making it more suitable for applications such as telephony or streaming media where a relatively smooth sending rate is of importance.

So, TFRC may be used as a congestion control mechanism for applications that traditionally run over RTP/UDP. The basic idea is to enable non-TCP based applications to also behave in a 'friendly' manner from the network congestion and fairness viewpoint. It is indeed a very useful feature as the share of non-TCP traffic in the Internet seems to increase regularly with the passage of time. However, it should be noted that in order to enable good end-user experience, some optimisations or adaptations for wireless networks are likely to be needed (as pointed out, e.g., in [49]).

References

[1] T. Halonen, J. Romero and J. Melero (eds), *GSM, GPRS and EDGE Performance*, John Wiley & Sons, 2nd edition, 2003, 615 pp.
[2] 3GPP, R6, TS 22.101, Service Aspects; Service Principles, v. 6.10.0.
[3] 3GPP, R5, TS 22.003, Circuit Teleservice Supported by a Public Land Mobile Network (PLMN), v. 5.2.0.
[4] ETSI, R98, TR 101 714, Performance Characterization of the GSM Adaptive Multi-Rate (AMR) Speech Codec, v. 7.2.0.
[5] 3GPP, R4, TR 23.972, Circuit Switched Multimedia Telephony, v. 4.0.0.
[6] H. Holma and A. Toskala (eds), *WCDMA for UMTS*, John Wiley & Sons, 3rd edition, 2004, 450 pp.
[7] 3GPP, R5, TS 23.228, IP Multimedia Subsystem (IMS), v. 5.7.0.
[8] M. Poikselkä, G. Mayer, H. Khartabil and A. Niemi, *The IMS: IP Multimedia Concepts and Services in the Mobile Domain*, John Wiley & Sons, 1st edition, 2004.
[9] IETF RFC 3261, SIP: Session Initiation Protocol, 2002.
[10] IETF RFC 2616, Hypertext Transfer Protocol – HTTP/1.1, 1999.
[11] O. Verscheure, X. Garcia, G. Karlsson and J. P. Hubaux, User-oriented QoS in packet video delivery, *IEEE Network Magazine*, 1998.
[12] WAP Forum, Wireless Session Protocol Specification.
[13] Open Mobile Alliance, Enabler Release Definition for MMS, candidate version 1.2.
[14] Open Mobile Alliance, MMS Architecture Overview, candidate version 1.2.
[15] Open Mobile Alliance, MMS Client Transactions, candidate version 1.2.
[16] Open Mobile Alliance, MMS Encapsulation Protocol, candidate version 1.2.
[17] Open Mobile Alliance, Generic Content Download over the Air Specification.
[18] 3GPP, R6, TS 23.233, Transparent End-to-end Packet-switched Streaming Service, Stage 1, v. 6.3.0.
[19] 3GPP, R5, TS 26.233, Transparent End-to-end Packet-switched Streaming Service, General Description, v. 5.0.0.

[20] 3GPP, R5, TS 26.234, Transparent End-to-end Packet-switched Streaming Service, Protocol and Codecs, v. 5.0.0.

[21] IETF, RFC 2326, Real Time Streaming Protocol (RTSP), 1998.

[22] IETF, RFC 2327, SDP: Session Description Protocol, 1998.

[23] Nokia, *Multiplayer Game Performance Over Cellular Networks*, White Paper, 2004, v. 1.0.

[24] Open Mobile Alliance, Gaming Architecture Overview, August 2004, draft version 0.11.

[25] Open Mobile Alliance, Gaming Platform Version 2.0 Server Framework, 2004, draft version 0.0.5.

[26] Open Mobile Alliance, OMA GP2.0 Client/Server Protocol Specification, 2004, draft version 1.0.

[27] IETF, RFC 2401, Security Architecture for the Internet Protocol, 1998.

[28] IETF, RFC 2409, The Internet Key Exchange (IKE), 1998.

[29] IETF, RFC 2406, IP Encapsulating Security Payload, 1998.

[30] IETF, RFC 2402, IP Authentication Header, 1998.

[31] IETF, RFC 2246, The TLS Protocol Version 1.0, 1999.

[32] Open Mobile Alliance, Push to Talk over Cellular (PoC) – Architecture, 2004, v 1.0.

[33] IETF, RFC 3311, The Session Initiation Protocol (SIP) UPDATE Method, 2002.

[34] IETF, RFC 3262, Reliability of Provisional Responses in the Session Initiation Protocol (SIP), 2002.

[35] R. Cuny and A. Lakaniemi, VoIP in 3G Networks: An End-to-end Quality of Service Analysis, *VTC 2003, Spring, April 2003*, Vol. 2, pp. 930–934.

[36] 3GPP, R6, TS 22.141, Presence Service, v. 6.4.1

[37] R. Cuny, End-to-end performance analysis of Push to Talk over Cellular (PoC) in WCDMA, *IASTED, Communication Systems and Networks, September 1–3, 2004*, pp. 1–5.

[38] Ericsson, Motorola, Nokia and Siemens, Push-to-Talk over Cellular (PoC) User Plane; (E)GPRS/UMTS Specification, Release 1.0, 2003.

[39] IETF, RFC 3150, End-to-end Performance Implications of Slow Links, 2001.

[40] IETF, RFC 3155, End-to-end Performance Implications of Links with Errors, 2001.

[41] IETF, RFC 3481, TCP over Second (2.5G) and Third (3G) Generation Wireless Networks, 2003.

[42] IETF, RFC 2414, Increasing TCP's Initial Window, 1998.

[43] IETF, RFC 3168, The Addition of Explicit Congestion Notification (ECN) to IP, 2001.

[44] IETF, RFC 2988, Computing TCP's Retransmission Timer, 2000.

[45] IETF, RFC 4015, The Eifel Response Algorithm for TCP, 2005.

[46] Y. Swami *et al.*, Decorrelated Loss Recovery (DCLOR) Using SACK Option for Spurious Timeouts, Internet draft, 2005.

[47] Y. Swami *et al.*, Lightweight Mobility Detection and Response (LMDR) Algorithm for TCP, Internet draft, 2005.

[48] IETF, RFC 3448, TCP Friendly Rate Control (TFRC): Protocol Specification, 2003.

[49] D. Beaufort, L. Fay, C. Samson and A. Teil, Measured performance of TCP Friendly Rate Control Protocol over 2.5G network, *IEEE Vehicular Technology Conference Proceedings, September 2002*, Vol. 1, pp. 563–567.

3

QoS in 3GPP Releases 97/98, 99, 5, 6 and 7

Anna Sillanpää and David Soldani

The content of this chapter is prevalently based on 3rd Generation Partnership Project (3GPP) specifications and it is mainly aimed at giving an overview of the Universal Mobile Telecommunication System (UMTS) quality of service (QoS) concept and architecture. The roles and importance of protocols used to control the execution of network functions in a co-ordinated manner across system interfaces are described in Chapter 4. The implementation aspects of the QoS management functions in network elements needed to handle bearer services with specific QoS and the relation between the functions internal to nodes are presented in Chapters 5 and 6.

3.1   Where does QoS come from?

A new QoS paradigm emerged with the packet-based networks. In the traditional circuit-switched environment, such as the fixed-line telephone network, QoS is inherently present as the circuit is reserved throughout the telephone networks for each call. Resources are reserved when the call is established, and kept so until the call participants hang up. On the other hand, in the packet-switched (PS) networks, a fixed circuit is not reserved to enable the efficient use of resources. The efficiency gain results from the flexibility in allocating resources in the PS domain, and from the fact that the reserved fixed-line circuits are not used to their full capacity in the circuit-switched environment. For example, in a traditional phone call between two people the two persons are not speaking and listening to each other all the time, but mostly one is speaking and the other is listening. Similarly, an interactive game application does not require the exchange of continuous bidirectional packet flow at a certain bit rate, but rather an exchange of various amounts of data with pauses in-between. Service applications are also different in tolerating packet delays. Some are delay-sensitive and require a steady flow of packets, such as traditional telephony calls; some other services can tolerate packet delays – for

QoS and QoE Management in UMTS Cellular Systems
Edited by David Soldani, Man Li and Renaud Cuny   © 2006 John Wiley & Sons, Ltd

example, file transfers. Furthermore, some service applications cannot tolerate transmission errors while others can. For example, file transfer is error-sensitive, and if all the packets cannot be correctly received, the entire file may be useless to the recipient. Lost or incorrectly received packets need to be retransmitted for such applications. Some other applications can tolerate minor transmission errors and retransmission may not even be feasible due to the time delay resulting from that. Bandwidth requirements can also vary between service applications. Real time video streaming requires a certain amount of bandwidth and file transfer benefits from larger bit rates but a voice conversation does not gain from having double the amount of resources in the networks.

While PS networks can take advantage of unused bearer periods they need to be able to correctly determine when a particular application service (e.g., a call) needs resources as well as the quality and amount of resources. The QoS concept is created to do that and it is defined in such a way that various types of users and service applications with different bearer requirements can be supported in packet-based networks.

3.1.1 Application and bearer service categorisation

From the QoS requirements' perspective the different service applications and the bearer services carrying them need to be categorised. There are several ways to do this, and the European Telecommunications Standards Institute (ETSI) defined 3GPP Release 99 (R99) categorisation is one of the most widely used if not the most popular one. According to this model, both the application and the bearer services are first divided into two main classes – real time (RT) and non-real time (NRT) – and both are further split into two additional subclasses.

NRT requirements are the easiest ones to support in PS networks. Typical applications are of background or interactive type. The background class is used when a terminal is sending and receiving data files in the background. Examples of background applications are email, SMS, download of databases and reception of measurement records. The interactive class is applicable when a machine or a human is online requesting data from remote equipment; for example, for web browsing, data base retrieval, server access, polling for measurement records and automatic data base enquiries (tele-machines).

The RT class requires a short packet delay and consists of two basic types of applications: streaming and conversational. The term 'streaming' refers to an application playing unidirectional, synchronised (if several media streams are involved) and continuous media stream(s) while those streams are being transmitted over the data network. Examples of these media streams are audio and video. Streaming applications can also be further divided into 'on-demand' and 'live' information delivery. Examples of the first category are music and news-on-demand applications; the latter type can be live delivery of radio and television programmes. These applications require steady and time-sensitive delivery of user packet flows in duplex manner – that is, both to and from the UE – so that the user can speak and listen to the other party simultaneously. The conversational class is used to support conversational applications such as telephony-like speech over the PS domain.

In the 3GPP standards the terms RT and NRT are used to describe both the application and the bearer services. However, to make the distinction between whether an application or a bearer service is referred to, in this book RT/NRT are used for the service applications, and for the bearer services the RT bearer or guaranteed bit rate (GB or GBR) and the NRT bearer or non-GB (NGB or NGBR) are used instead of just RT and NRT, respectively.

More information on service applications can be found in Chapter 2.

3.1.2 GPRS network architecture

The Special Mobile Group (SMG) group of ETSI created the General Packet Radio Service (GPRS) system. It was a part of the standardisation work for the Global System for Mobile Communications (GSM) network. The GPRS was designed to offer the packet-based service Core Network (CN) to complement the GSM circuit-switched (CS) domain. 3GPP later adopted the GPRS for its third-generation (3G) network.

Figure 3.1 depicts the basic logical GPRS architecture with the home location register (HLR), base station system (BSS)/radio access network (RAN), UE – user equipment, consisting among other things of the terminal equipment (TE), and the mobile termination (MT) – and the interfaces between nodes. The complete GPRS architecture contains additional elements and interfaces as described in [1]. These elements and interfaces optimise the GPRS architecture and enable the use of other application and network services with the GPRS; for example, the Short Message Service (SMS) and customised applications for mobile network-enhanced logic (CAMEL).

The serving GPRS support node (SGSN) and gateway GPRS support node (GGSN) are the GPRS-specific nodes in the GSM and 3GPP 3G systems. The role of the SGSN is to support subscribers within its coverage area. It comprises the subscriber authorisation with the information fetched from the HLR, as well as authentication, registration and mobility management. As far as the UE is concerned, the SGSN handles protocol conversion and signalling as well as the routing of user plane traffic toward the GGSN the user is connected to. The SGSN may also be in contact with another SGSN when the UE is changing its location from the routing area of one SGSN to an area covered by a different SGSN. The SGSN is located in the same 3GPP network – that is, the public land

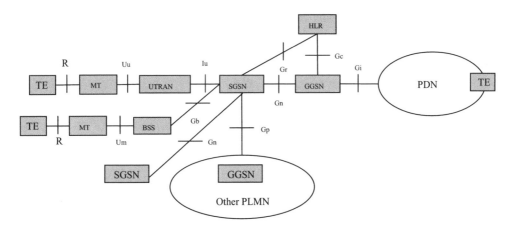

Figure 3.1 Basic GPRS architecture [1].

mobile network (PLMN) – as the UE is camped in. For example, if the subscriber has a subscription in Finland, and the subscriber is located in Finland, a SGSN of the Finnish operator is serving him. However, when the subscriber travels to another country – for example, to the UK – and uses the local network, a SGSN of the British operator is serving him.

The GGSN is the GPRS node that is connected to the external packet data network (PDN) – for example, the Internet or a virtual private network (VPN) intended for the use of a particular company. The GGSN routes the packets sent by the UE to the 'outside world' towards the other party in the communication – for example, another UE or an application server. The GGSN also routes the data destined for the UE from the external network. From the external network point of view, the GGSN is a gate to the mobile network and the UEs within. It hides GPRS network details from external networks. The GGSN may also allocate the user's dynamic Internet Protocol (IP) address to be used while the UE is connected to the GPRS network (i.e., to that GGSN). A home network GGSN often supports the subscriber regardless of his/her current location. However, depending on the subscription, a GGSN in the visited network can also be used.

As far as the radio access part is concerned, either the UMTS terrestrial radio access network (UTRAN) or the BSS is applicable. The UTRAN is the 3GPP radio network connected to the CN with the Iu interface and having the Uu radio interface. The UTRAN air interface is designed to have two modes, the frequency division duplex (FDD) and the time division duplex (TDD). The name 'wideband code division multiple access' (WCDMA) is used to cover both FDD and TDD operations. However, throughout this book, the term 'WCDMA' relates only to UTRA FDD, since TDD is not within the scope of the following chapters. The TDD is further divided into two variants: R99-defined time division–code division multiple access (TD-CDMA) and R4-standardised time division–synchronous code division multiple access (TD-SCDMA). The operator can select the air interface technology (e.g., FDD and/or TDD) to be used in his network according to each particular case.

The BSS is the GSM/EDGE radio access network (GERAN) based 3GPP radio network – that is, the GSM/enhanced data rates for global evolution (EDGE) RAN. The GERAN is based on the earlier SMG GSM air interface, which has later been enhanced to support EDGE. The enhanced GPRS (EGPRS) part of EDGE covers the PS end and it allows for more efficient modulation in the GPRS radio interface. The GERAN radio interface is denoted by Um and is based on time division multiple access (TDMA). The GERAN is connected to the PS CN with the Gb interface. In addition, from Release 5 (R5) onward, the GERAN can also be connected to the GPRS CN with the Iu interface; this particular scenario is not shown in Figure 3.1.

The UE, comprising the TE and MT, is also called the 'mobile station' (MS) in 3GPP specifications. In spoken language it is often referred to as a *mobile phone* or *terminal*. The MT contains the 3GPP protocol stack needed for signalling and user traffic transfer over the air interface to the UTRAN/GERAN and further to the CN. The TE contains the user application and user interface – for example, the keyboard and the display. The TE may be co-located with the MT in the UE, when the only UE is the mobile phone, or outside it, such as a laptop or personal computer (PC) connected to the UE. In addition, the 3GPP UE contains the universal subscriber identity module (USIM) holding key subscriber information and security keys needed to use the 3GPP networks.

3.1.3 A/Gb and Iu mode

The combined A and Gb interface architecture forms the A/Gb mode connecting the second-generation (2G) CN and BSS. The A interface is used to connect the CS CN to the BSS. Its PS end counterpart is the Gb interface. The Gb interface was originally developed as part of the 2G GSM GPRS system in R97. In R99, the GPRS domain has been enhanced, with EGPRS improving radio interface capacity.

The Iu mode and the corresponding UTRAN were first standardised in R99 as part of the first 3GPP 3G release. The Iu mode is divided into two functional parts: the Iu-ps interface supporting PS services and the Iu-cs interface supporting CS services. In R5, the Iu interface was defined to support the GERAN in addition to the UTRAN.

Standardisation has been continuing for both the A/Gb and Iu modes from one release to another. Thus, there are three variants of PS interface combinations in 3GPP standards:

- The Gb interface connecting the 2G-SGSN and the BSS based on the GPRS/EDGE radio network and the Um air interface. The Gb interface was first defined in R97.
- The Iu-ps interface connecting the 3G-SGSN and the UTRAN supporting the Uu interface. The Iu-ps interface for the UTRAN was first defined in R99.
- The Iu-ps interface connecting the 3G-SGSN and the BSS based on the GPRS/EDGE radio network and the Um air interface. The Iu-ps interface for the GERAN is defined from R5 onward. However, this option has not been used thus far in operating systems.

The main architectural differences between the Gb and Iu-ps interface systems are the location of the encryption, header compression, and cell level mobility management. In the Gb interface architecture, the 2G-SGSN carries out these functions. On the other hand, the radio network controller (RNC) is responsible for these in the Iu-ps interface architecture.

Regarding the QoS in the PS domain, both Gb and Iu-ps interfaces support NRT applications from their first releases. These applications are supported by NGB or NRT bearers. On the other hand, the biggest QoS-related difference between the Gb and Iu interface architecture is the timing of RT application support by the GB or RT bearer services. For the Gb interface, RT application support is gradual: in R97/98, streaming applications can be provided by NGB bearers. The 'simple streaming application' was standardised in R4 which also defined the codecs to be used. Streaming bearer service standardisation was mainly done in R5 with further enhancements in R6 and R7. R6 enhancements include PS handovers and R7 is likely to contain dedicated channels for the Gb interface architecture. Support for the bearer service by the Gb interface architecture will be finalised and explicitly done in R7. For the Iu-ps interface, both streaming and conversational service applications with GB were supported from R99 onward. However, some Iu-ps QoS aspects were added in R4 in support of RT applications, such as advanced QoS negotiation over the Iu interface between the RNC and SGSN, and an improved PS domain handover minimising packet delay.

The R99 term to indicate QoS categorisation is the 'UMTS traffic class' (or QoS class). This parameter has four values: two values for the GB bearer service, which are conversational and streaming, and two values for NGB bearer service, which are interactive and background. The R97/98 counterpart to the UMTS traffic class is the delay class. This parameter has four defined values, three predictive values and the 'best-effort' value being the lowest. Additional QoS parameters are used in both releases to further provide distinctive bearer services to fulfil various service application requirements – for example, regarding bit rate and reliability. More information on the topic is provided in Section 3.2.

3.1.4 QoS in transport network

The QoS handling described in this chapter mostly concentrates on the upper layers of the protocol stack, and in particular on Packet Data Protocol (PDP) context related procedures of the GPRS system. The PDP context provides a pipe between the UE and the GGSN enabling user packet transfer to and from the UE. However, the lower layers of the protocol stack also play a role in the realisation of the QoS in operating networks. These lower layers are also called the *transport network layer*. These layers and their QoS means are not fully standardised in 3GPP and, thus, many aspects are left for operators and manufacturers to decide.

The GPRS CN part of the system is always deployed over an Internet Engineering Task Force (IETF) defined IP network. Thus, the differentiated services (DiffServ) specified for the IP may be used for QoS handling (i.e., classification, metering, marking, shaping and dropping) [2]. At the same time, mapping of PDP context QoS parameter (negotiated or authorised QoS profile) values onto DiffServ code points (DSCPs), which are used for QoS handling in DiffServ, is not standardised. Additional methods such as multiprotocol label switching (MPLS) can also be used with the IP transport solution. The L2 and L1 below the IP layer are not defined in 3GPP. For example, the Ethernet and its virtual local area network (VLAN) feature can be used to prioritise packet streams differently.

The Gb interface was originally based on frame relay (FR) connections, but from R4 onward the IP-based Gb interface has been specified in 3GPP. FR may use its own QoS handling, but it may also be deployed without such QoS support. In the latter case, QoS can be provided over the FR-based Gb interface by implementing proprietary packet scheduling or other mechanisms to allow for prioritisation among different application packets. For the IP-based Gb interface, DiffServ and other IP QoS methods can be used as described for the GPRS CN.

The Iu-ps interface has been based on IP from the very start and, thus, IP QoS methods can be used.

In the radio network, the GSM BSS user packet exchange is mostly carried over time division multiplexing (TDM) based lines between the base station controller (BSC) and base stations (Abis interface). The E1 variant of the TDM is deployed in Europe and the T1 in North America. Proprietary scheduling and other QoS mechanisms can be used to provide QoS handling for these lines.

Regarding the UTRAN, the interface between the RNC and the UTRAN base stations is called 'Iub'. (The term 'Node B' is used in 3GPP specifications to indicate a logical node responsible for radio transmission/reception in one or more cells to/from the UE. The more generic term 'base station' used elsewhere in this book means exactly the same thing. In a Node B more cells can be set up; a 'cell' is defined by means of the following: a cell identification (C-ID), configuration generation ID, timing delay (T_Cell), UTRA absolute radio frequency channel number (UARFCN), maximum transmission power, closed-loop timing adjustment mode and primary scrambling code.) This interface was originally based on asynchronous transfer mode (ATM) transport, and ATM is deployed over the TDM. From R5 onward, IP transport was also possible in the Iub interface. In the IP transport solution, ATM or a packet-based transport solution such as the ethernet can be used below the IP layer. In ATM, QoS differentiation is possible – for example, through constant, variable and unspecified bit rate support. This can be used to prioritise packet transfer through ATM connections. As far as TDM and IP are concerned, QoS handling can be provided as described earlier.

More information on ATM (and MPLS) can be found at the MFA Forum's webpages – *http://www.mfaforum.org/* – and on Ethernet at the Institute of Electrical and Electronics Engineers' (IEEE) webpages – *http://www.ieee.org/portal/site*

More information on QoS in the transport network can be found in Chapters 6 and 8.

3.1.5   ETSI and 3GPP

The SMG of ETSI defined the GSM in the late 1980s and 1990s. As part of this work, the GPRS system was first defined in the R97. GPRS has been evolving ever since, and became in the late 1990s part of the 3GPP standardisation project along with other GSM specifications. 3GPP was founded by standardisation organisations from Europe (ETSI), Japan (Association of Radio Industries and Businesses, ARIB; and Telecommunication Technology Committee, TTC), South Korea (Telecommunications Technology Association, TTA), the USA (Alliance for Telecommunications Industry Solutions, ATIS) and China (China Communications Standards Association, CCSA). Individual companies can be members of 3GPP through these regional organisation partners, and most members are telephone operators and equipment manufacturers. The aim of 3GPP was to specify a 3G mobile system. The new mobile system was to comprise WCDMA FDD and TDD radio accesses and an evolved GSM CN inherited from the ETSI SMG group. The 3GPP technical specification group (TSG) organisation shown in Figure 3.2 still defines 3GPP specifications. The organisation has changed over the years and the last changes took place in spring 2005 when the TSGs responsible for the terminals (TSG-T) and CN (TSG-CN) were combined to form the Core Network and Terminals TSG (TSG-CT). Each group has a chairman and possibly one or more vice chairmen responsible for the group's work. Specifications are created in the working groups (WGs) which sometimes work together by holding joint meetings or by sending so-called 'liaison statements' (LSs) to each other. LSs can be used to inform other groups about decisions, put questions to other groups and so forth.

The most important WGs from the QoS perspective are:

- SA WG1 – this WG is responsible for defining user requirements, mainly from the operators' view point.
- SA WG2 – this WG defines the general 3GPP architecture as well as application and bearer service and system functionality. This is the most important group in defining 3GPP QoS functionality.

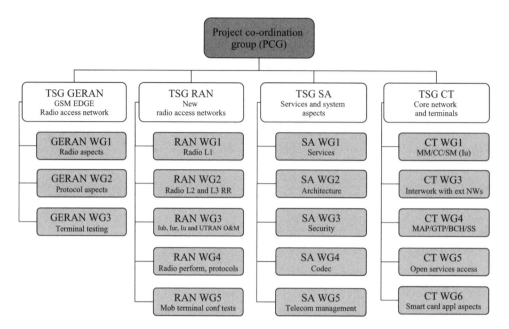

Figure 3.2 3GPP TSG organisation responsible for creating the technical specifications.

- CT groups, in particular CT WG1 and CT WG3. These groups define the details of GPRS and related QoS mechanisms of the CN and terminals (UEs).
- Various GERAN and RAN groups, mainly GERAN WG1 and WG2 and RAN WG1, WG2, WG3 and WG4. These groups define the air interface and RAN mechanisms for GERAN and UTRAN, respectively, including QoS handling.

The 3GPP standardisation work is contribution-driven and consensus-based. This means that in order to be approved a technical solution needs to be presented to the appropriate WG in the form of a contribution. Telephone manufacturers as well as the operators create and present contributions to the WG meetings that are held several times a year. If the contribution is accepted, the solution becomes part of the specification. Often many contributions are needed before a specification is completed and ready to be approved. A contribution can be accepted through consensus in the working group, and sometimes several meetings are needed to reach it. For very controversial issues an official vote can be used. Once a specification is ready, the WG submits it to its TSG for approval. At this point, the specification needs to be 80% complete. After the approval, the specification is frozen. It then enters the maintenance phase during which only so-called 'essential corrections' are allowed. These are changes that are seen as necessary from the functionality perspective. The meeting participants can propose essential corrections to be amended by using the change request (CR) procedure.

The GSM (GERAN) and 3G (UTRAN) specifications have a 3GPP specification number of four or five digits – for example, 09.02 or 29.002. The first two digits define the

Table 3.1   3GPP specification series numbering.

Subject of specification series	3G/GSM R99 and later	GSM-only (R4 and later)	GSM-only (before R4)
General information (long defunct)			00 series
Requirements	21 series	41 series	01 series
Service aspects (Stage 1)	22 series	42 series	02 series
Technical realisation (Stage 2)	23 series	43 series	03 series
Signalling protocol (Stage 3) – user equipment to network	24 series	44 series	04 series
Radio aspects	25 series	45 series	05 series
CODECs	26 series	46 series	06 series
Data	27 series	47 series (none exists)	07 series
Signalling protocols (Stage 3) – RSS – CN	28 series	48 series	08 series
Signalling protocols (Stage 3) – infra – fixed network	29 series	49 series	09 series
Programme management	30 series	50 series	10 series
Subscriber identity module (SIM/USIM), IC cards, test specs	31 series	51 series	11 series
OAM&P and charging	32 series	52 series	12 series
Access requirements and test specifications		13 series (1)	13 series (1)
Security aspects	33 series	(2)	(2)
UE and (U)SIM test specifications	34 series	(2)	*11 series*
Security algorithms (3)	35 series	55 series	(4)

[1] The 13 series GSM specifications relate to EU-specific regulatory standards. On the closure of the ETSI TC SMG, responsibility for these specifications was transferred to the ETSI TC MSG (Mobile Specification Group) and they do not appear on the 3GPP file server.
[2] The specifications of these aspects are spread throughout several series.
[3] Algorithms may be subject to export licensing conditions. See the relevant 3GPP and ETSI pages.
[4] The original GSM algorithms are not published and are controlled by the GSM Association.

series (see Table 3.1). These two digits are followed by either two more digits (01 to 13 series) or by three digits (21 to 55 series) identifying the various specifications in the series.

3GPP originally decided to prepare specifications on a yearly basis, and the first set of such specifications was called the R99 – as it was approved in 1999. The intention was to complete the next release in 2000. However, Release 2000 specifications were split into two consecutive releases: Release 4 (R4) and Release 5 (R5). Most of R4 was completed in 2001 but some specifications were completed as late as in September 2002. R5 was frozen in March 2002. R5 was followed by Release 6 (R6), which was officially completed in early 2005. R7 work was started in 2005 and will be approved in 2007 at the latest.

For each release, once the specification is approved, the specification version number indicates the release the specification belongs to. R97, R98 and R99 GSM-only specifications take four-digit numbers (e.g., 03.60) and have a version number of 6.x.x (R97), 7.x.x (R98) and 8.x.x (R99). For example, TS 03.60 version 6.11.0 is an R97 specification. On the other hand, the newer 3GPP specifications take a five-digit specification number and a version number, so that the 3GPP TS corresponding to GSM TS 03.60 is TS 23.060 and the version number for R99 is 3.x.x, for R4 4.x.x., for R5 5.x.x and for R6 6.x.x. The R7 version number will to bc 7.x.x and take a five-digit specification number. All specifications are not applicable for all releases, as new specifications may be created and old specifications discontinued for a particular release.

Generally, in the standardisation process a technical report (TR) may precede a technical specification (TS). A TR is created as a feasibility study to assess possible solutions. However, technical specifications form the core of a release and define the standardised solutions on which implementations are to be based. Technical specifications specify the technical aspects – in particular, the messages sent between the system nodes and the functionality needed in these nodes. The TS process contains three stages: 1, 2 and 3. Stage 1 defines high-level user requirements – in other words, the service aspects from the user's perspective. Stage 2 concentrates on the architecture, key functionality and overall message exchange. By using the standardisation language this is called the 'functional entity behaviour and information exchange description'. Stage 3 contains the protocol (i.e., the detailed message and procedure descriptions). International Telecommunication Union (ITU-T) Recommendation I.130 [3] describes the three-stage method, and ITU-T Recommendation Q.65 [4] defines Stage 2 of the method. All three stages are not always needed, and in many cases only Stage 2 and Stage 3 – or even only Stage 3 – may be sufficient. It is also possible that only amendments to one or more existing TSs are made and, thus, added by using the CR process. On the other hand, if a major new functionality is defined, new TSs are often required.

3GPP specifications use a special language. When something is mandatory the word *shall* is used and when something is recommended *should* is utilised. The fact that an action is optional is indicated by using *may*, and *can* is only used to describe a possible situation or action. Similarly, negative forms can be used for these verbs. In some cases the specifications do not follow these language rules very strictly, and the most common situation is that the normal present term is used to indicate mandatory (*shall*) behaviour. This may be due to the inexperience of the contributor or to the fact that more detailed specifications need to be created defining all the detailed aspects.

More information on 3GPP can be found under *www.3gpp.org*

3.1.6 *Internet Engineering Task Force (IETF)*

The organisation responsible for creating Internet standards is depicted in Figure 3.3. The IETF assumes the task of developing and evolving the Internet and its architecture. It also intends to ensure the Internet's smooth and secure operation. Unlike 3GPP, IETF members are individuals (and not companies) such as network designers, academics, engineers and researchers even if they are often from many companies and organisations creating IETF-compatible products.

The organizational home of the IETF is the Internet Society (ISOC), which has the Internet Architecture Board (IAB) under it. The IAB provides architectural guidance for Internet standardisation and comprises two subgroups: the IETF and the Internet Research Task Force (IRTF). The IETF carries out short-term or medium-term Internet research and standardisation work. The Internet Engineering Steering Group (IESG)

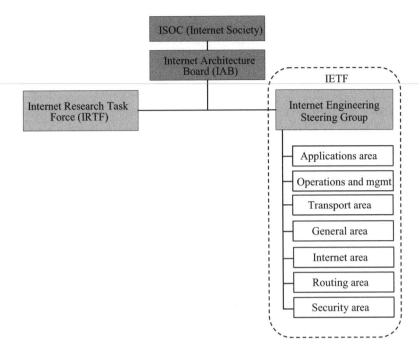

Figure 3.3   The IETF organisation.

ensures that solutions are sound enough from the security perspective and follow Internet methodologies. It is divided into areas that are managed by area directors. The area directors are members of the IESG. Each area is responsible for a specific topic and has several working groups under it further concentrating on smaller topics. The current areas are: applications, general, Internet, operations and management, routing, security, and transport. The WGs produce Internet drafts that after careful processing become numbered requests for comments (RFC).

The IRTF is headed by the Internet Research Steering Group (IRSG), which has research groups under it. The IRTF concentrates on long-term research activities, but is in practice less active than the IETF. Another important IETF standardisation-related group is the Internet Assigned Numbers Authority (IANA). It allocates unique number parameter names and values upon request needed for development and utilisation of the RFCs. The IANA is not shown in Figure 3.3.

3GPP and IETF work together. In particular, 3GPP adopts IETF-defined protocols, when one is available and suitable for a technical solution being standardised in 3GPP. The main reason for the reuse of existing solutions and protocols is to optimise 3GPP standardisation and implementation efforts. It is possible to define amendments within the IETF to satisfy 3GPP requirements. In some cases, 3GPP needs to create its own amendments to IETF protocols. These are used on top of or in conjunction with IETF-defined solutions. 3GPP has adopted many IETF-standardised protocols (e.g., IP, SIP, SDP, RTP, COPS and Diameter). Session Initiation Protocol (SIP), Session Description Protocol (SDP) and Real time Transport Protocol (RTP) are used for IP multimedia services (IMS) but can also be used for other applications. Common Open Policy Service (COPS) and Diameter are part of 3GPP R5 and R6 QoS solutions and will be further discussed in Section 3.2.

More information on the IETF is available at *www.ietf.org*

3.1.7 GSM Association (GSMA)

The GSM Association (GSMA) or the GSM Memorandum of Understanding (MoU) Association is an organisation set up specially for GSM family operators. The GSM family consists of the GSM, GPRS, EDGE and 3GSM (UMTS). The organisation was established in 1987 and is a global trade association that promotes, protects and enhances the interests of GSM mobile operators throughout the world. Its membership consists of more than 680 GSM family operators, and 160 manufacturers and suppliers (as of September 2005).

Among other things, the GSMA creates documents which specify technical aspects related to the GSM/3GPP systems. One example is the GSMA IR.34 Inter-PLMN Backbone Guidelines [5], describing the GPRS roaming network used to interconnect different operators' GPRS networks and their nodes.

More information on the GSMA is available at *http://www.gsmworld.com/index.shtml*

3.1.8 ITU-WARC and spectrum allocation

UMTS is one of the available 3G mobile technologies, which represents the European and Japanese answer to ITU international mobile telephony (IMT-2000) requirements for 3G mobile cellular radio systems.

CDMA-2000 1x-EVDO is an alternative to UMTS. EVDO stands for 'evolution data only' or 'evolution data optimised', often abbreviated as EV-DO, EvDO, 1xEV-DO or 1xEvDO. This technology is the wireless radio broadband data protocol being adopted by many CDMA mobile phone providers in USA, Brazil, Japan, Korea, Israel, Australia and Canada as part of the cdma2000 standard evolution. CDMA-2000 1x-EVDO has been deployed commercially nationwide in Japan since December 2003 with transfer rates of 2.4 Mb/s, which is about six times faster than UMTS in the first development stage in Europe. The personal handy phone system (PHS), which is widely deployed in Japan, China and other areas of Asia, also achieves similarly high data transmission rates and therefore is a strong competitor with UMTS networks in the market place. Data connections at wireless local area network (WLAN) locations (hot spots) also compete with UMTS in urban centres.

In order to differentiate it from competing technologies, UMTS is sometimes marketed as 3GSM (see Section 3.1.7), emphasising the combination of the 3G nature of its technology and the GSM standard, which it was designed to succeed. UMTS supports up to 1920-kb/s data transfer rates, although, at the time of writing, users in deployed networks can expect performance up to 384 kb/s. In 2006, UMTS networks will be upgraded to high-speed downlink packet access (HSDPA), known as 'evolved WCDMA' systems or 3.5G. This will provide a downlink transfer speed up to 14.4 Mb/s. Work is also progressing on improving the uplink transfer speed up to 5.76 Mb/s with the high-speed uplink packet access (HSUPA).

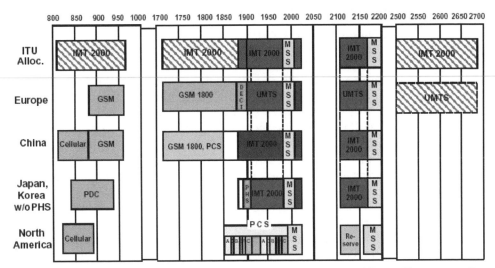

Figure 3.4 WRC-2000 IMT-2000 frequencies (MSS stands for 'mobile satellite spectrum').

The specific frequency bands originally defined at the ITU World Administrative Radio Conference in 1992 (WARC-92) for the IMT-2000 systems were 1885–2025 MHz for the uplink and 2110–2200 MHz for the downlink. The spectrum allocation in Europe, China, Japan, Korea and North America is illustrated in Figure 3.4 and in Table 3.2.

Table 3.2 3GPP-planned operating bands for UMTS and GSM [7], [8].

Operating band	Uplink (MHz)	Downlink (MHz)	Total	Channels
GSM400	478.8–486	488.8–496	2×7.2 MHz	35
GSM400	450.4–457.6	460.4–467.6	2×7.2 MHz	35
GSM850	824–849	869–894	2×25 MHz	124
RGSM	876–915	921–960	2×39 MHz	194
EGSM	880–915	925–960	2×35 MHz	173
GSM900	890–915	935–960	2×25 MHz	124
GSM1800 (DCS)	1710–1785	1805–1880	2×75 MHz	374
GSM1900 (PCS)	1850–1910	1930–1990	2×60 MHz	299
UMTS-FDD (I)[a]	1920–1980	2110–2170	2×60 MHz	12
UMTS-FDD (II)	1850–1910	1930–1990	2×60 MHz	12
UMTS-FDD (III)	1710–1785	1805–1880	2×75 MHz	15
UMTS-FDD (IV)	1710–1755	2110–2155	2×45 MHz	9
UMTS-FDD (V)	824–849	869–894	2×25 MHz	5
UMTS-FDD (VI)[b]	830–840	875–885	2×10 MHz	2
UMTS-FDD (VII)	2500–2570	2620–2690	2×70 MHz	14
UMTS-TDD	1900–1920 and	2010–2025	$20 + 15$ MHz	7

[a] Core band
[b] Japan

UMTS uses a pair of 5-MHz channels with a chip rate of 3.84 MHz, one in the 1900-MHz range for the uplink and one in the 2100-MHz range for the downlink. The nominal channel spacing for UMTS is 5 MHz [7]. As shown in Figure 3.4, the UMTS spectrum identified at WARC-92 and WARC-2000 is already used in North America for 2G personal communications system (PCS) and mobile satellite spectrum (MSS) communications. Regulators are trying to free up the 2100-MHz range for 3G services, although UMTS in North America will still have to share spectrum with existing 2G services in the 1900-MHz band. 2G GSM services elsewhere use 900 MHz and 1800 MHz and therefore do not share any spectrum with planned UMTS services.

In Europe and Japan, the actual number of 3G operators per country is between three and six, and the number of FDD carriers (2 × 5 MHz) per operator varies from two to three. The frequency reuse factor is one.

GSM technology uses a combination of time- and frequency-division multiple access (TDMA/FDMA). The FDMA part involves division of the available bandwidth into carrier frequencies. One or more carrier frequencies are assigned to each base station. The channel spacing is 200 kHz. Each of these carrier frequencies is then divided in time (eight time slots, denoted by TSL), using a TDMA scheme. The frequency reuse factor varies from 1 to 18.

GSM900 and GSM1800 are used in most of the world. GSM1800 is also called the digital communication system (DCS) in Hong Kong and the United Kingdom. In some countries the GSM900 band has been extended to cover a larger frequency range. Extended GSM (EGSM) uses frequency ranges of 880–915 MHz (uplink) and 925–960 MHz (downlink), adding 50 traffic channels to the original GSM900 band [8].

The GSM specifications also describe 'railways GSM' (RGSM), which uses frequency ranges of 876–915 MHz (uplink) and 921–960 MHz (downlink). RGSM provides additional channels and specialised services for use by railway personnel. All these variants are included in the GSM900 specification [8].

GSM850 and GSM1900 are used in the USA, Canada and many other countries in the Americas. GSM850 is also sometimes called 'GSM800'. 'PCS' merely represents the original name in North America for the 1900-MHz band. 'Cellular' is the term used to describe the 850-MHz band, as the original analogical cellular mobile communication system was allocated in this spectrum. Providers commonly operate in one or both frequency ranges.

Another less common GSM version is GSM400. It uses the same frequency as and can co-exist with old analogical Nordic mobile telephone (NMT) systems. NMT is a first-generation (1G) mobile phone system which was primarily used in Nordic countries, Eastern Europe and Russia prior to the introduction of GSM.

3.2 QoS concept and architecture

This section presents the QoS concept and architecture as defined in 3GPP R97/98 and later releases. Bearer service attributes and mapping thereof between different network releases are also described. More information on the covered topics may be found in the references provided at the end of this chapter.

3.2.1   Releases 97 and 98 (R97/98)

The user's communication in the GPRS network is based on the PDP context. The PDP context is established between the UE and GGSN. It is a logical pipe established between the UE and CN. User traffic is carried in it from the UE to the GGSN and towards external IP networks and onward, finally, to a peer entity with whom the user is communicating. The peer entity may be another mobile user, an application server hosting HTTP, gaming, etc. The PDP context established according to R97/98 specifications is called the 'primary PDP context'. When the PDP context is established a QoS profile is assigned to it and the data conveyed in the PDP context is handled according to that QoS profile.

The QoS attributes that are used as the *R97/98 QoS profile* for the PDP context are:

- *Precedence class*: indicates the priority for maintaining the bearer service. The possible values are high, normal and low level. High level guarantees a bearer service that is ahead of all other precedence levels. Normal level guarantees a bearer service that is ahead of low level. Low level is given to bearer services after high and normal priority commitments have been fulfilled.
- *Delay class*: the delay attribute refers to end-to-end transfer delay through the GPRS system. There are three predictive delay classes and one best-effort class.
- *Reliability class*: defines data reliability in terms of residual error rates for the probability of data loss, data delivery out of sequence, duplicate data delivery and corrupted data. This is defined with different bearer service characteristics to be used for each class. For example, if acknowledged or unacknowledged mode is used in the link layer, there are five different reliability classes to fulfil the bearer service needs for different RT/NRT, error-sensitive/non-error-sensitive and data loss characteristics.
- *Throughput class*: this is user data throughput specified in terms of peak and mean classes that characterise the expected bandwidth required for a PDP context. Peak throughput specifies the maximum rate at which data are expected to be transferred across the network for an individual PDP context. There is no guarantee that the level can be sustained, especially as UE capability and radio resource availability have an impact on throughput. Mean throughput specifies the average rate at which data are expected to be transferred across the GPRS network for the PDP context. In addition to various exact values, a 'best-effort' mean throughput class may be negotiated. This means that throughput shall be made available to the UE on a per need and availability basis.

The PDP context can be modified during its lifetime. For example, a UE may decide to request an update due to a change in the application session, or the SGSN may initiate the procedure based on an HLR update on the subscription information. When the PDP context is not needed any more, it can be torn down. It is typically the UE that initiates PDP context deactivation, but the SGSN or GGSN may also initiate the tear-down.

Depending on the UE and network capabilities, one or more primary PDP contexts can be established for the UE. If several PDP contexts are supported, then each of them can be used for different purposes. For example, one could be established as a general

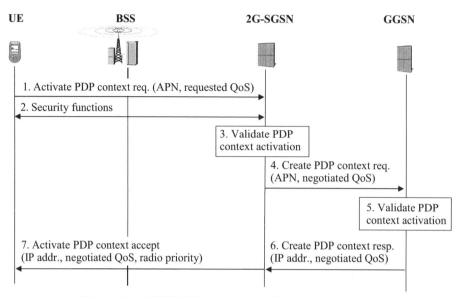

Figure 3.5   R97/98 PDP context establishment procedure.

purpose PDP context to be used to exchange control messages between the UE and network. Another could be created for gaming, and a third one for Internet browsing.

Figure 3.5 describes the R97/98-defined PDP context establishment procedure. It has the following steps:

1. The UE sends the activate PDP context request via the BSS to the SGSN. In the request, the UE includes information on the PDP context and may include the requested QoS profile (i.e., precedence class, delay class, reliability class, peak through-put class and/or mean throughput class as described previously), the Access Point Name (APN) and the UE IP address, or the network may provide them. The QoS parameters are requested based on the service application for which the PDP context is to be used. The APN identifies the external network to which the user wishes to be connected and possibly the application/application server – for example, the Internet or a private intranet, and maybe a WAP, gaming, banking or other server. In addition, the APN may define the GPRS network of the GGSN, if the subscriber is roaming in a visited network. The UE IP address may either be in IPv4 or in IPv6 format.
2. The SGSN may initiate the security procedures. There are different procedures to prevent unauthorised access through authentication, and to provide user, user data and signalling confidentiality by using ciphering and temporary identification. The security procedures are not related to QoS handling.
3. The SGSN validates the PDP context request regarding its capabilities and current load based on the earlier fetched subscriber HLR record containing the subscribed QoS profile(s). For example, the SGSN verifies that the requested APN and QoS parameters (QoS profile) are valid and allowed for the subscriber. Also, if the UE did not provide an APN or the requested QoS, the SGSN selects the default values for these from the subscriber's HLR record.

4. The SGSN determines the GGSN address based on the APN and sends the create PDP context message to the GGSN for its validation. The message contains the negotiated QoS profile (processed by the SGSN) and the APN among other parameters. The GGSN may also be requested to provide a dynamic IP address for the user.
5. The GGSN performs further admission control and validation for the request – for example, whether the requested PDP context parameters are compatible with the service application for which the PDP context is to be used. The GGSN also provides the dynamic IP address, if needed.
6. The GGSN accepts the PDP context establishment and provides the necessary information, such as the negotiated QoS profile (processed by the GGSN) and IP address – to the SGSN.
7. The SGSN may need to process the negotiated QoS based on the GGSN response. The SGSN returns a positive acknowledgement to the UE with the negotiated QoS profile, IP address (if a dynamic address is used) and the radio priority, which the SGSN assigns based on the negotiated QoS. The UE uses the radio priority in radio link contro/medium access control (RLC/MAC) for uplink user data transmission. The UE saves the received information and uses them for data conveyed in the PDP context.

According to the R97/98 specifications, the BSS is not involved in the QoS negotiation of PDP context establishment. However, when the application packets are conveyed in the PDP context through the GPRS CN, the BSS and the air interface, the packets are marked and handled with the QoS profile of the PDP context.

In practice, the R97/98 QoS functionality is most suitable for service applications with NRT requirements – that is, those of interactive and background type. Examples of such service applications are file transfer and gaming. For RT streaming service applications, radio interface capacity may or may not suffice. This depends on the particular application service (i.e., how much capacity it requires), on the load situation over the air interface and other resources as well as on the UE capabilities. If the UE can support it and if the load situation allows this, additional air interface resources (bandwidth) can be allocated for the UE. However, this may not always be possible. Thus, the streaming service applications that require a relatively high bit rate (e.g., video clips) are the most challenging. Generally, the conversational type of service applications are considered to be unsupported in the R97/98 standards, as they require timely and stringent packet delivery which cannot be guaranteed in various situations.

3.2.2   Release 99 (R99)

R99 is fundamental as it defines the first 3GPP 3G system as well as many important amendments to the 2G GSM system. R99 also introduced an end-to-end QoS architecture. This end-to-end QoS architecture covers the entire path from the UE through the RAN/BSS and the PS CN to the external network(s). Still, R99 QoS specifications define QoS functionality for the UE, the RAN/BSS and the GPRS CN nodes, and the interfaces between these elements, but neither the MT/TE interface nor the interface towards the external networks. These QoS aspects are further evolved in R5 specifications (see Section 3.2.3).

As part of the R99 end-to-end QoS model the traffic classification concept was created specifying RT services (i.e., conversational and streaming classes) as well as NRT services (i.e., interactive and background classes). This classification is used to characterise different types of applications and their bearer requirements.

Radio access bearer (RAB) reservation was defined at the outset for the Iu interface. RAB reservation contains explicit QoS support for the RNC, base stations and RAN interfaces. The R99 Gb interface architecture air interface was amended to include EGPRS and the packet flow context (PFC) for the Gb interface. These RAN and BSS amendments support, in particular, RT application and bearer services (i.e., streaming and conversational types). However, standardisation of Gb interface conversational bearer services will be finalised and explicitly done in R7. Note that some of the RAN and BSS QoS related amendments are briefly defined in this section rather than in subsequent ones in order to keep these aspects together.

In addition, the secondary PDP context is specified as part of R99.

3.2.2.1 End-to-end QoS concept and architecture

The purpose of the R99 end-to-end QoS architecture is to ensure that the bearer services are established and modified in the UE, RAN/BSS and PS CN according to UE-requested QoS and APN, the user's subscription rights, the current network load and the operator's QoS policies. This R99 functionality is defined in 3GPP TS 23.107 [9] and is applicable to both Iu and Gb interface architectures, even though for simplicity, in accordance with 3GPP specifications, the terms 'UMTS' and 'RAN' are used below instead of UMTS/GERAN, RAN/BSS and RNC/BSC.

The R99 end-to-end QoS concept complements the earlier defined GPRS CN bearer services and QoS concepts by adding new functionality to various entities in the system.

3.2.2.2 QoS architecture

Figure 3.6 depicts the R99 QoS architecture. It comprises the following functionality:

- *End-to-end service*: the end-to-end service depicted at the top is provided by the underlying services. This covers the path from the UE to the external networks.
- *UMTS bearer service*: this service is the R99-defined QoS functionality and it covers the various standardised QoS aspects of the UMTS network. It is composed of the RAB service and the CN bearer service. The RAB service is provided in the RAN containing the RNCs and the base stations as well as the interface between them, between the RNC and the CN, and the air interface from the base station to the UE. The GPRS packet core offers the CN bearer service, which is implemented in the SGSN and the GGSN, as well as in the interface between them.
- *CN bearer service*: this service is applicable in the GPRS CN node and its interfaces. It controls and utilises lower level backbone network services to provide the contracted UMTS bearer service. The backbone services route the traffic according to the negotiated QoS.
- *Backbone network service*: the backbone covers the L1 and L2 of the OSI model. The backbone supports different bearer services to supply the QoS needed for the various

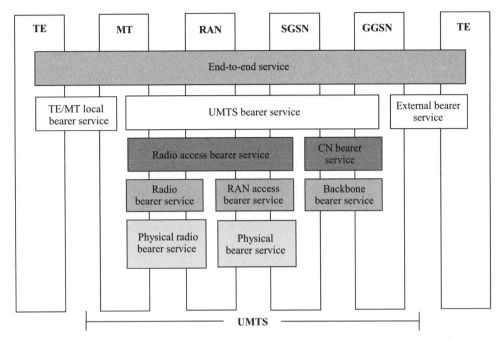

Figure 3.6   R99 end-to-end QoS architecture [9].

user applications. The exact method of providing QoS functionality in the backbone is left for the operators and manufacturers to choose. Backbone network services are implemented with generic (non-UMTS specific) methods.

- *Radio access bearer service*: the service comprises QoS support for the entire RAN and its interfaces. It is realised by two distinct services: the radio bearer service and the Iu bearer service.
- *Radio bearer service*: the service covers all the aspects of radio interface transport. The UTRA FDD/TDD service provides the radio bearer service.
- *UTRA FDD/TDD service*: this service provides the transport between the UE and the RAN by means of the radio bearer service. For A/Gb mode, there are corresponding GSM/EGPRS physical radio bearer services.
- *Iu bearer service*: the service provides QoS handling by means of the physical bearer service between the RAN and the CN. Iu bearer services provide different bearer services with a variety of QoS characteristics. For A/Gb mode, the corresponding services are offered over the Gb interface.
- *Physical bearer service*: this service provides the transport between the RAN and the CN.
- *Terminal equipment/mobile termination local bearer service*: this is responsible for the local QoS between the TE and MT. The TE and MT may be collocated in the UE, or the TE can be a separate device (e.g., laptop computer) that is connected to the MT (e.g., UE) providing the connection to the UMTS network. This QoS functionality is not further elaborated in R99 QoS specifications.

- *External bearer service*: the external bearer service is not defined in R99 specifications. The architecture configurations providing the service may vary and different QoS solutions may be offered; for example, the interface may be a direct interface from the UMTS network to an application server, or it may include one or more intermediate internets or intranets and at the other end another UMTS subscriber located in another UMTS network.

3.2.2.3 QoS functions for UMTS bearer service in control plane

Figure 3.7 shows the QoS functions in the UMTS bearer service architecture in the control plane. The *translation functions* in the MT and the GGSN convert the signalling between the external services and the internal UMTS services. This includes translation of the service attributes. The *UMTS bearer service managers* in the MT, SGSN and the GGSN signal between each other and with the associated *admission/capability control* at a bearer service request (initiation or modification) to verify that the requested service is supported and that the resources for it are available. The SGSN UMTS bearer service manager also contacts the *subscription control* to find out if the user is allowed to use the indicated bearer service characteristics. The UMTS bearer service manager of the MT translates the UMTS bearer service attributes to local bearer service attributes, and also requests that the radio bearer service manager does the same. The UMTS bearer service manager in the SGSN translates the UMTS bearer service attributes to RAB, RAN access bearer and CN bearer service attributes. It also requests the RAN access bearer service, CN bearer service and the RAB managers to provide the specified service. Furthermore, the GGSN UMTS bearer service manager translates the UMTS bearer service attributes to the CN bearer service attributes and requests the CN bearer service manager to provide the service. In addition, it translates the UMTS bearer service attributes to external bearer service attributes and requests the service from the external bearer service manger. The *RAB manager* verifies with its *admission/capability control* whether the RAN supports the requested service and whether the needed resources are available. It also carries out the attribute translation from the RAB service to the radio and RAN access bearer services, and *vice versa*, as well as requesting the radio bearer service and the RAN access bearer service managers to provide the bearer services with the indicated attribute values. The radio, Iu bearer service and CN bearer service managers use services provided by the lower layers (as shown in Figure 3.7). The *local bearer service manager* handles the local bearer service between the MT and TE, and the *external bearer service manager* the bearer service toward the external network; these are not defined in R99 specifications.

3.2.2.4 QoS functions for UMTS bearer service in user plane

The QoS functions of the UMTS bearer service for the user plane are shown in Figure 3.8. The *classification function* in the GGSN and in the MT assigns user data units received from the external bearer service or the local bearer service to the appropriate UMTS bearer service according to the QoS requirements of each user data unit. The *traffic conditioner* in the MT provides conformance of the uplink user data traffic with the QoS attributes of the relevant UMTS bearer service. In the GGSN a traffic conditioner

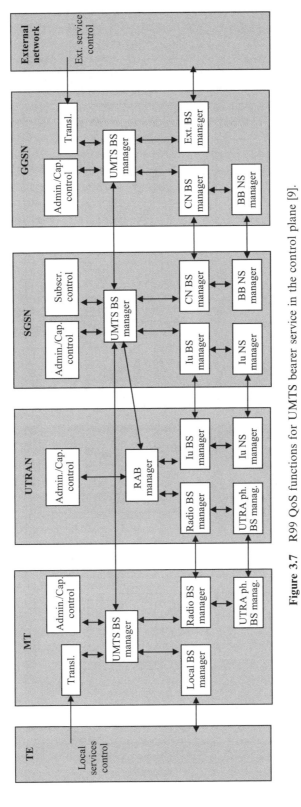

Figure 3.7 R99 QoS functions for UMTS bearer service in the control plane [9].

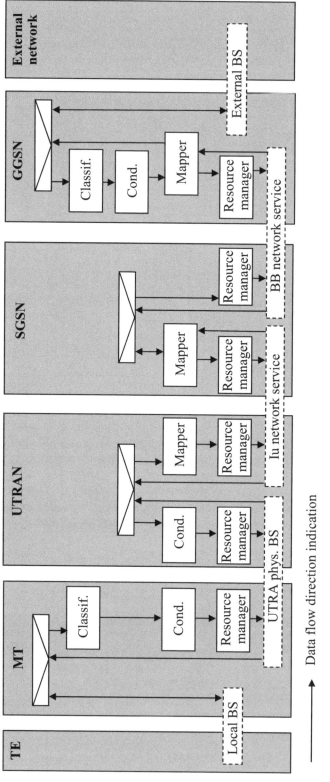

Data flow direction indication

Figure 3.8 R99 QoS functions for UMTS bearer service in the user plane [9].

may provide conformance of the downlink user data traffic with the QoS attributes of the relevant UMTS bearer service (i.e., per PDP context). The packet-oriented transport of the downlink data units from the external bearer service to the UTRAN and the buffering in the UTRAN may result in bursts of downlink data units not conformant with UMTS bearer service QoS attributes. A traffic conditioner in the UTRAN forms this downlink data unit traffic according to the relevant QoS attributes. The traffic conditioners are not necessarily separate functions. For example, a resource manager may also provide conformance with the relevant QoS attributes by appropriate data unit scheduling. The *mapping function* marks each data unit with the specific QoS indication related to the bearer service performing the transfer of the data unit. Each resource manager of a network entity is responsible for a specific resource. The resource manager distributes its resources between all bearer services requesting transfer of data units. Thereby, the resource manager attempts to provide the QoS attributes required for each individual bearer service.

3.2.2.5 UMTS bearer and RAB service attributes

UMTS bearer and RAB service attributes describe the bearer service provided by the UMTS network. These attributes form the *R99 QoS profile* for a PDP context and they are:

- *Traffic class*: this attribute has four possible values.
  - *Conversational*: this value is intended for voice applications. The key characteristic for this class is the stringent real time requirement. Transport delays cannot be long and vary between the data packets; otherwise, it becomes difficult for the receiving user to follow the conversation. In R99, this value is only supported for the Iu interface.
  - *Streaming*: when the user is looking at real time video or listening to real time audio, this class is used. The real time data flow is always aiming at a live (human) destination. It is a one-way transport.
  - *Interactive*: for this class, it is important that the payload content be preserved during the transport (low bit error rate) and that the time delay be reasonable as the recipient is waiting for the response.
  - *Background*: the value is applicable for the cases where the end-user is a computer sending and receiving data files in the background. The destination is not expecting the data within a specific time, so the timing requirements are not stringent. Nevertheless, it is important that the contents of the transported data be unchanged.
- *Maximum bit rate* (kb/s): this is the maximum number of bits delivered by UMTS and to UMTS within (divided by) a time period.
- *Guaranteed bit rate* (kb/s): this is the guaranteed number of bits delivered by UMTS and to UMTS within (divided by) a time period, provided that there are data to deliver. For the traffic exceeding this limit, the specified QoS is not guaranteed. The GBR is only used for streaming and conversational classes. For NRT traffic it is not so important to guarantee a specific bit rate for a particular time period.
- *Delivery order* (yes/no): indicates whether the in-sequence delivery of packets is needed – that is, whether the order of the data packets needs to be kept or not.

- *Maximum service data unit (SDU) size* (octets): this indicates the maximum SDU size for which the network needs to satisfy the negotiated QoS. It is used for admission control, policing and, especially in the RAN, for optimising transport.
- *SDU format information* (bits): the value indicates a list of the possible exact sizes of SDUs. It can be used to gain spectral efficiency for the so-called transparent RLC mode in the RAN, in which packet retransmissions are not used in the RLC layer to recover from errors.
- *SDU error ratio*: this indicates the fraction of SDUs lost or detected as erroneous. It may be used to configure the protocols, algorithms and error detection schemes, primarily in the RAN.
- *Residual bit error ratio*: this attribute indicates the undetected bit error ratio in the delivered SDUs. It may be used to configure the radio interface, algorithms and error detection coding.
- *Delivery of erroneous SDUs* (yes/no/–): this parameter is used to indicate whether SDUs detected as erroneous are delivered or discarded. For values 'yes' and 'no' error detection is applied, but for the value '–' it is not done.
- *Transfer delay* (ms): this indicates the maximum delay for the 95th percentile of the delay distribution for all delivered SDUs during the lifetime of a bearer service. It is used to specify the delay tolerated by the application, and it allows the UTRAN to set up the correct bearer service parameters.
- *Traffic handling priority*: specifies the relative importance of the SDUs belonging to the UMTS bearer compared with the SDUs of other bearers. It is a relative value (as opposed to an absolute one), and it is used especially within the interactive class. The purpose is to distinguish QoS handling within the interactive traffic class.
- *Allocation/Retention priority*: specifies the relative importance of a UMTS bearer compared with other UMTS bearers for allocation and retention of the UMTS bearer. In situations where resources are scarce, the attribute can be used to prioritise bearers with a high allocation/retention priority over bearers with a low allocation/retention priority when performing admission control.
- *Source statistics descriptor* (for RAB service only): the parameter specifies the characteristics of the source application service of submitted SDUs and has two values: speech and unknown. The RAN can use the value to calculate statistical multiplex gain for admission control on the radio and Iu interfaces.

In R99, the mapping of QoS attributes between the application attributes and the UMTS bearer service attributes is not standardised, and is thus left for the operators. In similar manner, the mapping from UMTS bearer service attributes to CN bearer service attributes in not standardised. On the other hand, the mapping between UMTS bearer and RAB service attributes and their values is specified. The relationship is mostly one-to-one as the attributes are the same.

3.2.2.6 Radio access bearer (RAB) reservation for the Iu interface

As part of the first UMTS release (i.e., R99) the RAB reservation procedure is defined for the Iu interface as well as the radio resource control (RRC) and lower layer procedures to support various types of QoS characteristics in the RAN (e.g., both GBR and NGBR bearer services in the UTRAN). These procedures enable resource reservation according to the negotiated QoS in the different parts of the RAN.

RAB reservation is described in Section 3.2.2.7 and as part of the R99 end-to-end QoS concept and functionality in this chapter. RAB reservation corresponds to the Gb interface PFC procedure described above.

3.2.2.7 PDP context activation for the Iu interface

Figure 3.9 describes the R99 PDP context activation for the Iu interface. The scenario contains the following steps:

1. The UE sends the activate PDP context request via the RAN to the SGSN. In the request, the UE includes information on the PDP context and may include the requested QoS profile (e.g., traffic class, maximum bit rate, GBR, etc. – see Section 3.2.2.5) and the APN and UE IP address, or the network may provide them. The QoS parameters are requested based on the application service for which the PDP context is to be used. The APN identifies the external network to which the

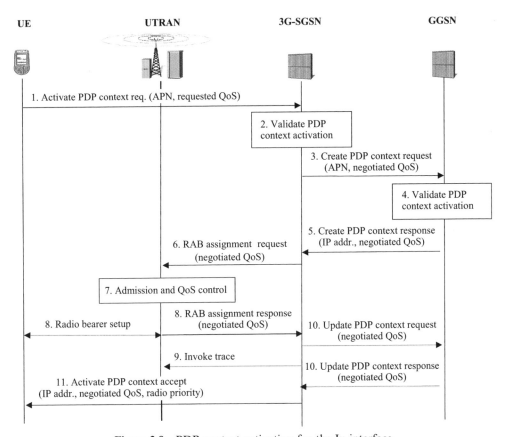

Figure 3.9 PDP context activation for the Iu interface.

user wishes to be connected and, possibly, the application/application server – for example, the Internet or a private intranet, and maybe a WAP, gaming, banking or other server. In addition, the APN may define the GPRS network of the GGSN, if the subscriber is roaming in a visited network. The UE IP address may either be in IPv4 or in IPv6 format.

2. The SGSN validates the PDP context request regarding its capabilities and current load as well as the earlier fetched subscriber HLR record containing the subscriber's QoS profile(s). For example, the SGSN verifies that the requested APN and QoS parameters (QoS profile) are allowed for the subscriber. Also, if the UE did not provide the APN or requested QoS, the SGSN selects the default values for these from the subscriber HLR record.

3. The SGSN determines the GGSN address based on the APN and sends the create PDP context message to the GGSN for validation. The message contains the SGSN-processed negotiated QoS and the APN among other parameters. The GGSN may also be requested to provide a dynamic IP address for the user.

4. The GGSN performs further admission control and validation for the request – for example, whether the requested PDP context parameters are compatible with the related application service. The GGSN also provides the IP address, if needed.

5. The GGSN accepts the PDP context establishment and provides the necessary information – such as the GGSN-processed negotiated QoS and IP address – to the SGSN.

6. The SGSN may need to process the negotiated QoS based on the GGSN response. The SGSN sends the RAB request to the RNC with the negotiated QoS (QoS profile). Based on the negotiated QoS it receives from the SGSN, the RNC carries out admission control and produces the appropriate L1 and L2 parameters to establish and maintain the radio bearer characteristics. The radio bearer configuration encompasses, among other things, the definition of the RLC protocol transmission mode – that is, transparent (TM), acknowledged (AM) or unacknowledged mode (UM) – the quality targets (BLER and SIR), selection of transport channels, and derivation of the corresponding codes of the physical channels employed at the radio interface.

7. The radio bearers are established according to the request over the air interface, and the RNC returns a positive acknowledgement to the SGSN.

8. The trace function may be invoked to track the subscriber. This functionality is not related to the QoS but is part of the PDP context establishment procedure.

9. Depending on the RAN's response, the SGSN may first initiate a new request towards the RNC with lowered QoS values and then to the GGSN – for example, if the RAN was not able to support the original SGSN-requested negotiated QoS values in Step 6. The GGSN replies to the received request.

10. The SGSN returns a positive acknowledgement to the UE with the negotiated QoS profile and IP address (if a dynamic address is used). The message also includes the SGSN-generated radio priority for R97/98 backward compatibility purposes. The UE saves the negotiated QoS profile and other information to be used for the PDP context.

The A/Gb mode PDP context activation procedure is the same as the Iu mode's apart from the fact that the RAB procedure is replaced by the BSS PFC procedure. The PFC is described in Section 3.2.2.9.

3.2.2.8 Air interface improvements for the Gb interface in R99 and later

As part of the R99, the EGPRS standardisation defined new modulation and coding schemes (MCS) for the GSM-based GPRS air interface. Octagonal phase shift keying (8-PSK) modulation was selected for EGPRS instead of the previously used Gaussian minimum shift keying (GMSK). This new modulation and coding scheme allows an optimised use of the air interface bandwidth enabling increased user data rates. In fact, with a single GSM data channel the bit rate is 9.6 kb/s or 14.4 kb/s, if high-speed CS data (HSCSD) are available. GPRS supports up to a theoretical maximum data rate of 140.8 kb/s (though typical rates are closer to 56 kb/s) with four different coding schemes (CS). With EGPRS the actual packet data rates can theoretically reach 384 kb/s with 48 kb/s per timeslot and up to 69.2 kb/s per timeslot in good radio conditions [6].

With these improvements there were two main advantages: more users could be served and users could be provided with higher data rates. This is especially useful for the packet-based service applications, which often require high bit rates due to the amount of data transfer involved – for example, for downloading video clips, large files or webpages with figures, and, in particular, for RT service applications, such as streaming services.

Additional Gb interface related air interface improvements were defined after R99. The temporary block flow (TBF) procedure was enhanced in R4 for the Gb interface architecture. The TBF is the L2 (link layer) used in the Gb interface architecture to transport data over the air interface between the UE and the base station. The TBF is a temporal, unidirectional connection established when either the UE or the base station has data to send to the other. As part of R4 the extended TBF is defined. This is a procedure where the TBF release is delayed so that the number of TBF establishments is reduced. Thus, user-experienced delay is decreased as the packets are transported over the air interface faster without a new TBF setup. In R4, network-assisted cell change (NACC) was also standardised, reducing the cell reselection time while the subscriber is moving from one cell to another.

In addition, the conversational types of bearer services were not supported until R7. The reason being that as the original TBF functionality supports neither dedicated traffic channels nor handovers, it is not so well-suited for the conversational type of handling. These additional TBF features are likely to be completed in the R7 time frame for the Gb interface architecture.

3.2.2.9 Packet flow context (PFC) for the Gb interface

In addition to EGPRS, another important feature was defined for the Gb interface architecture in R99: the PFC. It is used to provide QoS handling between the SGSN and BSS. The SGSN requests the PFC to be established as part of PDP context activation, and at that time the QoS parameters are also negotiated. In the process, the SGSN indicates the maximum timer value for the BSS to be used for the PFC. The BSS may use this or a lower value for the timer, and once the timer expires the BSS removes the PFC.

Subsequently, the BSS or SGSN can establish a new PFC as needed; for example, if the UE establishes a new application service and PDP context with different QoS requirements. The BSS and SGSN can also modify the PFC or remove it by specific procedures between the two. The 3GPP specifications are not very clear on whether one PFC can be shared by several UEs or only used by one UE. At least some companies have interpreted the specifications in such a way that a single PFC is used for only one UE.

A given PFC can be shared by one or more activated PDP contexts with identical or similar negotiated QoS requirements. The PDP contexts that share the same PFC constitute one *aggregated packet flow*, and the related QoS requirements are called the 'aggregate BSS QoS profile'. The aggregate BSS QoS profiles specify the QoS that the BSS needs to provide for a given packet flow between the UE and SGSN including the radio and Gb interfaces.

In R5, the PFC was further evolved by adding UE-specific flow control functionality. In this solution, the sending of packets over the Gb interface from the SGSN to the BSC is based on the BSC being given flow control information. The PFC allows the flow control to be defined between the UE PFCs. Thus, if the UE has two or more packet flows using different PFCs, the BSC may indicate flow control specific QoS information to the SGSN regarding the PFCs. This enables prioritisation among the user's bearer (and application) service flows. More information on packet data transfer across the EGPRS network is given in Chapter 4.

3.2.2.10 Secondary PDP context

In addition to the primary PDP context defined in R97, the R99 defined the secondary PDP context functionality. A secondary PDP context is established according to the related primary PDP context, but the QoS handling can be different. Thus, the secondary PDP context uses the same APN and IP addresses as the associated primary PDP context, but it typically has a different QoS profile.

3.2.2.11 QoS mapping between R97 and R99 onward

As the defined QoS parameters are different between R97 and R99, the mapping between the different values has been standardised to support smooth handovers between newer and older releases as well as operations between nodes supporting different releases. For example, it is possible to have an R99 SGSN connected to a R97 GGSN. Table 3.3 shows the mapping between these two releases when R99 attributes are derived from R97 attributes. Similar mapping is defined when R99 attributes are used for generating R97 attributes. R97 QoS attributes are also used in R98. By the same token, R99 QoS attributes are also used in later releases (e.g. R4, R5 and R6).

3.2.3 Release 5 (R5)

This section presents the main functionality of 3GPP R5 QoS amendments and the end-to-end QoS architecture.

Table 3.3 Rules for determining R99 attributes from R97/98 attributes [9].

Resulting R99 attribute		Derived from R97/98 attribute	
Name	Value	Value	Name
Traffic class	Interactive	1, 2, 3	Delay class
	Background	4	
Traffic handling priority	1	1	Delay class
	2	2	
	3	3	
SDU error ratio	10^{-6}	1, 2	Reliability class
	10^{-4}	3	
	10^{-3}	4, 5	
Residual bit error ratio	10^{-5}	1, 2, 3, 4	Reliability class
	$4 * 10^{-3}$	5	
Delivery of erroneous SDUs	No	1, 2, 3, 4	Reliability class
	Yes	5	
Maximum bit rate (kb/s)	8	1	Peak throughput class
	16	2	
	32	3	
	64	4	
	128	5	
	256	6	
	512	7	
	1024	8	
	2048	9	
Allocation/retention priority	1	1	Precedence class
(not relevant for the UE)	2	2	
	3	3	
Delivery order	Yes	Yes	Reordering required
	No	No	(information in the SGSN and the GGSN PDP contexts)
Maximum SDU size	1500 octets		(Fixed value)

3.2.3.1 The main functionality of R5 QoS amendments

In earlier releases, QoS allocation was carried out for bearer services based on the UE-requested QoS profile, the APN, subscriber HLR QoS profiles and the operator policies in the SGSN and GGSN. Different QoS characteristics can be used for different applications by defining a different APN for each of them, but dynamic service specific information is not used in a standardised manner. In R5 this changed. A new functionality, called the 'service-based local policy' (SBLP) was defined. This functionality allows the proxy-call session control function (P-CSCF) to dynamically provide the IMS application session characteristics to be used for bearer service QoS authorisation. In particular, SDP attributes are useful in gaining QoS authorisation. SDP is used as the application session protocol for 3GPP IMS services. After the initial bearer setup, QoS information can be updated due to changes in the IMS session. In order to support the

SBLP, the policy decision function (PDF) of the P-CSCF and the policy enforcement point (PEP) of the GGSN are defined. The PDF gains QoS authorisation, and the PEP enforces the decision in the GGSN. In R5, the IMS is the only application for which the SBLP was defined, because it was seen as the most important.

As part of R5 QoS amendments, QoS parameter mapping is defined for the UE, PEP/GGSN and PDF/P-CSCF. R5 QoS amendments also cover the dynamic provision of *gating information* based on IMS session information. Gating allows user packets to flow through or to be discarded. The gating filters are set in the GGSN for the PDP context. The filters ensure that the GGSN only allows authorised traffic to pass through to and from the UE. The gate is opened and closed depending on the application session status. In addition, the GPRS and IMS domains are able to exchange *charging identifiers* as part of QoS authorisation. The identifiers can be added to the charging records created in each domain. They allow the operator to correlate the records afterward and, thus, use them for charging purposes.

Despite the addition of dynamic QoS handling in R5, the GGSN, SGSN and UE QoS functionality of earlier releases is largely preserved; for example, the admission, subscription and APN-related control and verification remain in the SGSN and the GGSN. The RAN/BSS and Gb/Iu interface QoS mechanisms are also preserved.

3.2.3.2   End-to-end R5 QoS architecture

In R5, the R99 end-to-end QoS architecture is amended, as defined in TS 23.207 [10]. The R5 QoS architecture control plain is depicted in Figure 3.10. The tinted boxes are described as part of the R99 end-to-end functionality from where they were inherited. Thus, the most significant changes introduced in R5 are the PDF co-located with the P-CSCF and the associated Go interface between the PDF/P-CSCF and GGSN. In addition, the UMTS bearer service (BS) managers located in the GGSN and the UE are added. These new functions are responsible for end-to-end QoS handling. New mapping function is also added into the translation function in the UE and GGSN. The P-CSCF is part of the 3GPP IMS domain and it acts as a particular SIP proxy. It is situated in the same operator network as the GGSN used for subscriber bearer services. This is normally the home network, but it can also be the visited GGSN.

GGSN end-to-end QoS functionality
The *IP bearer service manager* in the GGSN uses standard IP mechanisms to manage IP bearer services. The IETF-defined differentiated services or, more commonly, the *DiffServ Edge function* (see [2]) is mandatory for R5 end-to-end QoS-compliant GGSN. It is used to provide QoS for the external bearer service. Parameters for the DiffServ Edge function (i.e., classifiers, meters, markers, and shapers/droppers) may be statically configured in the GGSN or derived from PDP context parameters. Although the R5 end-to-end QoS specification includes the IETF-defined Reservation Protocol (RSVP) functionality (see [11]) it is optional, and it also seems that RSVP is not widely implemented. Thus, it is not further evolved in this chapter.

In addition, the *SBLP enforcement point* is a mandatory function of the R5 GGSN, even though it is the operator choice whether to use it or not. The functionality controls the QoS for a set of IP flows carried within the same PDP context. It contains policy-

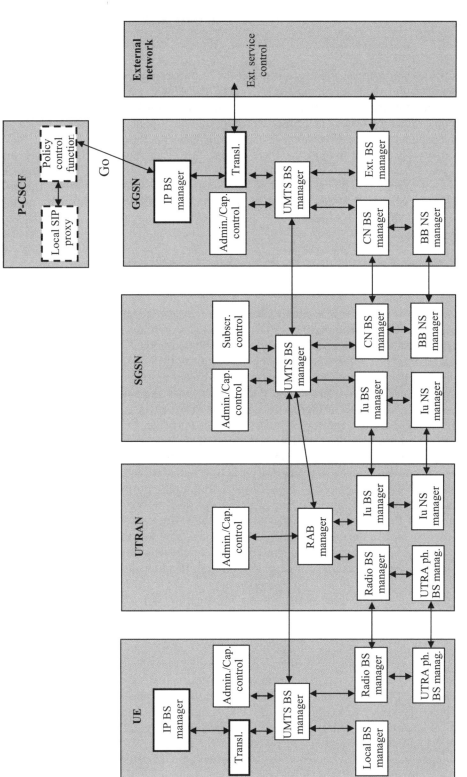

Figure 3.10 Release 5 QoS architecture for the control plane [10].

based admission control, which ensures that the resources used for a particular set of IP flows are within the limits indicated by the PDF/P-CSCF via the Go interface. The authorisation is expressed in terms of maximum bit rate (for NGB traffic) or GBR (for GB traffic) and QoS class. The translation/mapping function in the GGSN provides the mapping between these IP bearer service parameters and those used for the UMTS bearer services. In the user plane, there is also the gating function. It controls and manages the user packet flows as instructed by the PDF/P-CSCF. Each gate controls either an upstream or a downstream flow and consists of a packet classifier and the gate status, open or closed. When the gate is open the packet flow is let through and the appropriate DiffServ Edge functions are applied to them. If the gate is closed, the packets are dropped.

Key information exchanged over the Go interface

The most important information exchanged through the Go interface is:

- *Binding information*: this is used to bind or correlate together the IMS application session and GPRS bearer session (PDP context). The binding information consists of an authorisation token and the flow identifiers related to the IP flows within the same IMS session. The *authorisation token* consists of the IMS session identifier and PDF identifier. The PDF/P-CSCF allocates the authorisation token for each IMS application session at session setup. The P-CSCF provides the token to the UE in SIP signalling, and the UE further to the GGSN at PDP context activation (or modification). For the GGSN, the authorisation token also provides the correct PDF/P-CSCF address to authorise the QoS for the session. The flow identifiers are used for the identification of the IP flows within a media component associated with an IMS (SIP) session. A *media component* is a part of an SDP session description conveying information about media (e.g., media type, format, IP address, port(s), transport protocol, bandwidth and direction). A *flow identifier* consists of two parts: (1) the ordinal number of the position of the 'm=' lines in the SDP session description (see RFC 2327 [12]); and (2) the ordinal number of the IP flow(s) within the 'm=' line assigned in the order of increasing port numbers. (An IP flow is by definition a unidirectional flow of IP packets with the same source and destination IP addresses and port numbers and the same transport protocol. Port numbers are only applicable if used by the transport protocol.) The PDF receives the binding information as part of a request from the GGSN. The authorisation token is applied by the PDF to identify the IMS session. The flow identifier(s) is used to select the available information on the IP flows of the IMS session. If the binding information consists of more than one flow identifier, the PDF also verifies that the media components identified by the flow identifiers are allowed to be transferred in the same PDP context. For valid binding information consisting of more than one flow identifier, the information sent back to the GGSN includes the *aggregated QoS* for all the IP flows and the gate description (i.e., suitable *packet classifiers* (or filters) for these IP flows) and gate status to be applied. The filter(s) and associated gate(s) relate to the PDP contexts where the SBLP applies.
- *QoS authorisation*: also known as the 'authorised QoS' parameters for the bearer, this is the combination of the 'authorised QoS' of the individual IP flows of the media components of an IMS session. The 'authorised QoS' information consists of: (1) the

data rate (i.e., the maximum value of the *maximum bit rate* parameter for the NGB bearer services or *guaranteed maximum bit rate* for GB bearer services), and (2) the *maximum QoS class* (i.e., the highest class that can be used for the bearer). The 'authorised QoS' information for IP flows of a media component is extracted from the media-type information and bandwidth parameters of the SDP. The PDF maps the media-type information onto the QoS class that is the highest class that can be used for the media in the uplink and downlink directions, when both directions are used. In case of an aggregation of multiple media components within one PDP context, the PDF provides the 'authorised QoS' for the bearer as the combination of the 'authorised QoS' information of the individual IP flows of the media components. The QoS class in the 'authorised QoS' for the bearer contains the highest QoS class of those applied to individual media component IP flows and indicates the highest UMTS traffic class that can be applied to the PDP context. The data rate of the 'authorised QoS' for the bearer is the sum of the data rate values of the individual media IP flows of components and is used as the maximum data rate value for the PDP context. Detailed rules for calculating the 'authorised QoS' are specified in [13]. In the GGSN, the translation/mapping function performs the mapping of 'authorised QoS' information for the PDP context onto authorised UMTS QoS information. The GGSN derives the highest allowed UMTS traffic class for the PDP context from the QoS class in the 'authorised QoS' according to the rules specified in [10]. For example, QoS Class 'A' is mapped onto UMTS traffic class 'conversational'. In the case of RT (or GB) UMTS bearers (conversational and streaming classes), the GGSN considers the data rate value of the 'authorised QoS' information as the maximum value of the 'GBR' UMTS QoS parameter, whereas the 'maximum bit rate' UMTS QoS parameter is limited by the subscriber and service-specific setting in the HLR/Home Subscriber Server (HSS) (SGSN) and by the capacity/capabilities/service configuration of the network (GGSN, SGSN). In the case of NRT (or NGB) bearers (interactive and background classes) the GGSN considers the data rate value of the 'authorised QoS' information as the maximum value of the 'maximum bit rate' UMTS QoS parameter. The UMTS BS manager receives the authorised UMTS QoS information for the PDP context from the translation/mapping function. If the requested QoS exceeds the authorised QoS, the UMTS BS manager downgrades the requested UMTS QoS information to the authorised UMTS QoS information. The GGSN may store the authorised QoS for the binding information of an active PDP context in order to be able to make local decisions, when the UE requests a PDP context modification.

- *Gate description*: this includes the *packet classifiers* or *filter specification* parameters (source IP address, destination IP address, source ports, destination ports, protocol ID) explicitly describing a unidirectional IP flow, and gate status (opened/closed). The PDF derives the packet classifier from the connection information parameter of the SDP used with SIP for the IMS application session. The GGSN receives the packet classifier from the PDF in an authorisation decision. The packet classifier(s) are used to identify the allowed packets for the bearer service. In the GGSN, the packet filters are installed correspondingly to the received packet classifiers. IP packets matching an SBLP-supplied filter are subject to the gate associated with that packet filter. In the uplink direction, IP packets which do not match any SBLP-supplied filter are silently discarded. In the downlink direction, IP packets which do not match any

SBLP-supplied filter are matched against UE traffic flow template (TFT) supplied filters (installed for PDP contexts where SBLP is not applied). The commands to open or close the gate lead to the enabling or disabling of the passage of packets through the GGSN from the external network towards the UE and from the UE towards the external network. If the gate is closed all packets of related IP flows are dropped. If the gate is opened the packets of related IP flows are allowed to be forwarded.

• *Charging identifiers*: the GPRS and IMS charging identifiers (GCID and ICID, respectively) are provided from the GGSN to the PDF/P-CSCF and *vice versa*. They can be added to the charging records generated in the IMS and GPRS domains so that the operator can correlate the corresponding IMS and GPRS charging records to each other.

COPS and COPS-PR

The IETF-defined COPS protocol, its provisioning amendments called 'COPS–policy provisioning' (COPS-PR), and 3GPP-specific Go definitions are used over the Go interface to provide the R5 Go functionality. COPS has an advantage over the IETF-defined Remote Authentication Dial In User Service (RADIUS) protocol as it can hold state information. COPS is run over TCP/IP and its use offers a fairly simple authorisation functionality using client–server architecture. Architecturally, it contains the PEP, which is the client located in the GGSN and which enforces the QoS and authorisation decision made by the server – that is, the PDF (a.k.a. the policy control function or the PCF), co-located with the P-CSCF. COPS has two sets of messages: the first is used for setting up, maintaining and terminating the transport connection between two nodes – that is, the PDF/P-CSCF and the GGSN. The second set is used for QoS authorisation.

3.2.4   Release 6 (R6)

This section describes R6 QoS functionality. The combined R5 and R6 signalling flows are provided in the following section. The R5 Go interface is not changed in R6, but the R6 Gq interface is added between the PDF and the P-CSCF.

3.2.4.1   Main QoS amendments in R6

In R6, the key QoS amendment is the Gq interface, which is depicted in Figure 3.11. The Gq interface is a new interface between the PDF and the application function (AF) (e.g., the P-CSCF). In R5, the PDF is co-located with the P-CSCF and, thus, only supports the IMS. However, the R6 PDF is standardised as a stand-alone element and can be used to authorise other session-based applications than the IMS. In addition, other application session protocols than the SDP and SIP can be used, even though R6 supports these two at a more detailed level than other protocols. The R6 PDF maintains the same role it has in R5, and it is the key QoS and policy control entity in the architecture. The R5 Go interface between the PDF and GGSN is not changed in R6, but it is used as such in conjunction with R6 amendments. By the same token, most of the QoS functionality defined in R5 and before is kept in R6.

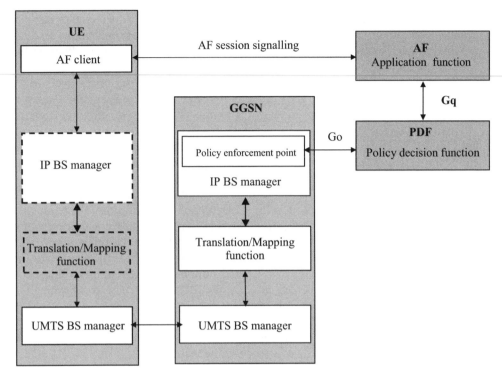

Figure 3.11   Release 6 QoS architecture amendments. The changed elements are shown shaded and the new Gq interface is in a bold font.

Another change in R6 is the possibility to multiplex several service applications' (application sessions or media components or multimedia sessions) data flows into one PDP context. In this case, there can be one or more AFs offering the application sessions (i.e., an AF session description can consist of more than one media component), but only one PDF to authorise the QoS for the related bearer services. In addition, all the application sessions receive the same QoS class treatment as is given for the PDP context. Typically, the application requiring the highest QoS class defines the value for the PDP context. The bit rates can vary for each application session within the PDP context as long as the value authorised for the PDP context is not exceeded. Typically, when defining the allowed bit rate for the PDP context the requirements from all the multiplexed applications are added together and used as the allowed maximum bit rate.

3.2.4.2   Gq interface

Key information exchanged over the Gq interface
The key information exchanged over the Gq interface is the following:

• *Authorisation token*: the Gq authorisation token is used in accordance with the R5-defined authorisation token and is used as binding information to correlate an application session (e.g., the IMS) and the related bearer session (PDP context), and to provide the GGSN with the PDF address. In R6, the PDF allocates the authorisation token and provides it to the P-CSCF. If several application sessions are multiplexed into one PDP context, a distinct authorisation token is allocated for each of the application sessions.

- *Media component description* is a part of an AF session description (e.g., SDP) conveying information about the media of the application session; for example, bandwidth (kb/s), media type (audio, video, etc.), *IP flow information* and related gating (gate open or closed). SDP is a commonly used application session protocol and, thus, this information is normally derived from the corresponding SDP parameters characterising the application session. The PDF uses this information for the QoS authorisation and gating decision.
- *AF-Application Identifier*: the AF-Application Identifier enables the AF to provide additional application information to the PDF for QoS authorisation. The parameter values and the logic that AF uses to determine the application identifier are not standardised but are left for the AF and PDF operator(s) and manufacturer(s). The decision may be based on a pre-defined server address used as the 'called party' for the application service if that server has a fixed address or a set of fixed addresses permanently assigned to it. The AF can also use SDP parameters to determine the AF-Application Identifier value to be used.
- *Charging identifiers*: the GPRS and IMS charging identifiers (GCID and ICID, respectively), as described in R5 for the Go interface, may also be exchanged over the Gq interface to correlate the related charging records.

Diameter and NASREQ

For the Gq interface, the IETF-defined Diameter is used. Generally, Diameter offers secure and scalable roaming support that is well-suited for single interfaces as well as for large networks. Diameter is based on the older RADIUS protocol but also contains new important features:

- Diameter is based on the client–server model but it can also be used in a peer-to-peer manner. In the latter, so-called 'Diameter agents' are used between the Diameter server and the client. In the Gq interface architecture, the AF (e.g., P-CSCF) acts as the client and the PDF as the server.
- Diameter has agent features that significantly reduce complexity in larger networks with many nodes and interfaces between them. For example, relay, proxy and redirect agents can be used to forward messages to the correct end-points, avoiding the necessity for senders to be aware of all configuration details and updates to them. The proxy agent can contain additional features, as needed in a particular environment. Furthermore, Diameter has a translation agent, which can be used to convert RADIUS messages to Diameter ones, and *vice versa*, between RADIUS and Diameter based interfaces.
- Diameter is stateful – that is, it contains state-information regarding each established Gq session.
- Diameter provides transport level security mechanisms; these are discussed further below.

- Diameter provides reliable transport as it uses TCP or the newer Stream Control Transmission Protocol (SCTP) instead of the User Datagram Protocol (UDP). The TCP, SCTP and UDP are defined in the IETF.
- Diameter has the capability negotiation function with which the neighbouring nodes can exchange capability support information.
- Peer entities can be easily found with the widely used IETF-defined domain name system (DNS) query mechanism.
- Diameter provides support for fail-over situations by providing access to a new node if another node fails.

The Diameter network access server application (NASREQ) is used for the Gq interface. It is an authentication, authorisation and accounting (AAA) application, from which only two messages are reused for 3GPP Gq interface purposes.

Diameter also has a set of connection setup, maintenance and disconnection messages used to manage the interface connection between the PDF and the AF. There is also negotiation capability between the nodes, but these negotiation messages cannot be forwarded – instead, they can only be delivered between two neighbouring nodes. Thus, Diameter's negotiation capability is not exchanged between the AF and the PDF if a proxy is located between the two. However, it is assumed that such capabilities are known in the 3GPP environment, as the interface is intra-operator or, if inter-operator, additional pre-agreements are needed due to other reasons such as security and charging. The Diameter and NASREQ protocols also define messages (called 'commands' in Diameter) and their parameters (called 'attribute value pairs' or AVPs in Diameter), which are used for QoS authorisation.

The 3GPP Gq interface has been defined as a vendor-specific Diameter application, where the Internet Assigned Numbers Authority (IANA)-assigned vendor identifier is '3GPP'. This is needed due to the fact that NASREQ and Diameter messages have been altered for Gq interface purposes in 3GPP. In addition, a specific IANA-allocated Diameter application identifier is needed for the Gq interface. This is used for Diameter purposes, and is separate from the Gq AF-Application identifier presented earlier in this section.

When the Gq interface is used between two operators the 3GPP-defined TS 33.210 is used to ensure security at the interface [14]. The specification defines the common security mechanisms for IP-based inter-operator interfaces in 3GPP. The specification assumes the use of the IETF-defined IP security (IPsec) (see IETF, RFC 2401 [15]). IPsec is a collection of protocols and algorithms including key management. In addition, the encapsulating security payload (ESP) of IPsec is to be used for these 3GPP interfaces. The provided security support includes data integrity (ensuring that the original content has not been altered), data origin authentication (the sender is who he claims to be), anti-replay protection (the message is not an earlier message that is being repeated maliciously) and, optionally, confidentiality (not revealing the contents to others). In 3GPP, the security mechanisms for intra-operator interfaces are left for operators and manufacturers to decide. However, in some cases the IETF-defined protocols mandate certain security mechanisms and those should then be used. For example, the Diameter specification requires that IPsec at least should be used for intra-domain traffic, but also transport layer security (TLS) could be used.

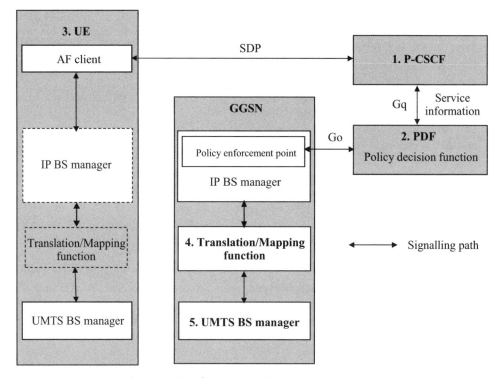

Figure 3.12   QoS mapping functions in the UE, GGSN, P-CSCF and PDF.

3.2.4.3   Combined R5 and R6 QoS parameter mapping

QoS parameter mapping is needed in several nodes of the QoS architecture (see Figure 3.12). As part of R6 the following mapping functionality is defined for IMS sessions using the SDP and PDF QoS authorisation:

- The P-CSCF maps the SDP media components onto the service information used over the Gq interface. This mapping is done at IMS session initiation or modification. The QoS information contains among other parameters the media type (e.g., audio, video) and maximum requested bit rate. In addition, the AF-Application Identifier can be employed to provide additional application information to be used for QoS authorisation in the PDF. The mapping is done as defined in 3GPP TS 29.208 [16] and those rules need to be provisioned in the P-CSCF.
- The PDF maps the service information received from the P-CSCF onto the 'authorised QoS' parameters provided to the GGSN. The maximum bit rate values received from the P-CSCF for the separate flows are added together to form the PDP context specific maximum allowed bit rate. For the QoS class, the highest of the various QoS classes is normally selected for the PDP context. The mapping is done as defined in 3GPP TS 29.208 [16] and those rules need to be provisioned in the PDF.
- At PDP context activation and modification, the UE maps the SDP QoS requirements to the requested UMTS QoS parameters. They are sent to the network to acquire the

negotiated or *authorised UMTS QoS information*, which is then used for the application service flow(s) conveyed in the PDP context. If several IP flows are multiplexed into one PDP context, the maximum bit rate should not exceed the PDP context specific maximum bit rate of the authorised UMTS QoS information. The mapping should be done as defined in 3GPP TS 29.208 [16] and those rules need to be provisioned in the UE.

- When PDP context activation and modification impacts QoS values, the GGSN maps the upper layer 'authorised QoS' information for the PDP context (e.g., the QoS class) received from the PDF onto the authorised UMTS QoS parameters (e.g., traffic class, indicating conversational). The mapping is done as defined in 3GPP TS 29.208 [16] and those rules need to be provisioned in the GGSN. After mapping, the GGSN verifies that the requested UMTS QoS values received from the SGSN do not exceed the PDF-provided authorised UMTS QoS parameters. If the requested values exceed the authorised ones, the GGSN downgrades the authorised values it returns to the SGSN. The SGSN provides these values to the UE as part of PDP context activation or modification.

- In addition to the above-defined mapping rules, the mapping between PDP context related authorised UMTS QoS parameters and DSCP values is not defined in 3GPP specifications. This mapping needs to be provisioned in the GGSN (e.g., based on operator preferences) and in the UE, if supported in it.

The specification 3GPP TS 29.208 [16] contains a more detailed description of standardised mapping rules and principles.

3.2.4.4   R6 QoS authorisation signalling flows for IMS

This section describes the combined R6 and R5 QoS authorisation signalling flows based on R6 amendments and the R5 Go interface. The IMS is used as the application, and the Iu interface (instead of Gb) to connect the GPRS network to the RAN. The complete signalling flows can be found in [16] and [1], which contain the QoS authorisation flows and the GPRS system flows, respectively.

Network element distribution between the home and visited networks is not shown in the following figures. However, the following scenarios are applicable for the IMS depending on subscriber location:

- The subscriber is located in his home network: all elements – RAN, SGSN, GGSN, and P-CSCF (3GPP SIP proxy) – are located in the home network.
- The subscriber is roaming in a visited network:
  - typically, GGSN and P-CSCF in home network, and RAN and SGSN in the visited network;
  - possibly, RAN, SGSN, GGSN and P-CSCF in the visited network.

In the following figures, in accordance with 3GPP specifications, a solid-line arrow indicates a mandatory flow and a dotted-line arrow an optional one.

Resource authorisation and reservation at the originating end

The resource authorisation signalling flow at the originating end is depicted in Figure 3.13. In the first step, the UE initiates the IMS service by sending a 'SIP invite' carrying among other things the SDP parameters containing the characteristics of the new application service. The UE normally sends the request using a general purpose PDP context but could also use a specific signalling PDP context. When the P-CSCF receives the request, it extracts the IP address and port numbers of the initiating UE. The P-CSCF verifies that the application service is supported in the IMS system, and the serving CSCF (S-CSCF) controlling the IMS session verifies that the subscriber is allowed to use the requested service. In addition, SDP parameters are negotiated between the terminating and originating ends. The P-CSCF provides the PDF with the application session characteristics used for QoS authorisation and the PDF can make the QoS authorisation decision – that is, the maximum bit rate and QoS class for the bearer. (The session characteristics and the QoS authorisation can also be provided later, if the session characteristics are not available at this point. The timing depends on the particular application.) The PDF generates the authorisation token and provides it to the P-CSCF. The token is used to identify and bind the particular GPRS bearer session (PDP context) and the application (IMS) sessions together. It also contains the PDF address that the GGSN uses to contact the correct PDF at PDP context activation that carries the requested application service. Corresponding connection information identification and QoS authorisation take place at the terminating end too.

At the initiating end, the scenario continues by the P-CSCF sending the authorisation token and negotiated SDP parameters to the UE in a SIP message. Then, the UE initiates a new PDP context for the application service. (This can be a secondary PDP context with specific QoS requirements, if an already existing primary PDP context has a suitable APN and IP address for reuse. If the UE has a suitable PDP context setup, it could also use that as such or by modifying it.) The UE sends the activate PDP context request with the requested QoS, binding information (i.e., the authorisation token, flow identifiers) and possibly the APN to the SGSN. The SGSN carries out similar checks to those in earlier releases; for example, it verifies that requested values are within the subscriber maximum authorised QoS and allowed APN information as indicated by the information provided earlier by the HLR. After that, the SGSN initiates PDP context establishment toward the GGSN. The request contains among other things the SGSN-processed negotiated QoS and APN. The GGSN validates the request and sends the authorisation request with the binding information to the PDF using the Go interface.

The PDF identifies the appropriate IMS session by means of the authorisation token and provides the GGSN with the QoS authorisation decision. The PDF may have already received the IMS session information or it may need to request that information from the P-CSCF. It may also need to contact the P-CSCF if the P-CSCF has requested in the first message to be contacted at this point.

In addition to QoS authorisation, the PDF may provide the gating decision to the GGSN. The gating decision indicates either 'gate open' or 'gate closed'. The value 'gate open' allows the GGSN to pass user traffic through in the GGSN right away, whereas the decision 'gate closed' does not permit such packet transfer.

The GGSN enforces the PDF-provided QoS (and gating) decision, and acknowledges successful PDP context establishment to the SGSN, including the negotiated QoS. The

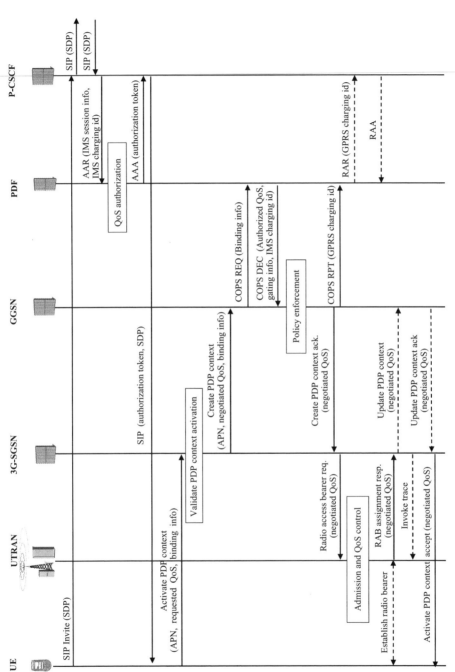

Figure 3.13 Resource authorisation and reservation at the originating end.

GGSN may also need to allocate a new IP address, unless one is already available for the PDP context. The process continues with the GGSN reporting the positive outcome to the PDF.

The SGSN requests the RAN to verify QoS values. RAN checking mainly concentrates on whether it supports the requested bearer service characteristics and whether it has resources available for it. In addition, the RNC and the base station reserve resources according to the request; for example, typically a dedicated traffic channel for RT bearer services. If the RAN cannot support the SGSN-requested QoS, the SGSN may initiate a new RAN bearer request toward the RNC to establish the RAN bearers. If the QoS is changed due to SGSN and RAN negotiation, the SGSN needs to carry out PDP context modification towards the GGSN to provide the updated QoS values for it. Finally, the SGSN returns a positive response to the UE for the PDP context establishment containing the negotiated QoS. The UE can either accept the context or reject it. The UE could reject the context if it felt the QoS is not sufficient for the application service, and it could initiate a new PDP context activation.

During PDP context activation, GPRS and IMS charging identifiers may be exchanged for charging correlation purposes (i.e., to be included in the generated charging data records (CDRs) created in the GPRS and IMS domains). The GPRS charging identifier (GCID) is sent from the GGSN to the PDF, and then forwarded to the P-CSCF; the ICID is provided from the P-CSCF to the GGSN via the PDF.

3GPP specifications also allow the CN to initiate PDP context activation. However, this appears not to be implemented.

Opening and closing the gate
The P-CSCF may also open or close the gate after the PDP context has been established and QoS authorisation granted. This gate action is depicted in Figure 3.14. If the P-CSCF wishes to open the gate to let the user traffic flow through, the P-CSCF sends the 'gate open' instructions to the PDF. Subsequently, the PDF forwards the 'gate open' instructions to the GGSN. The GGSN opens the relevant ports (i.e., letting the traffic pass through both from and to the UE).

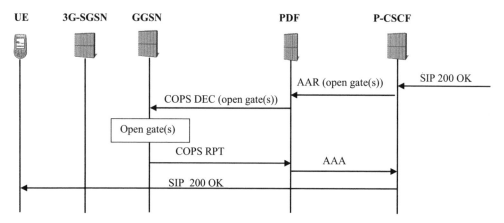

Figure 3.14   Opening and closing the gate.

Similarly, the gate can be closed as a separate action during an ongoing bearer service session. In this case, the message flow sequence is equal to the one shown to open the gate, but the 'gate open' message is replaced by the instruction to close the gate.

An exemplary situation of the use case is putting media on hold: when the media stream is put on hold the 'gate close' flow takes place, and when the flow is resumed again after a while the 'gate open' is applicable.

UE-initiated PDP context modification due to IMS session modification

A change in the IMS session may require a new QoS authorisation to replace the existing one for the PDP context. Figure 3.15 shows this scenario. There are many reasons for such a change; for example, if the user has first accessed a webpage and then started some other action on the webpage, which could be an image download requiring an increase in the bit rate.

An example scenario starts by a SIP signalling exchange in which SDP characteristics are modified. That modification is followed by the P-CSCF pushing the new IMS service information (e.g., based on SDP parameters) to the PDF. The PDF makes a new QoS authorisation for the PDP context. The UE also detects the need for the QoS change and sends the PDP context modification with the appropriate QoS parameters. The request is first checked by the SGSN and then forwarded to the GGSN for its validation. If the requested QoS is within the already authorised QoS, the GGSN does not need to contact the PDF. However, if that is not the case, then the GGSN sends the 'authorisation request with the binding information' to the PDF, which sends the new QoS and gating authorisation to the GGSN. The PDF may also inform the P-CSCF of the new authorisation; this is done if the P-CSCF has requested to be informed about such changes in the initial authorisation. Once the GGSN has received the new authorisation from the PDF it enforces it and sends the negotiated QoS back to the SGSN. At that point, the radio bearers may need to be modified. In this case, the RNC verifies the request before accepting the updated values. It is also possible that the radio bearers are re-established at this point. Re-establishment is needed if any of the timers supervising the radio resources has expired. The timers depend on the particular bearer services supported in the RAN, and those depend on the service applications and operator settings. Finally, the new QoS settings are in place for the PDP context, and the SGSN sends an acknowledgment to the UE containing the new negotiated QoS to be used for the PDP context.

In practice, the UE may also initiate a new PDP context when it invokes a new application service, instead of modifying an existing PDP context as shown above. This depends on UE implementation and capabilities, and, among other things, the number of PDP contexts the UE can support at one time.

Then, in addition to the UE, 3GPP specifications also allow the various network entities – that is, the GGSN, SGSN, and RAN – to initiate PDP context modification for various reasons. Regarding the QoS, the GGSN is in a good place to detect incompatibility between the current PDP context QoS authorisation and the QoS needed by the service application (e.g., the need to upgrade or downgrade the QoS). However, the network should not request the QoS to be upgraded beyond the original UE-requested QoS for the PDP context. In addition, if an application session service flow is terminated,

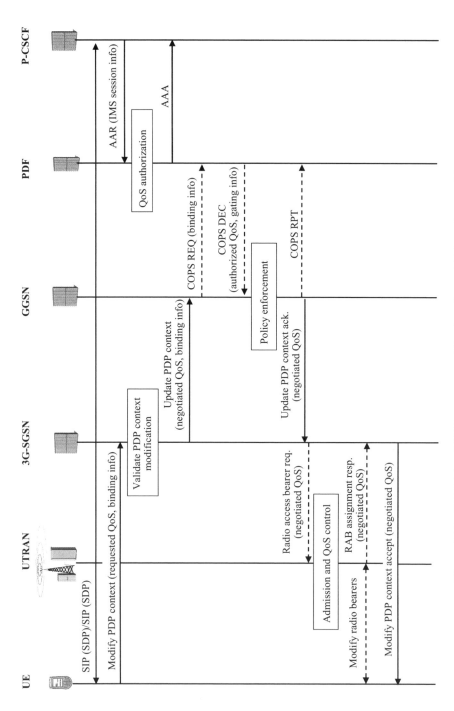

Figure 3.15 UE-initiated PDP context modification due to IMS session modification.

the P-CSCF pushes that information to the PDF, and the PDF makes a new QoS authorisation. This new authorisation may impact the filters and gating (the flows allowed to be transported over the PDP context) and also the QoS characteristics (maximum QoS class and bit rate) may need to be downgraded for the PDP context.

In operating networks it is possible that all system entities may not support all the standardised modification procedures, in which case they cannot be carried out. In addition, in these modifications the UE can only accept or reject network-indicated QoS values, but it is not able to negotiate them. This fact poses some practical limitations to the use of the various network-initiated modification procedures because if the UE rejects the request the PDP context is released.

PDP context deactivation due to IMS session release

At some point, the IMS session is released. This occurs when the application service is consumed; for example, the video stream the user has been downloading ends, the user finishes viewing a webpage or one party in a voice communication disconnects the call. The signalling flow depicting the termination scenario is shown in Figure 3.16.

In this scenario, the UE initiates the SIP session termination towards the P-CSCF and peer entity. At the receipt of session termination information, the P-CSCF removes the session information and pushes the termination information to the PDF as well. The PDF notices that no other application service flows or sessions are using the PDP context and so starts a timer to supervise PDP context removal. The UE also notices that it does not need the PDP context any more, and starts the PDP context deactivation towards the SGSN after receiving the termination acknowledgement from the P-CSCF. The SGSN forwards the request to the GGSN. The GGSN also releases the associated dynamic IP address (if used), if the terminated PDP context is the last PDP context using the address. Then, the GGSN informs the PDF about the termination. In this case, the P-CSCF has already initiated the SIP session termination sequence towards the PDF. When the GGSN acknowledges PDP context deactivation to the SGSN, the SGSN acknowledges it to the UE. Finally, PDP context deactivation is also indicated to the RAN. PDP context related resources and information are removed in the various nodes throughout the process.

Note that it is assumed above that the PDP context is no longer used for another application session or flow and, thus, due to session termination, the PDP context is no longer needed and can be removed. Of course, if this is not the case, the PDP context should be kept even if its QoS authorisation could be impacted.

In the scenario described above, the UE initiates the SIP session and the PDP context release. However, the SIP session could just as well be terminated by the peer entity (e.g., by another UE or an application server). In addition, network entities may initiate PDP context termination. For example, the GGSN, SGSN or RNC can initiate PDP context termination for various reasons, such as the subscriber running out of credit, resource prioritisation (needed for other higher priority bearer services) and error situations. In addition, the PDF or P-CSCF may trigger a GGSN-initiated PDP context deactivation, if an application session is terminated or QoS authorisation removed.

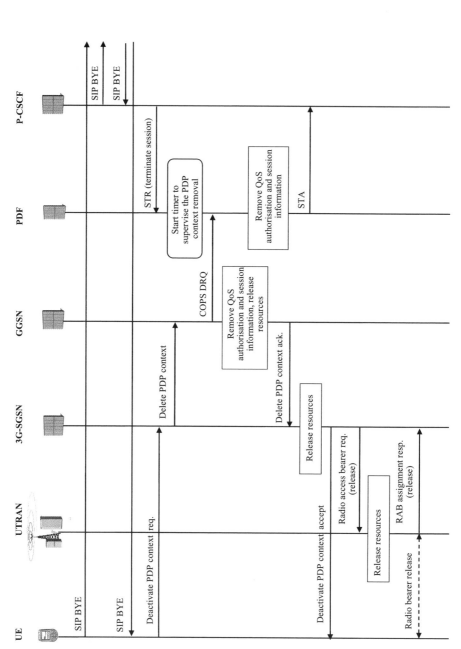

Figure 3.16 PDP context deactivation due to IMS session release.

3.2.5 Release 7 (R7)

In R7 a new work item called 'policy and charging control' (PCC) is established which concentrates on combining R5 and R6 QoS and policy control functionality with R6's flow-based charging (FBC) functionality. Whereas QoS and policy control provides QoS authorisation for the bearer based on dynamic service requirements, the FBC provides the possibility to give rules for differentiated charging.

The draft 3GPP R7 PCC architecture is depicted in Figure 3.17. This standardisation effort was in its early stages when this book was being written (end of 2005). A feasibility study, called a 'technical report' (TR) in the 3GPP language, has been approved, and Stage 2 TS 23.203 has been started. As R7 brings some important changes to the QoS architecture, this chapter gives a short description of the technical solutions agreed and discussed thus far. Additional changes are possible – if not likely – as the work evolves in 3GPP until R7 standardisation is completed at the end of 2006, or in 2007.

The PCC standardisation effort concentrates on merging the R5 and R6 QoS and policy control Go and Gq interfaces with the R6 FBC Rx and Gx interfaces. No major changes are foreseen for the already existing charging interfaces Gy and Gz in R7. As a result, the R7 architecture contains the following:

- A gateway with a policy and charging enforcement function (PCEF): R6 GGSN QoS and R6 gateway/traffic plane function (GW/TPF) functionality;
- Policy and Charging Rules Function (PCRF): R6 PDF and R6 CRF functionality,
- AF: R6 QoS and policy control AF and R6 FBC AF functionality;

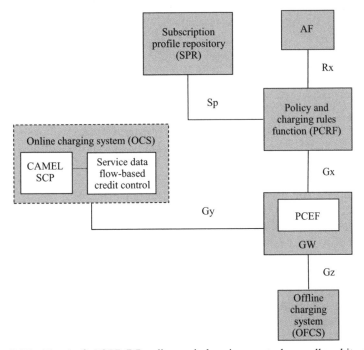

Figure 3.17 The draft 3GPP R7 policy and charging control overall architecture.

- Gx: R6 Gx and R5 Go functionality; and
- Rx: R6 Rx and R6 Gq functionality.

A new interface providing the PCRF with subscriber-related information is added in R7 to the architecture. This new interface is called 'Sp' and is also shown in Figure 3.17. The subscriber profile repository (SPR) contains subscriber information used for QoS and service authorisation, as well as charging information. Basically, the PCRF fetches the needed information when the subscriber establishes his first bearer (e.g., a PDP context for GPRS access). The PCRF could cache the information for further use, but it could access the SPR later, if needed.

In addition, the possibility to retrieve the SPR and PCRF information from the roaming subscriber (located in the visited PLMN) home network is being discussed. The needed information could be fairly generic – for example, operator-specific (the same for all the operator's subscribers) – and preconfigured in the visited PLMN gateway or PCRF. It could also be obtained from the home network either through the 'visited PCRF–home PCRF' interface or the 'visited PCRF–home SPR' interface.

There are some differences between R5 and R6 QoS and policy control and R6-based FBC solutions, which need to be addressed as these two are combined:

- The R6 FBC architecture is different from the R6 QoS architecture regarding WLAN interworking support (for more information see Chapter 5). R6 FBC functionality supports 3GPP WLAN access (WLAN interworking) and, thus, charging rules can be provided for WLAN bearers. On the other hand, R6 QoS and policy control is intended for GPRS access only. It appears that R7 PCC will support both GPRS and WLAN access.
- R6 FBC functionality allows the charging rules to be provided to the gateway without dynamic application service information. The rules can be provided to the gateway at the subscriber's first PDP context establishment. The charging rules include filters or references to pre-configured filters (or other information) in the gateway identifying service applications and their charging rules. In R7 PCC, this functionality can be used to allow specific service applications at bearer establishment without further interaction with the PCRF. QoS information can also be associated with this authorisation information provided from the PCRF to the PCEF. For example, this can be used for non-session based and some session-based service applications whereas AF involvement is used for specific session-based services (e.g., the IMS).
- The handling of user plane traffic is flow-based in the gateway for R6 charging control in the sense that the charging rules are given at the flow level. On the other hand, the QoS and policy control authorisation is at the PDP context level. Consequently, the same QoS handling is carried out for all the flows in one PDP context whereas charging can be done separately for each flow.
- The R6 Rx interface allows the radio access type (RAT), the APN and the subscriber home PLMN information (also indicating if the subscriber is roaming or not) to be provided to the CRF for charging rules authorisation. In R7 PCC this information can be used to determine the allowed service applications and bearer service QoS characteristics in addition to charging rules determination.

- The R5 Go is based on COPS, whereas the FBC Gx and Rx are based on Diameter as well as the R6 Gq interface. Then, as the R7 PCC Gx and Rx interfaces are to be based on the R6 Gx and Rx, the R7 Gx and Rx are to be based on the Diameter protocol. By the same token, as the Diameter application is NASREQ for the R6 Rx (as well as the R6 Gq) and the Diameter credit control application (DCCA) for the Gx, the R7 Rx will use NASREQ but the R7 Gx the DCCA.

- R5 and R6 QoS and policy control use the authorisation token as the bearer and application session binding method. The advantage of the authorisation token is that it is very precise and can easily be used even when the subscriber has several application and bearer sessions simultaneously ongoing. It is also dynamically allocated and indicates the PDF address to the GGSN. However, the disadvantage of the authorisation token based binding is the fact that it requires explicit support from the UE. Then, if a UE does not support the method, it cannot be used; for example, this is the case for the older (legacy) mobiles complying with pre-R5 specifications.

- R6 FBC uses the subscriber identity (number) both to find the correct CRF and as binding information. Optionally, additional information (e.g., QoS information) can be used as the binding mechanism. The main advantage of the subscriber identifier based binding solution is that it does not require any additional support from the UE. Thus, it can be used as long as it is supported in the infrastructure, where it is only needed in the gateway, PCRF and AF. This makes its use simpler in real-life systems.

- In the R7 PCC architecture the valid PCRF is found by the subscriber identity – for example, the mobile subscriber integrated services digital network number (MSISDN) – and the main binding mechanism is likely to be the subscriber identity (e.g., UE IP address) and additional information such as QoS information; the authorisation token is only used for backward compatibility.

References

[1] 3GPP, R6, TS 23.060, General Packet Radio Service (GPRS); Service Description; Stage 2, v. 6.10.0.

[2] IETF, RFC 2475, An Architecture for Differentiated Services, 1998.

[3] ITU-T Recommendations I.130, Method for the Characterization of Telecommunication Services Supported by an ISDN and Network Capabilities of an ISDN.

[4] ITU-T Recommendation Q.65, The Unified Functional Methodology for the Characterization of Services and Network Capabilities.

[5] GSM Association, Official Document, IR.34, Inter-PLMN Backbone Guidelines.

[6] T. Halonen, J. Romero and J. Melero (eds), *GSM, GPRS and EDGE Performance*, John Wiley & Sons, Second Edition, April 2003, 615 pp.

[7] 3GPP, R6, TS 25.101, User Equipment (UE) Radio Transmission and Reception (FDD), v. 6.8.0.

[8] 3GPP, R6, TS 45.005, Radio Transmission and Reception, v. 6.11.0.

[9] 3GPP, R99, TS 23.107, Quality of Service (QoS) Concept and Architecture, v. 3.9.0.

[10] 3GPP, R5, TS 23.207, End-to-end Quality of Service (QoS) Concept and Architecture, v. 5.10.0.

[11] IETF, RFC 2205, Resource ReSerVation Protocol (RSVP), 1997.

[12] IETF, RFC 2327, SDP: Session Description Protocol.

[13] 3GPP, R5, TS 29.208, End-to-end Quality of Service (QoS) Signalling Flows, v. 5.9.0.
[14] 3GPP, R6, TS 33.210, 3G Security; Network Domain Security; IP Network Layer Security, v. 6.5.0.
[15] IETF, RFC 2401, Security Architecture for the Internet Protocol, 1998.
[16] 3GPP, R6, TS 29.208, End-to-end Quality of Service (QoS) Signalling Flows, v. 6.5.0.

4

Packet Data Transfer in UMTS Cellular Networks

David Soldani and Paolo Zanier

This chapter introduces packet data transmission and the protocols used for carrying user plane application data across Universal Mobile Telecommunication System (UMTS) cellular networks. Our target is to give an overview of the roles of the supported protocols and explain the mapping between the service access points (SAPs) of the protocols in questions and the information available in the network elements to classify and collect performance measurements. As explained in Section 9.3.1, such identifiers will allow the network management system (NMS) to monitor the performance of service applications based on a particular subset of quality of service (QoS) profile attributes.

The high level functional grouping of Layer 1–3 protocols described in this chapter allows the reader to get a clear view of the protocol architecture and the transfer of a specific type of information over the air (radio) interface on common, shared or dedicated resources. More information on the topics covered in Section 4.1 for enhanced General Packet Radio Service (EGPRS) and Section 4.2 for wideband code division multiple access (WCDMA) can be found in [1]–[11] and in [12]–[26], respectively.

4.1 Packet data transfer across EGPRS networks

This section introduces the user and control plane protocol stacks implemented in EGPRS networks. Besides this, the radio channels and frame structure are also presented. The radio channels include traffic, control and packet data channels.

4.1.1 User plane protocols

The user plane protocol stack for EGPRS networks is depicted in Figure 4.1. The numbers in the figure define the SAPs between the protocol layers where the

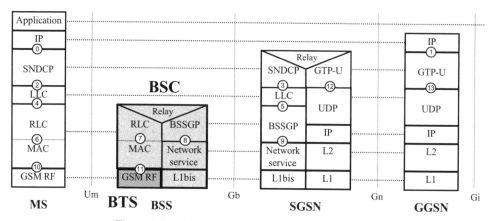

Figure 4.1   EGPRS user plane protocol stacks [1].

corresponding performance can be assessed. In the base station system (BSS), we have broken the protocol stack to show how the different entities may be deployed in the radio access network (RAN). The channel codec unit (CCU) – that is, Layer 1 functions – is in the base transceiver station (BTS), and the packet control unit (PCU) – that is, medium access control (MAC) and radio link control (RLC) functions – is deployed in the base station controller (BSC) [1].

The Packet Data Protocol (PDP) context between the 2nd-generation (2G) Serving GPRS Support Node (SGSN) and the mobile station (MS) is uniquely addressed with a temporary logical link identity (TLLI) and a network layer SAP identifier (NSAPI) pair. The NSAPI and TLLI are used for network layer routing. An NSAPI/TLLI pair is unambiguous within a routing area (RA).

The NSAPI identifies the PDP context associated with a PDP address; it is represented by a transaction identifier (TI) in some session management (SM) signalling messages. The MS produces an unused NSAPI any time it requests the activation of a PDP context. The TI is dynamically allocated by the MS for MS-requested PDP context activation, and by the network for network-requested PDP context activation. The TI is de-allocated when a PDP context has been deactivated.

The TLLI unambiguously identifies the logical link between the MS and SGSN. Within an RA, there is a one-to-one correspondence between the international mobile subscriber identity (IMSI) and TLLI, which is only known in the MS and SGSN. The TLLI is derived from the packet temporary mobile subscriber identity (PTMSI), which is allocated by the SGSN, and is valid only in the RA associated with the PTMSI.

Sub-network Dependent Convergence Protocol (SNDCP) minimises the transfer of redundant control information (e.g., TCP/IP header) and user data between the SGSN and MS through compression techniques. In addition, the SNDCP multiplexes N protocol data units (N-PDUs) from one or several NSAPIs onto one Logical Link Control (LLC) protocol SAP identifier (SAPI). The NSAPI multiplexed onto the same SAPI must use the same radio priority level, QoS traffic handling priority and traffic class. The output of the compression subfunctions are segmented (reassembled) to LLC frames of maximum length (to SNDCP packets) [2].

The relationship between TLLI/NSAPI and LLC/SNDCP is illustrated in Figure 4.2. The figure shows the end-to-end packet data transfer across the EGPRS network and relevant information stored and available in the MS, BSS and packet core network (CN). Besides this, Figure 4.2 depicts how the network layer uniquely identifies the ongoing packets belonging to distinct communications. As described in Section 9.3.1, by means of such identifiers, it is possible to classify measurements to assess the performance of the carried service applications based on a particular combination of QoS attributes.

The LLC permits information transfer between the SGSN and one or more MSs using the same physical radio resources with different service criteria. An LLC connection is identified by a data link connection identifier (DLCI), which consists of a SAPI, which identifies the SAP at the SGSN end and the MS end of the LLC interface, and the TLLI, which represents a specific MS. LLC protocol supports acknowledged, unacknowledged and ciphering types of operations. The LLC frames are multiplexed onto BSS GPRS protocol (BSSGP) virtual connections (BVCs) [3].

RLC functions are: segmentation of LLC PDUs into RLC data blocks and reassembly of RLC data blocks into LLC PDUs; segmentation of RLC/MAC control messages into RLC/MAC control blocks and reassembly of RLC/MAC control messages from RLC/MAC control blocks; backward error correction (BEC) enabling the selective retransmission of RLC data blocks. An RLC data block transfer has a unique temporary flow identity (TFI), a set of physical data channels (PDCHs) to be used for downlink transfer; and, optionally, a temporary block flow (TBF) starting time indication. A TBF is comprised of two peer entities, which are the RLC endpoints [4].

MAC enables multiple MSs to share a common transmission medium, which may consist of several physical channels. MAC may allow an MS to use several physical channels in parallel – that is, use several time slots within the time division multiple access (TDMA) frame [4].

The interface between the SGSN and BSS (denoted by Gb in Figure 4.1) allows many users to be multiplexed over a common physical resource. The communication between BSSGP entities is based on BVCs. Each BVC is used in the transport of BSSGP PDUs between: peer point-to-point (PTP) functional entities; peer point-to-multipoint (PTM) functional entities and peer signalling functional entities. There is a one PTP functional entity per cell, which is identified by a BSSGP virtual connection identifier (BVCI). Each BVC is identified by means of a BVCI, which has end-to-end significance across the Gb interface. Each BVCI is unique between two peer network service entities (NSEs). The identifier of the network service entity (NSEI), together with the BVCI, uniquely identifies a BVC (e.g., a PTP functional entity) within an SGSN. The NSEI is used by the BSS and the SGSN to determine the network service virtual connections that provide the service to a BVCI. In the downlink, the SGSN includes the IMSI (or TLLI) in the PDU and makes the following available to the BSS: MS radio access capability; QoS profile (peak bit rate), type of BSSGP SDU (signalling or data), type of LLC frame (ACK, NACK or not), precedence class (1, 2, 3) and transmission mode to be used when transmitting the LLC-PDU across the radio interface – that is, acknowledged mode (AM) (using RLC/MAC ARQ functionality) or unacknowledged mode (UM) (using RLC/MAC unit data functionality). In the uplink the BSS provides the SGSN with the following: TLLI, received from the MS, and the negotiated QoS profile (peak bit rate, the

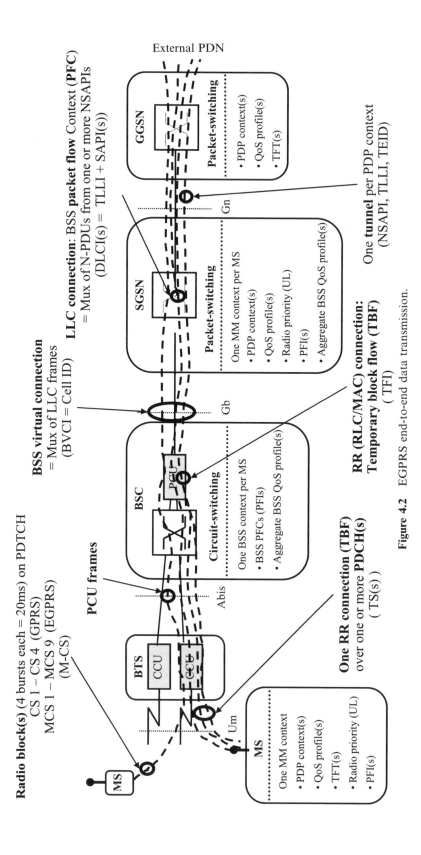

Figure 4.2 EGPRS end-to-end data transmission.

precedence used at radio access and the Tx mode used across the radio path). The SGSN obtains the BVCI and NSEI from the underlying network service [5], [6].

In packet idle mode, no temporary block flows exist; the upper layers can require the transfer of an LLC PDU that, implicitly, may trigger the establishment of a TBF and transition to packet transfer mode. The MS listens to the broadcast channel and to the paging subchannel for the paging group the MS belongs to in idle mode. In packet transfer mode, the MS is allocated radio resources providing a TBF on one or more physical channels. Concurrent TBFs may be established in opposite directions so that the transfer of LLC PDUs in RLC AM or RLC UM is provided. When selecting a new cell, the MS leaves packet transfer mode, reads the system information and enters packet idle mode where it switches to the new cell [7].

One radio block is by definition carried by four normal bursts (20 ms). The MAC header is of constant length, 8 bits, whereas the RLC header is variable in length. The RLC data field contains octets from one or more LLC PDUs. The block check sequence (BCS) is used for error detection [8]. More information on frame structure and radio channels is provided in Section 4.1.3.

In R98, four different coding schemes (i.e., CS-1, CS-2, CS-3 and CS-4) are defined for the radio blocks carrying RLC data blocks. CS-1 is used for the slow associated control channel (SACCH) – that is, 1/2 convolutional code for forward error correction (FEC) and a 40-bit FIRE code for the BCS. CS-2 and CS-3 are punctured versions of CS-1 for FEC. CS-4 has no FEC. CS-2 to CS-4 use the same 16-bit CRC for the BCS over the whole uncoded RLC data block. Table 4.1 summarises the channel coding for the packet data traffic channel (PDTCH) [8].

In R99, the radio block structure for user data transfer is different for GPRS and EGPRS. For EGPRS, a radio block for data transfer consists of one RLC/MAC header and one or two RLC data blocks. Interleaving depends on the modulation and coding scheme (MCS) used. Nine different modulation and coding schemes, MCS-1 to MCS-9, for the EGPRS radio blocks carrying RLC data blocks are defined. For all EGPRS packet control channels the corresponding GPRS control channel coding is used. Details of the EGPRS coding schemes are shown in Table 4.2. Transmission and reception data flows are the same for GPRS and EGPRS, except for EGPRS MCS-9, 8 and 7, where four normal bursts are used for carrying two RLC blocks (one RLC block within two bursts for MCS-9 and 8) [8].

Table 4.1 GPRS channel coding for PDTCH [8].

Scheme	Code rate	Radio block size (bytes)	Modulation	Data rate (kb/s)	Data rate excluding RLC/MAC headers (kb/s)
CS-1	1/2	23	GMSK	9.05	8
CS-2	≈2/3	34	GMSK	13.4	12
CS-3	≈3/4	39	GMSK	15.6	14.4
CS-4	1	54	GMSK	21.4	20

Table 4.2 EGPRS channel coding for PDTCH [8].

Scheme	Code rate	Header code rate	Modulation	RLC blocks per radio block (20 ms)	Raw data within one radio block	Data rate (kb/s)
MCS-9	1.0	0.36	8-PSK	2	2×592	59.2
MCS-8	0.92	0.36		2	2×544	54.4
MCS-7	0.76	0.36		2	2×448	44.8
MCS-6	0.49	1/3		1	592	29.6
					48 + 544	27.2
MCS-5	0.37	1/3		1	448	22.4
MCS-4	1.0	0.53	GMSK	1	352	17.6
MCS-3	0.85	0.53		1	296	14.8
					48 + 248 and 296	13.6
MCS-2	0.66	0.53		1	224	11.2
MCS-1	0.53	0.53		1	176	8.8

Note: The italic captions indicate the 6 octets (48 bits) of padding when retransmitting an MCS-8 block with MCS-3 or MCS-6. For MCS-3, the 6 octets of padding are sent every second block.

4.1.2 Control plane protocols

The user plane protocol stack for EGPRS networks is depicted in Figure 4.3. The control plane consists of protocols for control and support of user plane functions [1]:

- Controlling the packet domain network access connections, such as attaching to and detaching from the packet domain network.
- Controlling the attributes of an established network access connection, such as activation of a PDP address.
- Controlling the routing path of an established network connection in order to support user mobility.
- Controlling the assignment of network resources to meet changing user demands.

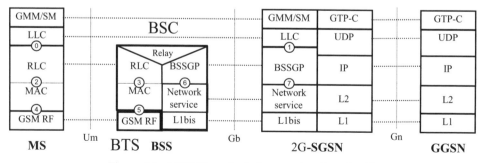

Figure 4.3 EGPRS control plane protocol stacks [1].

GPRS Mobility Management and Session Management (GMM/SM) protocol supports mobility management functionality such as GPRS attach, GPRS detach, security, RA update, location update, PDP context activation and PDP context deactivation.

GPRS Tunnelling Protocol for the control plane (GTP C) tunnels signalling messages between SGSNs and gateway GPRS support nodes (GGSNs) (Gn), and between SGSNs in the backbone network (Gp).

User Datagram Protocol (UDP) transfers signalling messages between GSNs. UDP is defined in RFC 768.

The remaining protocols in the stacks are the same as in the user plane (see Section 4.1.1).

4.1.3 Radio channels and frame structure

Each base station (BS) is assigned one or more carrier frequencies. A carrier frequency is also referred to as a TRX (transmitter–receiver, or better, transceiver), which is a combination of two frequencies (uplink and downlink frequency) carrying certain channel definitions. Each of these carrier frequencies is then divided in time, using a TDMA scheme [9], [10].

The fundamental unit of time in this TDMA scheme is called a *burst period*. Each burst period lasts 15/26 ms (or approx. 0.577 ms). A burst is a period of RF carriers modulated by a data stream, and therefore represents the physical content of a *time slot* (TS). A time slot is divided into 156.25 symbol periods. For Gaussian minimum shift keying (GMSK) modulation a symbol is equivalent to a bit. A particular bit period within a time slot is referenced by a *bit number* (BN), with the first bit period being numbered 0, and the last (1/4) bit period being numbered 156. For octagonal phase shift keying (8-PSK) modulation one symbol corresponds to three bits. In this case, the last (3/4) bit is numbered 468. The bits are mapped to symbols in ascending order according to [11]. Eight burst periods are grouped into a TDMA *frame* (120/26 ms, or approx. 4.615 ms), which forms the basic unit for the definition of logical channels. The time slots within a TDMA frame are numbered from 0 to 7 and are called the *time slot number* (TN) [10].

Radio channels implementing traffic in the radio interface are called *physical channels*. A physical channel uses a combination of frequency and time division multiplexing and it is defined as a sequence of *radio frequency channels* (RFCHs) and time slots. The RFCH sequence is determined by a function that, in a given cell with a given set of general parameters – time slot number, *mobile radio frequency channel allocation* (MA) and *mobile allocation index offset* (MAIO) – maps the TDMA frame number onto a radio frequency channel. Therefore, in a cell there is, for a physical channel assigned to a particular mobile, a unique correspondence between radio frequency channel and TDMA frame number. A given physical channel always uses the same time slot number in every TDMA frame. Hence, a *time slot sequence* is defined by a time slot number and a TDMA *frame number* (FN) sequence. A *physical channel* is therefore defined as a sequence of TDMA frames, a time slot number (modulo 8) and a *frequency hopping sequence* (FHS).

Logical channels are defined based on the type of information carried over the air interface. They can be divided into dedicated channels, which are allocated to an MS, and common channels, which are used by MSs in idle mode.

The detailed mapping of logical channels onto physical channels, the mapping of physical channels onto TDMA frame numbers, the permitted channel combinations and the operation of channels and channel combinations can be found in [10].

4.1.3.1 Hyperframes, superframes and multiframes

The organisation of burst, TDMA fames and multiframes for speech and data is illustrated in Figure 4.4. The multiframe structure for packet data channels is shown in Figure 4.5. The longest recurrent time period of the structure is called a *hyperframe* and has a duration of 3 h 28 min 53 s 760 ms (or 12 533.76 s). The TDMA frames are numbered modulo this hyperframe (TDMA frame number from 0 to 2 715 647). One hyperframe is subdivided into 2048 *superframes* which have a duration of 6.12 s. The superframe is itself subdivided in four types of *multiframes*:

- A 26-multiframe (51 per superframe) with a duration of 120 ms, comprising 26 TDMA frames. This multiframe is used to carry the TCH (and SACCH/T) and FACCH (see the following sections for a description of logical channels).
- A 51-multiframe (26 per superframe) with a duration of ≈235.4 ms (3060/13 ms), comprising 51 TDMA frames. This multiframe is used to carry the BCCH, CCCH (NCH, AGCH, PCH and RACH) and SDCCH (and SACCH/C), or PBCCH and PCCCH.
- A 52-multiframe (25.5 per superframe) with duration of 240 ms, comprising 52 TDMA frames. This multiframe is used to carry the PBCCH, PCCCH (PNCH, PAGCH, PPCH and PRACH), PACCH, PDTCH, and PTCCH.

4.1.3.2 Time slots and bursts

Four different types of bursts exist in the system:

- *Normal burst*: this burst is used to carry information on traffic and control channels, except for the RACH, PRACH and CPRACH. It contains 116 encrypted symbols and includes a guard time of 8.25 symbol duration (≈ 30.46 µs).
- *Frequency correction burst*: this burst is used for frequency synchronisation of the mobile. It is equivalent to an unmodulated carrier, shifted in frequency, with the same guard time as the normal burst. It is broadcast together with the broadcast control channel (BCCH). The repetition of frequency correction bursts is also named *frequency correction channel* (FCCH).
- *Synchronisation burst*: this burst is used for time synchronisation of the mobile. It contains a long training sequence and carries the information of the TDMA frame number and *BS identity code*. It is broadcast together with the frequency correction burst. The repetition of synchronisation bursts is also named the *synchronisation channel* (SCH).
- *Access burst*: this burst is used for random access and is characterised by a longer guard period (68.25 bit duration or 252 µs) to cater for burst transmission from a mobile which does not know the timing advance at the first access (or after handover).

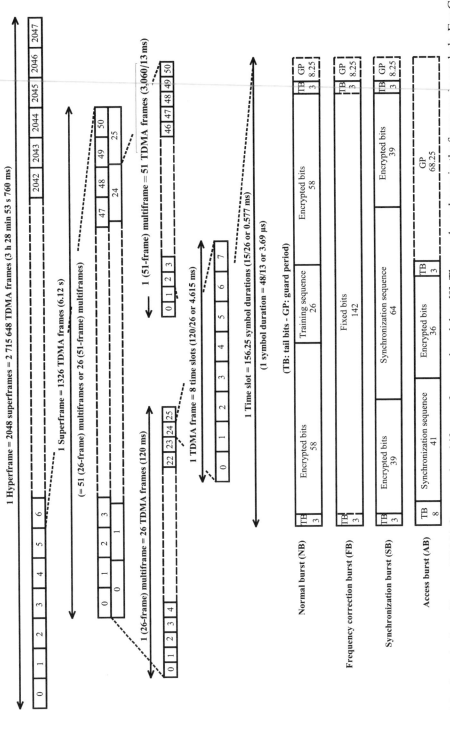

Figure 4.4 Organisation of bursts, TDMA frames and multiframes for speech and data [9]. The numbers shown in the figure are in symbols. For GMSK modulation, one symbol is one bit. For 8PSK modulation, one symbol is three bits.

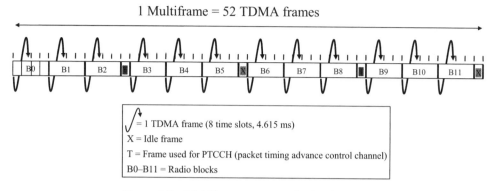

Figure 4.5 Multiframe structure for PDCH [10].

4.1.3.3 Traffic channels

Traffic channels are used for carrying either encoded speech or user data in circuit-switched (CS) mode. Traffic channels for the uplink and downlink are separated in time by three burst periods, so that the MS does not have to transmit and receive simultaneously, thus simplifying the electronics. The traffic channels are:

- *Full-rate traffic channel* (TCH/F): this channel carries information at a gross rate of 22.8 kb/s. TCH/Fs are defined using a 26-frame multiframe, or group of 26 TDMA frames. Of the 26 frames, 24 are used for traffic, one is used for the SACCH and one is currently unused (see Figure 4.4).
- *Half-rate traffic channel* (TCH/H): this channel effectively doubles the capacity of a system (i.e., speech coding at 7 kb/s, instead of 13 kb/s).
- *Enhanced full-rate traffic channel* (E-TCH/F): this channel carries information at a gross rate of 69.6 kb/s including the stealing symbols.

4.1.3.4 Common control channels

Common control channels can be accessed both by idle mode and dedicated mode mobiles. They are used by idle mode mobiles to exchange the signalling information required to change to dedicated mode. Mobiles already in dedicated mode monitor the surrounding BSs for handover and other information. The common control channels are defined within the 51-frame multiframe, so that dedicated mobiles, using the 26-frame multiframe TCH structure, can still monitor control channels. Common control channels include:

- *Broadcast control channel* (BCCH): this channel continually broadcasts, in the downlink, information including BS identity, frequency allocations and frequency-hopping sequences.
- *Frequency correction channel* (FCCH) and *synchronisation channel* (SCH): these channels are used to synchronise the mobile to the time slot structure of a cell by

defining the boundaries of burst periods, and the time slot numbering. Every cell in a GSM network broadcasts exactly one FCCH and one SCH, which are by definition on time slot number 0 (within a TDMA frame).

- *Random access channel* (RACH): this channel is the Slotted ALOHA channel used by the mobile to request access to the network.
- *Paging channel* (PCH): this channel is used to alert the MS of an incoming call.
- *Notification channel* (NCH): this channel exists in the downlink only; it is used to notify MSs of voice group and voice broadcast calls.
- *Access grant channel* (AGCH): this channel is used to allocate an SDCCH to a mobile for signalling (in order to obtain a dedicated channel), following a request on the RACH.

4.1.3.5   Dedicated control channels

Dedicated control channels are used for signalling between the network and the MS. They comprise:

- *Stand-alone dedicated control channel* (SDCCH): this channel is used to provide a reliable connection for signalling and short message services (SMS); it may be combined with CCCH (SDCCH/4). The SACCH/C is used to support this channel.
- *Slow associated control channel* (SACCH): this channel provides a relatively slow signalling connection. The SACCH is associated with either a TCH (SACCH/TH or SACCH/TF) or SDCCH (SACCH/C4 or SACCH/C8). The SACCH can also be used to transfer SMS messages if associated with a TCH.
- *Fast associated control channel* (FACCH): the FACH/F or FACH/H appears in place of the TCH/F or TCH/H when lengthy signalling is required between a GSM mobile and the network while the mobile is in call. The channel is indicated by the use of stealing flags in the normal burst. Typical signalling where this may be employed is during call handover.

All associated control channels have the same direction (bidirectional or unidirectional) as the channels with which they are associated.

4.1.3.6   Packet data channels

Packet data channels are also defined for dedicated and common traffic. They include:

- *Packet random access channel* (PRACH): the uplink PRACH is used by the MS to initiate uplink transfer (signalling) or for sending data; it is mapped onto one or several physical channels.
- *Packet paging channel* (PPCH): the downlink PPCH is used to page an MS prior to downlink packet transfer; it can be used for paging of both CS (Class A and CB GPRS MSs) and packet-switched (PS) data services; it is mapped onto one or several physical channels in the same way as done for the PCH.

- *Packet access grant channel* (PAGCH): the downlink PAGCH is used in the packet transfer establishment phase to send the resource assignment to an MS prior to packet transfer; it is mapped onto one or several physical channels.
- *Packet notification channel* (PNCH): the downlink PNCH is used to send a PTM-M (Point To Multipoint–Multicast) notification to a group of MSs prior to a PTM-M packet transfer. The PNCH is mapped onto one or several blocks on the PCCCH.
- *Packet broadcast control channel* (PBCCH): the downlink PBCCH is for system information. If not allocated in the cell, the packet data specific system information is broadcast on the BCCH. It is mapped onto one or several physical channels in the same way as done for the BCCH. The existence of the PCCCH, and consequently the existence of the PBCCH, is indicated on the BCCH.
- *Packet timing advance control channel, uplink* (PTCCH/U): this channel is used to transmit random access bursts to allow estimation of the timing advance for one MS in packet transfer mode. Two defined frames of a multiframe are used to carry the PTCCH.
- *Packet timing advance control channel, downlink* (PTCCH/D): this channel is used to transmit timing advance information updates to several MSs. One PTCCH/D is paired with several PTCCH/Us. Two defined frames of a multiframe are used to carry the PTCCH. Four normal bursts comprising a radio block are used for carrying the channels.
- *Packet data traffic channel* (PDTCH): this channel is allocated for unidirectional data transfer, either uplink (PDTCH/U) or downlink (PDTCH/D). It is temporarily dedicated to one MS or to a group of MSs in the PTM-M case. One MS may use multiple PDTCHs in parallel for individual packet transfer in multislot operation, and all packet data traffic channels may be used for mobile-terminated packet transfer. Up to eight PDTCHs with different time slots but with the same frequency parameters may be allocated to one MS at the same time. One PDTCH is mapped onto one physical channel.
- *Packet associated control channel* (PACCH): this channel is of a bidirectional nature and conveys signalling information (e.g., acknowledgements and power control information) related to a given MS. It carries resource assignment and reassignment messages, comprising the assignment of capacity for PDTCH(s) and for further occurrences of the PACCH. The PACCH shares resources with PDTCHs, which are currently assigned to one MS. An MS that is currently involved in packet transfer can be paged for CS services on the PACCH. This channel is dynamically allocated on the block basis on the same physical channel as carrying PDTCHs.

4.1.4 Mapping of packet data channels

As illustrated in Figure 4.5, mapping of logical channels in time is defined by a multiframe structure of 52 TDMA frames, divided into 12 blocks (of 4 frames), 2 idle frames and 2 frames used for the PTCCH. B0 is used as the PBCCH when allocated, and if required up to 3 more blocks on the same PDCH can be used as additional PBCCHs. On any PDCH with a PCCCH (with or without PBCCH), up to the next 12 blocks in the ordered list of blocks are used for the PPCH, PAGCH, PNCH, PDTCH or PACCH in the downlink. On an uplink PDCH that contains a PCCCH, all blocks in the multiframe can be used as the PRACH, PDTCH or PACCH. The mapping of channels onto multiframes is controlled by several parameters broadcast on the PBCCH. On a PDCH that does not contain a PCCCH, all blocks can be used as the PDTCH or PACCH. Two frames are used for the PTCCH and the two idle frames as well as the PTCCH frames can be used by the MS for signal measurements and BSIC identification. When no PCCCH is allocated, the MS camps on the CCCH and receives all system information on the BCCH. The MS monitors the uplink state flags on the allocated PDCHs and transmits radio blocks on those which currently bear the uplink state flag value reserved for the usage of the MS [8].

In short, PCCCHs are mapped together with the PBCCH (or BCCH) and PDTCH onto one or several physical channels according to the 52-multiframe. If the PCCCH (PNCH, PAGCH, PPCH and PRACH) is not allocated in the cell, the CCCH (PCH, RACH, AGCH and NCH) is used to initiate the packet data transfer. Possible channel combinations are:

- PBCCH + PCCCH + PDTCH + PACCH + PTCCH
- BCCH + PCCCH + PDTCH + PACCH + PTCCH
- BCCH + CCCH + PDTCH + PACCH + PTCCH.

A *multislot configuration* consists of multiple CS or PS traffic channels together with associated control channels, allocated to the same MS. The multislot configuration occupies up to eight basic physical channels, with different time slot numbers, but with the same frequency parameters – *absolute radio frequency channel number* (ARFCN) or MA, MAIO and *hopping sequence number* (HSN) – and the same *training sequence code* (TSC).

4.2   Packet data transfer across WCDMA networks

The section introduces end-to-end packet data transmission and the combined models for protocols used to control, support and carry user plane application data across WCDMA networks. Our target is to explain the mapping between bearer services and the SAPs of protocols, and the information available in the network elements in order to classify performance counters and indicators during measurements. As already pointed out, such identifiers will allow the NMS to measure the implemented service applications based on the corresponding PDP contexts. The concepts of high-speed downlink packet access (HSDPA), introduced in 3rd Generation Partnership Project (3GPP) R5 specifications, and high-speed uplink packet access (HSUPA), defined in 3GPP R6 specifications, are presented in Sections 4.3 and 4.4, respectively, where more details on adopted protocols and radio channels are given. The high-level functional grouping into the access stratum (AS) and non-access stratum (NAS) is defined in [12]. The AS is the functional grouping of protocols specific to the access technique (i.e., radio and Iu protocols). The NAS is the functional grouping of protocols aimed at: call control (CC) for CS voice and data; session management (SM), for PS data; mobility management for circuit-switched MM and PS domains (GMM); Short Message Services (SMS) for PS and CS domains; supplementary services (SS) and RAB management for re-establishment of radio access bearer (RABs) which still have active PDP contexts.

4.2.1 User plane protocol stack

The UMTS user plane protocol stack is depicted in Figure 4.6. The numbers in the figure define the SAPs between protocol layers where the performance of the related bearer and thus the corresponding offered QoS may be assessed. The mapping of bearer services onto protocol SAPs is reported in Table 4.3.

The GPRS Tunnelling Protocol (GTP) encapsulates all PDP PDUs – that is, tunnels user data between the RNC and SGSN, and between GPRS support nodes (GSNs) in the backbone network. UDP/IP is the backbone network protocol used for routing user data and control signalling [13].

The Packet Data Convergence Protocol (PDCP) exists only in the user plane and only for services from the PS domain. The main PDCP functions are: header compression and decompression of IP data streams (e.g., TCP/IP and RTP/UDP/IP headers) at the transmitting and receiving entity, respectively; transmission of user data means that PDCP receives a PDCP SDU from the NAS and forwards it to the RLC layer and *vice versa*; support for lossless serving radio network subsystem (SRNS) relocation or lossless downlink RLC PDU size change; and maintenance of PDCP sequence numbers for radio bearers that are configured to support lossless SRNS relocation or lossless downlink RLC PDU size change [14].

Broadcast Multicast Control (BMC) protocol provides a broadcast/multicast transmission service in the user plane on the radio interface for common user data in UM. The BMC functions are: storage of cell broadcast messages (CBMs); traffic volume monitoring and radio resource (CTCH/FACH) request for the cell broadcast service (CBS); scheduling and transmission of BMC messages to terminals; and delivery of CBMs to the upper layer (NAS) [15].

Radio Link Control (RLC) protocol provides segmentation/reassembly (payloads units, PUs) and retransmission services for both user (radio bearer) and control data (signalling radio bearer). Each RLC instance is configured by Radio Resource Control (RRC) protocol to operate in one of the three modes: transparent mode (TM), where no protocol overhead is added to higher layer data; unacknowledged mode (UM), where no retransmission protocol is in use and data delivery is not guaranteed; and acknowledged mode (AM), where the Automatic Repeat reQuest (ARQ) mechanism is used for error correction. For all RLC modes, CRC error detection is performed on the physical layer and the results of the CRC are delivered to the RLC together with the actual data. Other relevant functions of the RLC are: in-sequence delivery of upper layer PDUs; duplicate detection of RLC PDUs; flow (rate) control of the peer RLC transmitting entity; sequence number check in AM RLC to guarantee the integrity of reassembled PDUs and provide a mechanism for the detection of corrupted RLC SDUs through checking sequence numbers in RLC PDUs when they are reassembled into an RLC SDU; protocol error detection and recovery; ciphering in the RLC layer for non-transparent RLC mode; and service data unit (SDU) discard. RLC transfer mode indicates the data transfer mode supported by the RLC entity configured for that particular radio bearer. The transfer mode for a radio bearer is the same in both uplink and downlink directions; and it is

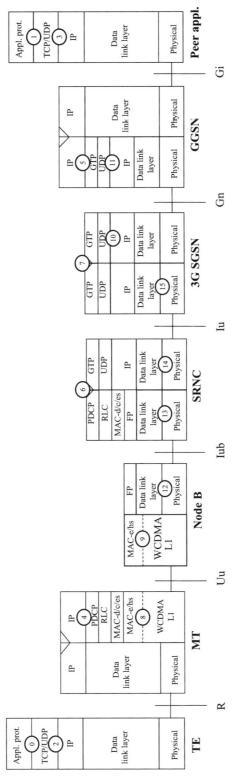

Figure 4.6 PS domain user plane protocol stack [12].

Table 4.3 Mapping of bearer services onto protocol service access points.

Bearer service (BS)	Service access point (SAP)	
Service applications	0	1
Network services	2	3
UMTS bearer service	4	5
Radio access bearer service	4	7
Core network bearer service	7	5
Radio bearer service	4	6
RAN access bearer service	6	7
Backbone network service	10	11
Physical bearer service	12 (14)	13 (15)
UTRA FDD	8	9

determined by admission control in the serving RNC (SRNC) from the RAB attributes and CN domain information. RLC transfer mode affects the configuration parameters of outer-loop power control in the RNC and the user bit rate. The quality target is not affected if TM or UM RLC is used, whilst the number of retransmissions should be taken into account if AM RLC is employed. The user bit rate is affected by the transfer mode of the RLC, since the length of Layer 2 headers is: 16 bits for AM; 8 bits for UM and 0 bits for TM. Hence, the user bit rate for network dimensioning is given by the Layer 1 bit rate reduced by the Layer 2 header bit rate. The RLC provides logical link control over the radio interface. There may be several simultaneous RLC links per UE and each link is identified with a bearer ID [16].

Medium Access Control (MAC) protocol controls the access (request and grant) procedures for the radio channel. The functionality of the MAC layer includes: mapping of logical channels onto the appropriate transport channels; selection of the appropriate transport format (TF) for each TCH depending on the current source rate; priority handling between data flows of one user equipment (UE), when selecting the transport format combination (TFC) in the given transport format combination set (TFCS); priority handling between UEs by means of dynamic scheduling of common transport channels, shared transport channels and for the dedicated E-DCH transport channel; identification of UEs on common transport channels; multiplexing/demultiplexing of upper layer PDUs onto/from transport blocks delivered to/from the physical layer on common transport channels (service multiplexing for common transport channels, since the physical layer does not support multiplexing of these channels); multiplexing/demultiplexing of upper layer PDUs onto/from transport block sets (TBSs) delivered to/from the physical layer on dedicated transport channels (service multiplexing for dedicated transport channels, this function can be utilised when several upper layer services (e.g., RLC instances) can be mapped efficiently onto the same transport channel); traffic volume measures on logical channels and reporting to RRC (based on the reported traffic volume information, RRC performs transport channel switching decisions); transport channel-type switching (execution of switching between common and dedicated transport channels based on a switching decision derived by RRC); ciphering (for transparent RLC mode); access service class (ASC) selection for RACH transmission; hybrid ARQ (HARQ) functionality for HS-DSCH and E-DCH transmission; in-sequence delivery and assembly/disassembly of higher layer PDUs on the HS-DSCH; and in-sequence delivery and assembly/disassembly of higher layer PDUs on the E-DCH [17]–[19].

The data stream(s) is/are characterised by one or more frame protocols (FPs) specified for that interface [20].

A PDP context is a virtual communication pipe established between the UE and the GGSN (SAPs 4 and 5 in Figure 4.6) for delivering the data traffic stream. The PDP context is defined in the UE, SGSN and GGSN by:

- A PDP context identifier (index of the PDP context).
- A PDP type (e.g., PPP or IP).
- A PDP address (e.g., an IP address).
- An access point name (APN) (label describing the access point to the packet data network).
- A QoS profile (bearer service attributes).

There is a one-to-one correspondence between the PDP context, UMTS bearer and RAB, as well as between the RAB and the radio bearer service, which, however, can be carried by more transport channels of the same type at the radio interface. A QoS profile is associated with each PDP context. The QoS profile is considered to be a single parameter with multiple data transfer attributes, as illustrated in Table 4.4 with RAB attributes.

4.2.2 Control plane protocol stack

The UMTS control plane protocol stack is depicted in Figure 4.7. The GMM protocol supports mobility management functionality such as attach, detach, security and RA update.

The SM protocol supports PDP context activation, modification, deactivation and preservation procedures.

The SMS protocol supports mobile-originated and mobile-terminated short messages.

RAN Application Part (RANAP) encapsulates and carries higher layer signalling, handles signalling between the 3G SGSN and Iu mode RAN, and manages the GTP connections on the Iu interface. RANAP is specified in 3GPP TS 25.413. The layers below RANAP are defined in 3GPP TS 25.412 and 3GPP TS 25.414.

The RRC protocol handles the signalling of Layer 3 between UEs and the UTRAN. RRC performs the following functions: broadcast of information provided by the NAS (CN); as an example RRC may broadcast CN location service area information related to some specific cells; broadcast of information related to the AS (typically, cell-specific information); establishment, re-establishment, maintenance and release of an RRC connection between the UE and UTRAN; establishment, reconfiguration and release of radio bearers; assignment, reconfiguration and release of radio resources for the RRC connection; evaluation, decision and execution of handover; cell reselection and cell/area update procedures; paging and notification to selected UEs; routing of higher layer

Table 4.4 Value ranges for RAB attributes for UTRAN and GERAN in 3GPP R6 [20].

Attribute/Traffic class	Conversational	Streaming	Interactive	Background
Maximum bit rate (kb/s)	$\Leftarrow 16\,000^{3}$	$\Leftarrow 16\,000^{3}$	$\Leftarrow 16\,000 - \text{overhead}^{3}$	$\Leftarrow 16\,000 - \text{overhead}^{3}$
Deliver order	Yes/No	Yes/No	Yes/No	Yes/No
Maximum SDU size (octets)	$\Leftarrow 1500$ or 1502^{2}	$\Leftarrow 1500$ or 1502^{2}	$\Leftarrow 1500$ or 1502^{2}	$\Leftarrow 1500$ or 1502^{2}
SDU format information[1]	See [20]	See [20]		
Delivery of erroneous SDUs	Yes/No/-	Yes/No/-	Yes/No/-	Yes/No/-
Residual BER	$5*10^{-2} - 10^{-6}$	$5*10^{-2}\,10^{-6}$	$4*10^{-3}, 10^{-5}, 6*10^{-8}$	$4*10^{-3}, 10^{-5}, 6*10^{-8}$
SDU error ratio	$10^{-2} - 10^{-5}$	$10^{-1} - 10^{-5}$	$10^{-3}, 10^{-4}, 10^{-6}$	$10^{-3}, 10^{-4}, 10^{-6}$
Transfer delay (ms)	$80 - \text{max value}$	$250 - \text{max value}$		
Guaranteed bit rate (kb/s)	$\Leftarrow 16\,000^{3}$	$\Leftarrow 16\,000^{3}$		
Traffic handling priority			1, 2, 3	
Allocation/Retention priority[1]	1, 2, ..., 15	1, 2, ..., 15	1, 2, ..., 15	1, 2, ..., 15
Source statistic descriptor	Speech/Unknown	Speech/Unknown		
Signalling indication			Yes/No	

[1] This parameter is limited to the values 1, 2 and 3 for GERAN when the Gb bearer service is used.
[2] Valid for PPP only.
[3] In case of GERAN the highest bit rate value is 473.6 kb/s.

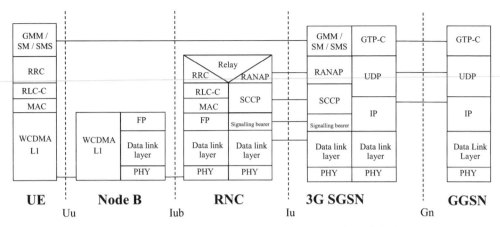

Figure 4.7 PS domain – control plane protocol stack [12].

PDUs; control of requested QoS (this includes the allocation of a sufficient number of radio resources); UE measurement reporting and control of reporting; outer-loop power control (the RRC layer controls setting of the target of the closed-loop power control); control of ciphering; arbitration of radio resources on uplink DCH (this function controls the allocation of radio resources on the uplink DCH on a fast basis, using a broadcast channel to send control information to all involved users); initial cell selection and reselection in idle mode; integrity protection; initial configuration for CBS; configuration for CBS discontinuous reception; timing advance control; Multimedia Broadcast Multicast Service (MBMS) control (the RRC controls the operation of MBMS point-to-point and point-to-multipoint radio bearers). The RRC is specified in 3GPP TS 25.331.

The RLC-C protocol offers logical link control over the radio interface for the transmission of higher layer signalling messages and SMS. RLC-C is defined in 3GPP TS 25.322.

GPRS Tunnelling Protocol for the control plane (GTP-C) is used for signalling messages between SGSNs and GGSNs (Gn), and between SGSNs in the backbone network (Gp).

User Datagram Protocol (UDP) is the transport protocol for signalling messages between GSNs. UDP is defined in RFC 768.

4.2.3   Radio interface protocol architecture and logical channels

The radio interface protocol architecture and the connections between protocols are shown in Figure 4.8. Each block represents an instance of the corresponding protocol. The dashed lines stand for interfaces through which the RRC protocol controls and configures the lower layers. The SAPs between MAC and the physical layer and between the RLC and MAC sublayers provide the *transport channels* (TrCHs) and the *logical channels* (LoCHs), respectively. TrCHs are specified for data transport between physical layer and Layer 2 peer entities, whereas logical channels define the transfer of a specific type of information over the radio interface.

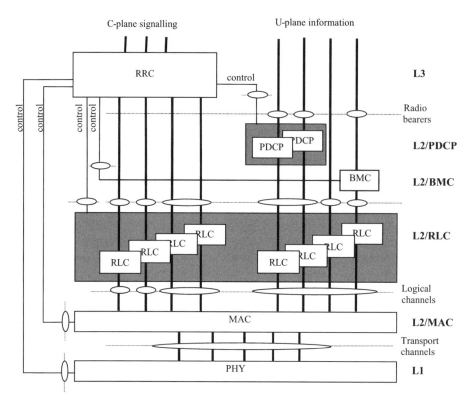

Figure 4.8 UTRA FDD radio interface protocol architecture [12].

The logical channels are divided into two groups: control channels and traffic channels. The control channels are used for transfer of control plane information and the traffic channels are used for the transfer of user plane information only [12].

The *control channels* are:

- *Broadcast control channel* (BCCH), for broadcasting system control information in the downlink.
- *Paging control channel* (PCCH), for transferring paging information in the downlink (used when the network does not know the cell location of the UE, or the UE is in cell-connected state).
- *Common control channel* (CCCH), for transmitting control information between the network and UEs in both directions (commonly used by UEs having no RRC connection with the network and by UEs using common transport channels when accessing a new cell after cell reselection).
- *Dedicated control channel* (DCCH). PTP bidirectional channel for transmitting dedicated control information between the network and a UE (established through a RRC connection setup procedure).
- *MBMS point-to-multipoint control channel* (MCCH). Point-to-multipoint downlink channel used for transmitting control information from the network to the UE. This

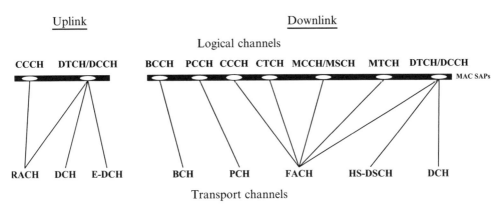

Figure 4.9 Mapping between logical channels and transport channels in the uplink and down-link directions [12].

channel is only used by UEs that receive the multimedia broadcast multicast service (MBMS).

- *MBMS point-to-multipoint scheduling channel* (MSCH). Point-to-multipoint downlink channel used for transmitting scheduling control information, from the network to the UE, for one or several MTCHs carried on a coded composite transport channel (CCTrCH). This channel is only used by UEs that receive an MBMS.

The *traffic channels* are:

- *Dedicated traffic channel* (DTCH). PTP channel, dedicated to one UE for the transfer of user information (a DTCH can exist in both uplink and downlink directions).
- *Common traffic channel* (CTCH). Point-to-multipoint unidirectional channel for transfer of dedicated user information for all or a group of specified UEs.
- *MBMS point-to-multipoint traffic channel* (MTCH). Point-to-multipoint downlink channel used for transmitting traffic data from the network to the UE. This channel is only used for an MBMS.

The mapping between logical and transport channels is depicted in Figure 4.9.

4.2.4   Radio Resource Control protocol states and state transitions

After power-on, terminals stay in idle mode until a request to establish an RRC connection is transmitted to the network. In idle mode the connection of the UE is closed on all layers of the AS. In idle mode the UE is identified by NAS identities such as the international mobile subscriber identity (IMSI), temporary mobile subscriber identity (TMSI) and packet TMSI (P-TMSI). The RNC has no information about any individual UE, and it can only address, for example, all UEs in a cell or all UEs monitoring a paging occasion [22].

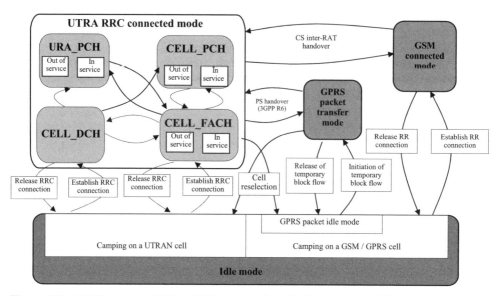

Figure 4.10   RRC states in UTRA RRC connected mode, including transitions between UTRA RRC connected mode and GSM connected mode for CS domain services, and between UTRA RRC connected mode and GSM/GPRS packet modes for PS domain services [22].

The transitions between idle mode and UTRA connected mode are illustrated in Figure 4.10. The UTRAN connected mode is entered when an RRC connection is established. The RRC connection is defined as a PTP bidirectional connection between RRC peer entities in the UE and UTRAN. A UE has either zero or one RRC connection. The RRC connection establishment procedure can only be initiated by the UE sending an RRC connection request message to the RAN. The event is triggered either by a paging request from the network or by a request from upper layers in the UE. The establishment of an RRC connection may include cell reselection, admission control, and Layer 2 signalling link establishment. When the RRC connection is established, the UE is assigned a radio network temporary identity (RNTI) to be used as its own identity on common transport channels. The release of an RRC connection can be initiated by a request from higher layers to release the last signalling connection for the UE or by the RRC layer itself in case of RRC connection failure. In case of connection loss, the UE requests re-establishment of the RRC connection. When the RRC connection is released, the signalling link and all radio bearers between the UE and the UTRAN are released [22].

As depicted in Figure 4.10, the RRC states are (a description of the physical channels can be found in Section 4.2.5.3):

- *Cell_PCH* or *URA_PCH*: in these states neither the DCCH nor DTCH are available. If the UE is 'inside the service area': it maintains up-to-date system information as broadcast by the serving cell on the BCH; performs a cell reselection process and periodic search for higher priority PLMNs; monitors the paging occasions and PICH monitoring occasions; receives paging information on the PCH; acts on RRC messages received on the PCCH and BCCH; performs a measurement process according to measurement control information; maintains up-to-date BMC data if it supports the CBS; acts on RRC messages received on the MCCH if it supports an MBMS and has activated an MBMS; runs the timer T305 for periodical UTRAN registration area (URA) update if the UE is in URA_PCH, or for periodical cell update if the UE is in CELL_PCH. If the UE is 'outside the service area', it performs the cell selection process to find a suitable cell.

- *Cell_FACH*: in this state the DCCH and DTCH are available. If the UE is 'inside the service area': it maintains up-to-date system information as broadcast by the serving cell on the BCH; performs the cell reselection process and measurement process according to measurement control information; runs the timer T305 for periodical cell update; acts on RRC messages received on the BCCH, CCCH and DCCH, and on the MCCH if it supports an MBMS and has activated an MBMS. If the UE is 'outside the service area', it performs the cell selection process to find a suitable cell.

- *Cell_DCH*: in this state the dedicated physical channel (DPCH), plus (if supported) the high-speed physical downlink shared channel (HS-PDSCH) and/or the enhanced dedicated physical data channel (E-DPDCH), is allocated to the UE. This state is entered from idle mode or from Cell_FACH state. In CELL_DCH the UE performs measurements according to the RRC's 'Measurement Control' message; acts on the RRC messages received on the DCCH or on the MCCH if it supports an MBMS and has activated an MBMS. The transition from Cell_DCH to Cell_FACH occurs either through the expiration of an inactivity timer or via explicit signalling.

4.2.5   Transport and physical channels

In UTRAN the data generated at higher layers is carried over the air interface using transport channels mapped onto different physical channels. The physical layer has been designed to support variable bit rate transport channels, to offer bandwidth-on-demand services and to be able to multiplex several services within the same RRC connection. The single output data stream from the coding and multiplexing unit is denoted by the CCTrCH, which may consist of one or more time-multiplexed transport channels. A CCTrCH is carried by one physical control channel and one or more physical data channels. In general, there can be more than one CCTrCH, but only one physical control channel is transmitted on a given connection or radio link (see [23] and [24]).

Two types of transport channels exist: dedicated channels and common channels. A common channel is a resource divided between all or a group of users in a cell, whereas a dedicated channel is by definition reserved for a single user. The connections and mapping between transport channels and physical channels are depicted in Figure 4.11.

4.2.5.1   Common transport channels

The common transport channels are resource-divided between all or a group of users in a cell (in-band identification of the users is needed). As depicted in Figures 4.9 and 4.11, the

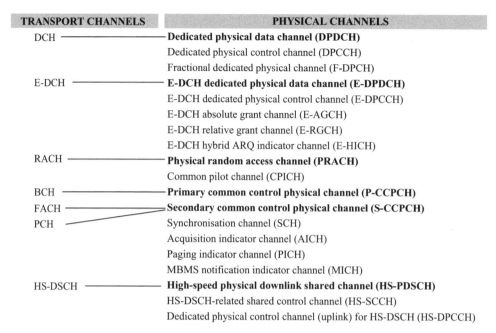

Figure 4.11 Mapping of transport channels onto physical channels [25].

common transport channels are [12]:

- *Random access channel* (RACH). The RACH carries uplink control information, such as a request to set up an RRC connection. It is further used to send small amounts of uplink packet data. It is characterised by collision risk and open-loop power control. It is mapped onto the physical random access channel (PRACH).
- *Broadcast channel* (BCH). The BCH is used to transmit information (e.g., random access codes, cell access slots, cell-type transmit diversity methods, etc.) specific to the UTRA network or to a given cell. It is mapped onto the primary common control physical channel (P-CCPCH), which is a downlink data channel only. It is broadcast in the entire coverage area of the cell.
- *Forward access channel* (FACH). The FACH carries downlink control information to terminals known to be located in a given cell. It is further used to transmit a small amount of downlink packet data with the possibility of changing rate fast (every 10 ms). There can be more than one FACH in a cell, even multiplexed onto the same secondary common control physical channel (S-CCPCH). The S-CCPCH may use different offsets between the control and data field at different symbol rates and may support only slow power control.
- *Paging channel* (PCH). The PCH carries data relevant to the paging procedure. The paging message can be transmitted in a single or several cells, according to the system configuration. It is broadcast in the entire coverage area of the cell. It is mapped onto the S-CCPCH associated with the page indicator channel (PICH).
- *High-speed downlink shared channel* (HS-DSCH). The HS-DSCH is available in 3GPP

R5 and later releases; it is a downlink transport channel shared by several UEs. It has no fast power control, but it may use link adaptation by varying modulation, coding and transmission power. It has the possibility of applying HARQ. It is always associated with a DPCH and one or several shared control channels (HS-SCCH). The HS-DSCH is transmitted over the entire cell or over only part of the cell using; for example, beam-forming antennas.

The common transport channels needed for basic cell operation are the BCH, RACH, FACH and PCH, while the use of the HS-DSCH may or may not be used by the operator.

4.2.5.2 Dedicated transport channels

In the dedicated transport channels the UEs are identified by the physical channel – that is, code and frequency for FDD and code, time slot and frequency for TDD. The dedicated transport channels are [12]:

- *Dedicated channel* (DCH). The DCH is a channel dedicated to one UE used in the uplink or downlink. It carries all user information coming from higher layers, including data for the actual service (speech frames, data, etc.) and control information (measurement control commands, UE measurement reports, etc.). It is mapped on the dedicated physical data channel (DPDCH). The DPCH is characterised by inner-loop power control and fast data rate change on a frame-by-frame basis; it can be transmitted to part of the cell using beam forming and supports soft/softer handover.
- *Enhanced dedicated channel* (E-DCH). The E-DCH is available in 3GPP R6 and later releases. It exists in the uplink only, with the possibility of changing rate each transmission time interval (TTI). It supports inner-loop power control and the possibility of applying HARQ and link adaptation by varying the coding, spreading factor and transmit power. It is mapped onto the E-DCH dedicated physical data channel (E-DPDCH).

To each transport channel, there is an associated transport format for a fixed or slow changing rate, or an associated transport format set for fast changing rate. A transport format is defined as a combination of encodings, interleaving, bit rate and mapping onto physical channels (see Section 4.2.5.5 and [23]–[24] for details). A transport format set is a set of transport formats. For example, a variable rate DCH has a transport format set (one transport format for each rate), whereas a fixed rate DCH has a single transport format.

4.2.5.3 Physical channels

Physical channels are defined by a carrier frequency, scrambling code, channelisation code (optional), time duration (start and stop instants) and, on the uplink, relative phase (0 or $\pi/2$). Scrambling and channelisation codes are specified in [26].

A *radio frame* is a processing duration which consists of 15 slots. The length of a radio frame corresponds to 38 400 chips (10 ms). A slot corresponds to 2560 chips (0.667 ms).

A *subframe* is the basic time interval for E-DCH and HS-DSCH transmission and related signalling at the physical layer. The length of a subframe corresponds to 3 slots (2 ms).

The physical channels are [25]:

- *Uplink dedicated physical channel* (UL DPCH). It consists of one dedicated physical control channel (DPCCH) and one or more dedicated physical data channels (DPDCH). Dedicated higher layer information, including user data and signalling, is carried by the DPDCH, and the control information generated at Layer 1 is mapped on the DPCCH. The DPCCH comprises: pre-defined pilot symbols (used for channel estimation and coherent detection/averaging); transmit power control (TPC) commands; feedback information (FBI) for closed-loop mode transmit diversity and the site selection diversity technique (SSDT); and, optionally, a transport format combination indicator (TFCI). There can be zero, one or several uplink DPDCHs (multi-code transmission) on each radio link, but only one uplink DPCCH is transmitted. DPDCH(s) and DPCCH are I/Q code-multiplexed with complex scrambling. Further, the uplink DPDCH can have a spreading factor from 256 (= 15 ks/s) down to 4 (= 960 ks/s), whereas the uplink DPCCH is always transmitted with a spreading factor of 256 (= 15 ks/s).

- *Uplink enhanced dedicated physical channel* (E-DPCH). This consists of the E-DCH dedicated physical data channel (E-DPDCH), the E-DCH dedicated physical control channel (uplink E-DPCCH), which are I/Q code-multiplexed with complex scrambling. The E-DPDCH is used to carry the E-DCH transport channel. There may be zero, one or several E-DPDCHs on each radio link. The E-DPCCH is a physical channel used to transmit control information associated with the E-DCH (see Section 4.2.5.6). There is at most one E-DPCCH on each radio link. The E-DPDCH can have a spreading factor from 256 (= 15 ks/s) down to 2 (= 1920 ks/s), whereas the E-DPCCH is always transmitted with a spreading factor of 256 (= 15 ks/s).

- *High-speed–dedicated physical control channel* (HS-DPCCH). The HS-DPCCH carries uplink feedback signalling related to downlink HS-DSCH transmission. The HS-DSCH-related feedback signalling consists of a HARQ acknowledgement (HARQ-ACK) and channel quality indication (CQI). There is at most one HS-DPCCH on each radio link. The HS-DPCCH can only exist together with an uplink DPCCH. The spreading factor of the HS-DPCCH is 256.

- *Physical random access channel* (PRACH). This is the only common uplink physical channel defined in 3GPP R6 specifications. Random access transmission is based on a Slotted ALOHA approach with fast acquisition indication. The UE can start random access transmission at the beginning of a number of well-defined time intervals, denoted access slots. There are 15 access slots per 2 frames. Information on what access slots are available for random access transmission is given on the BCH. Random access transmission consists of one or several preambles and a message. Each preamble consists of 256 repetitions of a signature of length 16 chips. There are a maximum of 16 available signatures. The length of the RACH message part can be 10 or 20 ms. The 10-ms message part radio frame is split into 15 slots. Each slot consists of two parts, a data part onto which the RACH transport channel is mapped and a control part that carries Layer 1 control information. The data and control parts are transmitted in parallel. A 10-ms message part consists of one message part radio frame, while a 20-ms message part consists of two consecutive 10-ms message part radio frames. The spreading factor of the data part can be 256, 128, 64 or 32. The control part consists of 8 known pilot bits to support channel estimation for coherent detection and 2 TFCI bits. The spreading factor for the message control part is 256. In case of a 20-ms PRACH message part, the TFCI is repeated in the second radio frame.

- *Downlink dedicated physical channel* (DL DPCH). This consists of a downlink DPDCH and a downlink DPCCH time-multiplexed with complex scrambling. The dedicated data generated at higher layers carried on the DPDCH is therefore time-multiplexed with pilot bits, TPC commands and TFCI bits (optional) generated by the physical layer. The DPCH may or may not include the TFCI; if the TFCI bits are not transmitted, then the DTX is used in the corresponding field. The I/Q branches have equal power and the spreading factors range from 512 (7.5 ks/s) down to 4 (960 ks/s).
- *E-DCH relative grant channel* (E-RGCH). This is a fixed-rate (60 kb/s, SF = 128) dedicated downlink physical channel carrying uplink E-DCH relative grants.
- *E-DCH absolute grant channel* (E-AGCH). The E-AGCH is a fixed-rate (30 kb/s, SF = 256) downlink physical channel carrying uplink E-DCH absolute grants.
- *E-DCH hybrid ARQ indicator channel* (E-HICH). This is a fixed-rate (its spreading factor is 128) dedicated downlink physical channel carrying the uplink E-DCH HARQ acknowledgement indicator.
- *Fractional dedicated physical channel* (F-DPCH). The F-DPCH carries only TPC commands (control information) at 1.5 ks/s (SF = 256). It is a special case of the downlink DPCCH, which is used in the case of HS-DSCH(s) without a DCH.
- *Common pilot channels* (CPICH). There are two types of common pilot channels, the primary and secondary CPICH. They are transmitted at fixed rate (15 kb/s, SF = 256) and carry a pre-defined symbol sequence. The primary common pilot channel (P-CPICH) is characterised by a fixed channelisation code and is always scrambled using a primary scrambling code. There is one P-CPICH per cell and it is broadcast over the entire cell. The P-CPICH is the phase reference for the SCH, P-CCPCH, AICH, PICH and the S-CCPCH carrying a PCH. The secondary common pilot channel (S-CPICH) is characterised by an arbitrary channelisation code of SF = 256 and is scrambled by either a primary or a secondary scrambling code. In a cell there may be zero, one or several S-CPICHs. Each S-CPICH may be transmitted over the entire cell or only over a part of the cell.
- *Primary common control physical channel* (P-CCPCH). The P-CCPCH is a fixed rate (15 ks/s, SF = 256) downlink physical channel used to carry the BCH. It is a pure data channel characterised by a fixed channelisation code. The P-CCPCH is broadcast over the entire cell and is not transmitted during the first 256 chips of each slot, where the primary SCH and secondary SCH are transmitted instead.
- *Secondary common control physical channel* (S-CCPCH). The S-CCPCH is used to carry the FACH and PCH, which can be mapped onto the same S-CCPCH (same frame) or onto separate S-CCPCHs. The S-CCPCH spreading factor ranges from 256 (15 ks/s) down to 4 (960 ks/s). Fast power control is not allowed, but the power of the S-CCPCH carrying the FACH only may be slowly controlled by the RNC. The S-CCPCH supports multiple transport format combinations (variable rate) using TFCI. It is on air only when there are data to transmit (available) and it may be

transmitted in a narrow lobe in the same way as a dedicated physical channel, if the PCH is not mapped onto the same S-CCPCH.

- *Synchronisation channel* (SCH). The SCH is a pure physical channel used in the cell search procedure. It consists of two subchannels transmitted in parallel, the primary SCH and the secondary SCH. The primary SCH is transmitted once every slot; it allows downlink slot synchronisation in the cell and is identical in every cell of the system. The secondary SCH allows downlink frame synchronisation and indicates which of the code groups the downlink primary scrambling code belongs to.

- *Acquisition indicator channel* (AICH). The AICH is a fixed-rate physical channel (SF = 256) used to indicate in a cell that the BS has received PRACH preambles (signatures). Once the BS has received a preamble, the same signature that has been detected on the PRACH preamble is then sent back to the UE using this channel. Higher layers are not involved in this procedure: a response from the RNC would be too slow to acknowledge a PRACH preamble. The AICH consists of a repeated sequence of 15 consecutive access slots of length 5120 chips.

- *Paging indicator channel* (PICH). The PICH is a physical channel used to carry paging indicators (PIs). This channel is transmitted at fixed rate (SF = 256) and is always associated with a S-CCPCH, where the PCH is mapped.

- *High-speed–shared control channel* (HS-SCCH). The HS-SCCH is a fixed-rate (60 kb/s, SF = 128) downlink physical channel used to carry downlink signalling related to HS-DSCH transmission (see Sections 4.2.5.5 and 4.2.5.6).

- *High-speed–physical downlink shared channel* (HS-PDSCH). The HS-PDSCH carries the HS-DSCH. A HS-PDSCH corresponds to one channelisation code of fixed spreading factor SF = 16 from the set of channelisation codes reserved for HS-DSCH transmission. Multicode transmission is allowed, which translates to the UE being assigned multiple channelisation codes in the same HS-PDSCH subframe, depending on its UE capability. An HS-PDSCH may use QPSK or 16QAM modulation. The channel symbol rate is 240 ks/s. All relevant Layer 1 information is transmitted in the associated HS-SCCH – that is, the HS-PDSCH does not carry any Layer 1 information.

- *MBMS indicator channel* (MICH). The MICH is a fixed-rate (SF = 256) physical channel for carrying MBMS notification indicators. The MICH is always associated with an S-CCPCH onto which a FACH transport channel is mapped.

4.2.5.4 Timing relationship between physical channels

The radio frame and access slot timing structure of downlink physical channels are illustrated in Figure 4.12. As shown in the figure, the cell system frame number (SFN) is transmitted on the P-CCPCH, which is used as a timing reference for all physical channels, since transmission timing in the uplink is derived from the timing of downlink physical channels [25].

The SCH, CPICH (primary and secondary) and P-CCPCH have identical frame timings. The S-CCPCH timing may be different for different S-CCPCHs, but the offset from the P-CCPCH frame timing is a multiple of 256 chips – that is, $\tau_{\text{S-CCPCH},k} = T_k \times 256$ chips, $T_k \in \{0, 1, ..., 149\}$. The PICH timing is $\tau_{\text{PICH}} = 7680$ chips prior to the timing of the S-CCPCH carrying the PCH with the corresponding

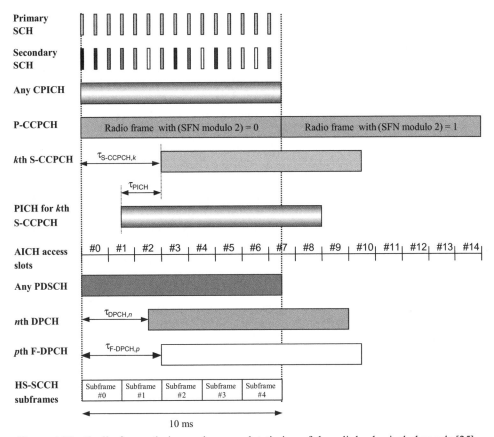

Figure 4.12 Radio frame timing and access slot timing of downlink physical channels [25].

paging information. The AICH access slot #0 starts at the same time as P-CCPCH frames with (SFN modulo 2) = 0. DPCH timing may be different for different DPCHs, but the offset from P-CCPCH frame timing is a multiple of 256 chips – that is, $\tau_{\text{DPCH},n} = T_n \times 256$ chips, $T_n \in \{0, 1, ..., 149\}$. F-DPCH timing may be different for different F-DPCHs, but the offset from P-CCPCH frame timing is a multiple of 256 chips – that is, $\tau_{\text{F-DPCH},p} = T_p \times 256$ chips, $T_p \in \{0, 1, ..., 149\}$. The start of HS-SCCH subframe #0 is aligned with the start of the P-CCPCH frames [25].

The timing relationship between the AICH and PRACH, downlink DPCCH/DPDCH and uplink DPCCH/DPDCH, F-DPCH and uplink DPCCH/DPDCH, HS-PDSCH and HS-SCCH, E-HICH, E-RGCH and E-AGCH, E-DPCCH and E-DPDCH uplink may be found in [25].

4.2.5.5   Formats and configurations

In order to describe how the mapping of TrCHs is performed and controlled by L1, some generic definitions and terms valid for all types of TrCH are introduced in this section. Further information can be found in [23] and [24].

- *Transport block* (TB) is the basic unit exchanged between L1 and MAC for L1 processing; a TB typically corresponds to an RLC PDU or corresponding unit; Layer 1 adds a CRC to each TB.
- *Transport block set* (TBS) is defined as a set of TBs, which are exchanged between L1 and MAC at the same time instance using the same transport channel. In case of HS-DSCH and E-DCH the TBS consists of one transport block only.
- *Transport block size* is defined as the number of bits in a TB. It is always fixed within a given TBS – that is, all TBs within a TBS are equally sized.
- *Transport block set size* is defined as the number of bits in a TBS.
- *Transmission time interval* (TTI) is defined as the inter-arrival time of TBSs and is equal to the periodicity at which a TBS is transferred by the physical layer on the radio interface. In 3GPP R99, it is always a multiple of the minimum interleaving period (i.e., 10ms, the length of one radio frame). In HSDPA, the TTI is 2 ms. In HSUPA both a TTI of 2 and 10 ms are supported. MAC delivers one TBS to the physical layer every TTI.
- *Transport format* (TF) is the format offered by L1 to MAC (and *vice versa*) for the delivery of a TBS during a TTI on a given TrCH. It consists of: one *dynamic part* (transport block size, transport block set size); and one *semi-static* part – TTI, type of error protection (turbo-code, convolutional code or no channel coding), coding rate, static RM parameter, size of CRC.
- *Transport format set* (TFS) is a set of TFs associated with a TrCH. The semi-static parts of all TFs are the same within a TFS. TB size, TBS size and TTI define the TrCH bit rate before L1 processing. As an example, for a DCH, assuming a TB size of 336 bits ($=$ 320 bit payload $+$ 16 RLC header), a TBS size of 2 TBs per TTI and a TTI of 10 ms, the DCH bit rate is given by $336 * 2/10 = 67.2$ kbit/s. Whereas the DCH user bit rate, which is defined as the DCH bit rate minus the RLC headers, is given by $320 * 2/10 = 64$ kbit/s. Depending on the type of service carried by the TrCH, the variable bit rate may be achieved by changing between TTIs either the TBS size only, or both the TBS and TBS size.
- *Transport format combination* (TFC), an authorised combination of the currently valid TFs that can be simultaneously submitted to Layer 1 on a CCTrCH of a UE (i.e., containing one TF from each TrCH that is a part of the combination).
- *Transport format combination set* (TFCS) is defined as a set of TFCs on a CCTrCH and a proprietary algorithm in the RNC that produces it. The TFCS is what is given to MAC by L3 for control. When mapping data onto L1, MAC chooses between the different TFCs specified in the TFCS. MAC has only control over the dynamic part of the TFC, since the semi-static part corresponds to the service attributes (quality, transfer delay) set by the admission control in the RNC. The selection of TFCs can be seen as the fast part of the RRC dedicated to MAC, close to L1. Thereby, the bit rate can be changed very fast, without any need of L3 signalling. An example of data exchange between MAC and the physical layer when two DCHs are multiplexed in the connection is illustrated in Figure 4.13. The TFCS may be produced as a Cartesian product between the TFSs of the TrCHs that are multiplexed onto a CCTrCH, every one of which is considered a vector. In theory, every TrCH can have any TF in the TFC, but in practice only a limited number of possible combinations are selected.

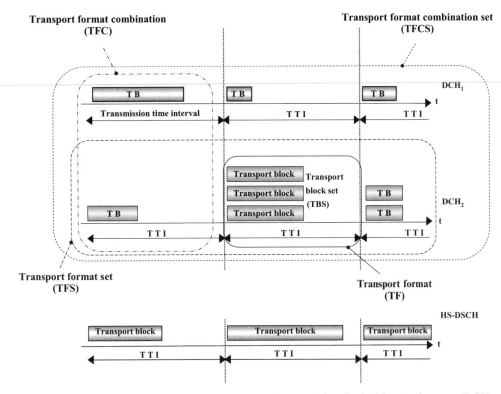

Figure 4.13 Example of data exchange between MAC and the physical layer when two DCHs and one HS-DSCH are employed [23].

- *Transport format indicator* (TFI), as pointed out in the introduction, is a label for a specific TF within a TFS. It is used in inter-layer communication between MAC and L1 each time a TBS is exchanged between the two layers on a transport channel.
- *Transport format combination indicator* (TFCI), as already explained, is used in order to inform the receiving end of the currently valid TFC and, hence, to decode, de-multiplex and transfer the received data to MAC on the appropriate TrCHs. MAC indicates the TFI to L1 at each delivery of TBSs on each TrCH. L1 then builds the TFCI from the TFIs of all parallel TrCHs of the UE, processes the TBs and appro-priately appends the TFCI to the physical control signalling (DPCCH). Through the detection of the TFCI the receiving end is able to identify the TFC.
- *Transport format for the HS-DSCH* consists of three parts – one *dynamic part*, one *semi-static part* and one *static part*. The transport format for the HS-DSCH is always explicitly signalled. There is no support of blind transport format detection. Attributes of the dynamic part are: transport block size (same as transport block set size); redundancy version/constellation; and modulation scheme. No semi-static attributes are defined. Attributes of the static part are: transmission time interval (fixed to 2 ms in FDD); error protection scheme to apply (turbo-coding, coding rate is 1/3); and size of CRC (24 bits).

- *HARQ information.* This is defined for the HS-DSCH and E-DCH. For the HS-DSCH, with the help of HARQ information the UE is able to identify the process being used for the transport block that is received on the HS-DSCH. For the E-DCH, the HARQ process is derived in an implicit way. For both the HS-DSCH and E-DCH, HARQ information also includes information that indicates whether a new data block is being transmitted for the first time or a retransmission. Furthermore, it is used to decode the received data correctly. The redundancy version is either explicitly indicated as part of HARQ information (for the HS-DSCH) or is derived from the retransmission sequence number (RSN), and the connection frame number (CFN), for the E-DCH.
- *Transport format and resource indication* (TFRI). The TFRI includes information about the dynamic part of the HS-DSCH transport format, including transport block set size and modulation scheme. The TFRI also includes information about the set of physical channels (channelisation codes) onto which the HS-DSCH is mapped in the corresponding HS-DSCH TTI.
- *Transport format for the E-DCH.* The transport format consists of three parts – one *dynamic part*, one *semi-static part* and *one static part*. The transport format for the E-DCH is always explicitly signalled. Attributes of the dynamic part are transport block size (same as TBS size) and redundancy version. The only attribute of the semi-static part is the TTI. Both TTIs of 2 ms (mandatory for certain UE categories) and 10 ms (mandatory for all terminals) are supported. Switching between the two values can be performed through L3 signalling. Attributes of the static part are error protection scheme to apply (turbo-coding 1/3) and size of CRC (24 bits).
- *E-DCH transport format combination indication* (E-TFCI). The E-TFCI includes information about the TBS size.

4.2.5.6 Physical layer models

In the uplink, when an E-DCH is not configured, the UE can use only one CCTrCH at the same time; otherwise, two CCTrCHs may be concurrently employed. Each CCTrCH has only zero (for blind transport format detection) or the corresponding TFCI. There is no physical layer multiplexing of RACHs, and there can only be one RACH and no other TrCH in a RACH CCTrCH. If the HS-DSCH is configured in the cell, only the HS-DPCCH is employed for reporting the HS-DSCH transport block acknowledgement (ACK/NACK) and channel quality indicator (CQI). The E-DCH CCTrCH consists only of one E-DCH TrCH, which is carried on the E-DPDCH(s) physical channel(s). E-DCH TFCI and E-DCH HARQ information are carried on a E-DPCCH physical channel [23].

In the downlink, multiple CCTrCHs can be transmitted simultaneously to one UE. The mapping between dedicated channels and physical channel data streams works in the same way as for the uplink. There can, however, be differences, which are mainly due to soft and softer handover. Further, pilot, TPC bits and TFCI are time-multiplexed onto the same physical channel(s). In the case of an HS-DSCH(s) without a DCH, the TPC bits are carried on F-DPCH(s). A PCH and one or several FACHs can be encoded and multiplexed together, forming a CCTrCH. Similarly, as in the DCH model there is one TFCI for each CCTrCH for indication of the transport formats used on each PCH and FACH. The PCH is associated with a separate physical channel carrying page indicators

(PIs) which are used to trigger UE reception of the physical channel that carries a PCH. A FACH or a PCH can also be individually mapped onto a separate physical channel. The BCH is always mapped onto one physical channel without any multiplexing with other transport channels, and there can only be one BCH and no other TrCH in a BCH CCTrCH. For each HS-DSCH TTI, each HS-SCCH carries HS-DSCH-related downlink signalling for one UE (i.e., TFRI, HARQ information and UE identity via a UE-specific CRC). The E-DCH active set can be identical or a subset of the DCH active set. E-DCH ACK/NACKs are transmitted by each cell of the E-DCH active set on an E-HICH. The E-HICHs of the cells belonging to the same radio link set (RLS) (i.e., same MAC-e entity and same Node B) have the same modulation and content, which is combined by the UE. The E-DCH absolute grant is transmitted by a single cell (i.e., the serving E-DCH cell) on the E-AGCH. E-DCH relative grants can be transmitted on the E-RGCH by each cell of the E-DCH active set. There is one serving E-DCH RLS (containing the serving E-DCH cell) and, optionally, one or several non-serving E-DCH radio link(s). For all UE categories, the uplink DCH capability is limited to 64 k/ps when the E-DCH is configured for the radio link [23].

4.2.5.7 Mapping of radio bearers onto transport channels

Figure 4.14 depicts a possible mapping of different radio bearer characteristics for CS and PS services onto RLC modes and types of transport channels between Layers 1 and 2. In the CS domain, guaranteed bit rate (GB) services are carried on conversational or streaming class using TM RLC and dedicated transport channels (DCHs). PS services with guaranteed bit rate run either on conversational class using UM RLC, or on streaming class using UM or AM RLC, depending on the transfer delay attribute value (see Table 4.4); for GB traffic DCHs are always used. Non-guaranteed bit rate (NGB) services are only PS, and are mapped onto interactive or background class using AM RLC – for example, PS services such as audio and video streaming, PoC, VS, WAP, MMS and Dialup may be offered using AM RLC). In this case, user data transmission is

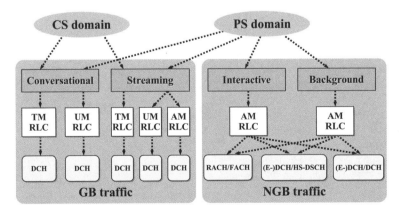

Figure 4.14 Mapping of bearer service characteristics onto RLC transmission modes and transport channels between Layers 2 and 1.

possible using the RACH and FACH in Cell_FACH state, or employing the combination of (E-)DCH with HS-DSCH or DCH in Cell_DCH state.

4.3 Introduction to high-speed downlink packet access (HSDPA)

The HSDPA concept relies on a new transport channel, the high-peed downlink shared channel (HS-DSCH), where a large amount of power and code resources are assigned to a single user during a certain TTI in a time- and/or code-multiplexed manner. The time-shared nature of the HS-DSCH provides significant trunking benefits over the DCH for bursts of high data rate traffic [27].

The following sections present the HSDPA concept, protocol architecture, radio channel structure, flow control and main functions of the physical layer. The text is based on [28]–[35], unless otherwise explicitly stated.

4.3.1 Concept description

The fundamental features of HSDPA are depicted in Figure 4.15.

In WCDMA, fast power control stabilises the received signal quality (or better E_s/N_0) by increasing the transmission power during fades in the received signal level. This causes peaks in ransmission power and subsequent power rise. Hence, there is a need to provide some headroom in total BS transmission power in order to accommodate its variations. Furthermore, WCDMA utilises variable spreading factors for long-term adjustment to average propagation conditions. Taking a different approach, a packet scheduler serves delay-tolerant traffic only under favourable radio channel conditions, avoiding transmission during inefficient, signal-fading periods. HSDPA does not use power control, so there is no power rise nor cell transmission power headroom. Instead, in order to adapt

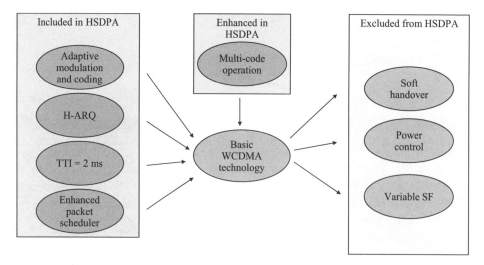

Figure 4.15 Fundamental features to be included and excluded in HSDPA.

transmission to the current channel quality, HSDPA is capable of varying modulation, coding rate (effective) and the number of used Walsh codes of fixed spreading factor (equal to 16).

Another consequence of not using closed power control is the need to minimise the channel quality variations across the TTI, which is accomplished by reducing its duration from 10 ms (minimum in R99) down to 2 ms. The fast HARQ technique is added, which rapidly retransmits the missing transport blocks and combines the soft information from the original transmission with any subsequent retransmission before the decoding process. The network may include additional redundant information that is incrementally transmitted in subsequent retransmissions (i.e., incremental redundancy).

Finally, HSDPA does not support soft handover due to the complexity of synchronising transmission from various cells. Thus, the HS-DSCH may provide full or partial coverage in the cell. Note, however, that the associated DPCH can still operate in soft handover mode, as further explained in Section 4.3.2.

4.3.2   Protocol architecture

All the HSDPA features shown in Figure 4.15 require the availability of recent channel quality information. This is why the MAC functionality in charge of the HS-DSCH channel (denote as MAC-hs) is implemented in the BS (see Figure 4.6). This is a major architecture modification compared with the R99 protocol stack.

More specifically, as illustrated in Figure 4.16, the MAC-hs is in charge of handling the HARQ functionality of every HSDPA user, distributing (scheduling) HS-DSCH

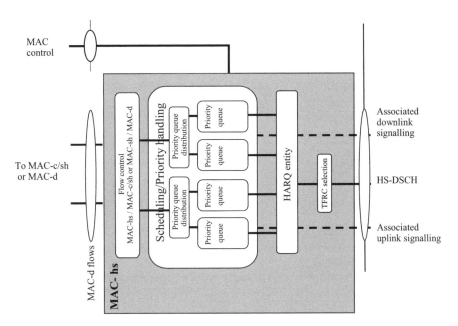

Figure 4.16   UTRAN-end MAC architecture/MAC-hs details [28], [29].

resources between all the users (MAC-d flows) according to their priority, and selecting the appropriate transport format for every TTI (link adaptation). The layers above MAC are not modified with respect to R99 architecture. Nonetheless, the RLC may only operate in either AM or UM, but not in TM due to ciphering. This is because (for TM) ciphering is done in the MAC-d, not in the RLC layer, and MAC-c/sh and MAC-hs do not support ciphering [28]–[29].

Also, MAC-hs stores the user data (MAC-d PDUs) to be transmitted across the air interface, which imposes some constraints on the minimum buffering capabilities of the BS. Movement of data queues from the CRNC to the BS creates the need for a flow control mechanism (HS-DSCH frame protocol) that aims at keeping the buffers full (see Section 4.3.7). The design of such flow control is a non-trivial task, because this functionality in co-operation with the packet scheduler is to ultimately regulate user-perceived service, which must fulfil QoS attributes according to the user subscription (e.g., the GB or the transfer delay for streaming bearers or traffic handling priority and the allocation/retention priority for interactive users).

4.3.3   Radio channel structure

The HS-DSCH transport channel can be seen as an evolution of the R99 DSCH. The HS-DSCH is mapped onto a pool of physical channels (i.e., channelisation codes), denoted as HS-PDSCHs, to be shared among all the HSDPA users in a time-multiplexed manner. The spreading factor of the HS-PDSCHs is fixed to 16, and the MAC-hs can use one or several codes (up to 15) simultaneously. Moreover, the scheduler may apply code-multiplexing by transmitting separate HS-PDSCHs to different users in the same TTI.

The uplink and downlink channel structure of HSDPA is described in Figure 4.17.

The downlink HS-SCCH is used to select users, when they are to be served and to signal the necessary information for the decoding process. As introduced in Section 4.2,

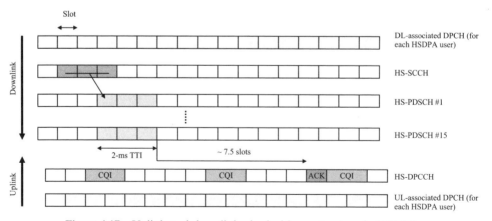

Figure 4.17   Uplink and downlink physical layer structure in HSDPA.

the HS-SCCH carries the following information [28]:

- UE ID mask, which identifies the user to be served in the next TTI.
- TFRI, which specifies the set of channelisation codes and the modulation scheme and transport block set size.
- HARQ-related information, which indicates whether the next transmission is a new data unit or a retransmission that should be combined, the associated ARQ process and information about the redundancy version.

The RNC can specify the recommended power of the HS-SCCH (offset relative to the pilot bits of the associated DPCH, [34]). HS-SCCH transmit power may be constant (possibly inefficient) or time varying according to a certain power control strategy, although the 3GPP specifications do not set any mandatory closed-loop power control modes for the HS-SCCH.

In the uplink, the HS-DPCCH carries the ARQ acknowledgements and CQI reports. Utilisation of this information is described in Section 4.3.5. In order to aid the power control operation of the HS-DPCCH, an associated DPCH is allocated for every user (radio link).

According to [34], the RNC may set the maximum transmission power on all the codes of the HS-PDSCHs and HS-SCCH in the cell. Otherwise, the BS may utilise all unused BS transmission power for these two channels; this option, though, will not be considered in the following sections. Likewise, the RNC determines the maximum number of channelisation codes to be used in the BS for HS-DSCH transmission.

4.3.4   Adaptive modulation and coding (AMC) and multicode transmission

As already mentioned, instead of using power control and variable spreading factors, in order to cope with the dynamic range of the E_s/N_0 at the UE, HSDPA adapts the modulation, the effective coding rate and number of channelisation codes to the current radio conditions. The combination of the first two mechanisms is denominated 'adaptive modulation and coding' (AMC).

Besides QPSK, HSDPA can optionally implement 16QAM modulation to increase the peak data rates for users served under favourable radio conditions. The inclusion of this high-order modulation introduces some complexity in the UE receiver, which needs to estimate the relative amplitude of the received symbols, whereas it only requires the detection of the signal phase in the QPSK case. A turbo-encoder is in charge of channel protection. The encoder is based on the R99 turbo-encoder with a rate of 1/3, although other effective coding rates within the range can be achieved by means of rate matching (i.e., puncturing and repetition). The resulting coding rate resolution has 64 steps, geometrically distributed. The combination of a modulation and a coding rate is abbreviated as a 'modulation and coding scheme' (MCS).

In addition to AMC, multicode transmission can also be considered as a tool for link adaptation purposes. If the user enjoys good channel conditions, the BS can exploit the situation by transmitting multiple parallel codes, reaching significant peak throughputs. For example, with MCS 5 and a set of 15 multicodes, a maximum peak data rate of 10.8 Mb/s can be obtained. With multicode transmission, the overall *dynamic range* of

AMC can be increased by up to $10 * \log_{10}(15) = 12\,\text{dB}$. The dynamic range of overall link adaptation combining AMC and multicode transmission is around 30 dB (i.e., 10 dB higher than that provided by variable spreading factors in WCDMA).

4.3.5 Link adaptation

The link adaptation functionality of the BS is in charge of adapting the modulation, coding format and number of codes to the current radio conditions. In order to understand the principles that should rule this functionality, let us first analyse the spectral efficiency of different modulation and coding schemes.

The bar graph of Figure 4.18 depicts the received E_b/N_0 per data bit per channelisation code, implementing the MCS of Table 4.5. Link level performance results are attained using the Pedestrian A channel profile, at 3 km/h, and 10% as the target block error rate (BLER) [37]. The figure adds the E_b/N_0 lower bound as a function of the peak data rate (PDR), which is the capacity of a channelisation code with a spreading factor of 16. This lower bound has been computed according to the channel capacity of a band-limited AWGN channel derived by Shannon [38]. Note that the theoretical link capacity represents the optimal performance limit (i.e., the lowest E_b/N_0 values). From Figure 4.18 it can be concluded that usage of the most protective transport formats (i.e., MCS 1 for the set described in Table 4.5) represents the lowest cost in terms of received E_b/N_0 per data bit.

Figure 4.19 plots the combination of the number of multicodes and MCSs that provides the highest first transmission throughput. The result assumes that the set of available MCSs is the one given in Table 4.5. The effects of different user positions are

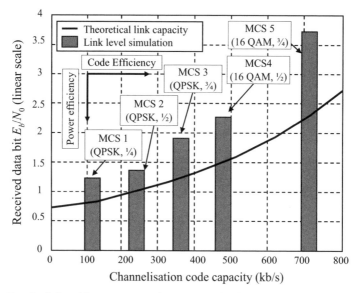

Figure 4.18 Received data bit energy to noise spectral density vs. the peak data rate (PDR) per code.

Table 4.5   Example of MCS set for HSDPA and available peak data rates.

MCS	Modulation	Effective coding rate	Bits per TTI	Peak rate with one code (kb/s)
1	QPSK	1/4	240	120
2		1/2	480	240
3		3/4	720	360
4	16QAM	1/2	960	480
5		3/4	1440	720

modelled using different geometry factor (G factor) values. The G factor is defined as the ratio between the total current own-cell power received by the UE, I_{or}, and the total current received interference from other cells, I_{oc}, plus noise [36]. For users employing low G factor values, who are typically very interference-limited with poorly received current signal quality, allocation of the most robust MCS appears as the most power-efficient solution. On the other hand, usage of higher order MCSs is attractive in code shortage allocations where the served user can afford a higher cost in terms of bit energy. With very fine resolution of the coding rate, the most spectrally efficient allocation would only resort to higher order MCSs when all the available multicodes are already used. However, in Figure 4.19 higher order MCSs (i.e., MCS 2 to 5) could achieve the highest first transmission throughput with a number of multicodes lower than the maximum available. This is due to the finite resolution of the coding rate in the MCS set of Table 4.5.

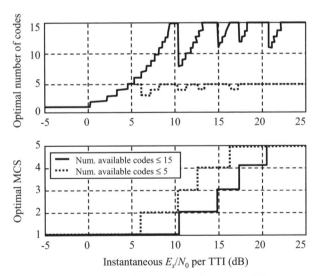

Figure 4.19   Optimal number of codes and MCSs as a function of the current E_s/N_0 per TTI. The results are attained using the Pedestrian A profile, at 3 km/h, assuming ideal channel quality estimation.

4.3.5.1   Methods for link adaptation

The selection criterion of the MCS to be employed can be based on various sources:

- *Channel quality indicator* (CQI): the CQI provides implicit information about the current signal quality received by the user. The CQI indicates the TBS size, number of codes and modulation from a set of reference values the UE is capable of supporting with a detection error no higher than 10% in the first transmission for a reference HS-PDSCH power. The RNC commands the UE to report the CQI with a certain periodicity from the set {2, 4, 8, 10, 20, 40, 80, 160 ms} (see [33]), with the possibility of disabling the reporting. The table including the set of reference CQI reports can be found in [32].
- *Power measurements on an associated DPCH*: every HS-DSCH runs a parallel DPCH for signalling purposes, whose transmission power can be used to gain knowledge about the current status of the user's channel quality. This information may be used for link adaptation and packet scheduling. The BS, at a given BLER target, may employ a table with relative E_b/N_0 offset between the DPCH and HS-DSCH for different MCSs. The advantages of using such information are that no additional signalling is required, and that it is available on a slot basis. However, this is limited to the case when the HS-DSCH and DPCH apply the same type of detector (e.g., a conventional RAKE), and cannot be used when the associated DPCH enters soft handover.
- *HARQ acknowledgements*: such an acknowledgement may provide an estimation of the user's channel quality too, although this information is expected to be less frequent than the previous information, because it is only received when the user is served. Hence, it does not provide current CQI. Note that it also lacks the channel quality resolution provided by the two above metrics, since a single information bit is reported.
- *Buffer size*: the amount of data in the MAC-hs buffer could also be applied in combination with previous information to select the transmission parameters.

For an optimal implementation of the link adaptation functionality, a combination of all the previous information sources is needed. If only one of them is to be selected, the CQI report appears the best choice due to its simplicity (as far as the network is concerned), its accuracy and its frequent report.

4.3.6   Fast hybrid ARQ

The retransmission protocol selected in HSDPA is the Stop And Wait (SAW) due to the simplicity of this form of ARQ. In SAW, the transmitter persists with transmission of the current TB until it has been successfully received before initiating transmission of the next one. Actually, up to eight SAW-ARQ processes may transmit in parallel over different TTIs for a UE [35]. The UE decoder combines the soft information of multiple transmissions of a TB at bit level. Note that this technique imposes some memory requirements on the mobile UE, which must store the soft information of unsuccessfully decoded transmissions. There exist different HARQ strategies:

- *Chase combining* (CC): every retransmission is simply a replica of the coded word employed for the first transmission. The decoder at the receiver combines these multiple copies of the transmitted packet weighted by the received SNR prior to decoding. This type of combining provides time diversity and soft combining gain at a low complexity cost and imposes the least demanding UE memory requirements of all HARQ strategies.
- *Incremental redundancy* (IR): retransmissions include additional redundant information that is incrementally transmitted if the decoding fails on the first attempt. This causes the effective coding rate to increase with the number of retransmissions. Incremental redundancy imposes demanding requirements on UE memory capabilities, and the standard only compels the UE soft memory to support the needs for chase combining [35].

4.3.7 Iub data transfer and flow control

As specified in [39], Frame Protocol (FP) is responsible for the transmission of HSDPA user data between the CRNC and BS. In principle, one UE may be associated with one or more MAC-d flows. In case of MAC-d multiplexing of different logical channels, each MAC-d flow contains HS-DSCH MAC-d PDUs for one or more priority queues (maximum 8 per MAC-d flow, maximum 8 per UE). In the MAC-d PDU header, the channel/type (C/T) field provides a means of identification of the logical channel instance when multiple logical channels (up to 15) are carried on the same MAC-d flow. Each logical channel is mapped onto a priority queue, characterised by a specific priority level called the 'scheduling priority indicator (SPI). The SPI, which ranges from 0 (low priority) to 15 (highest priority), is included only in the NBAP message used to set up the MAC-d flows (for more information see Section 5.3 and [34]). In FP, the relative priority of a HS-DSCH data frame is specified by the common channel priority indicator (CmCH-PI). The one-to-one mapping between the SPI and CmCH-PI for each user is the task of the RNC. Multiple MAC-d PDUs of the same length and same priority level (CmCH-PI) may be transmitted in one MAC-d flow in the same HS-DSCH data frame. On Iub the distinction between FP frames of different users is based on AAL2 channel identifiers (AAL2 CIDs).

When a new HSDPA connection (referred to as MAC-d flow) is set up, the CRNC can ask for capacity either using an FP HS-DSCH Capacity Request message or via the HS-DSCH initial capacity allocation as described in [34] (the latter is valid only for the first data frame transmission). When the RNC has data in its MAC-d buffers and the BS has granted capacity with a HS-DSCH Capacity Allocation message, a data frame is transmitted immediately according to the received allocation. The flow control interactions between the BS and RNC are illustrated in Figure 4.20.

The MAC-hs user plane interface towards the RNC is basically the user data buffer, which contains buffered MAC-d PDUs for the different HSDPA users that are allocated to a MAC-hs cell.

The MAC-hs flow control (FC) algorithm monitors the content of the MAC-hs user data buffer and also the rate at which data are removed from the buffer [36]. This is done using two thresholds, high and low. Comparing the buffer content with the thresholds, the MAC-hs FC algorithm allocates credits (CRs) to the user (priority queue), so that the

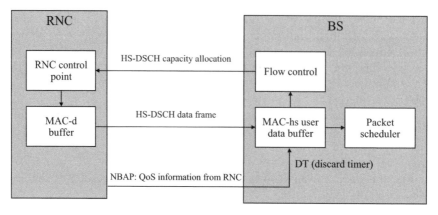

Figure 4.20 MAC-hs flow control interaction with RNC.

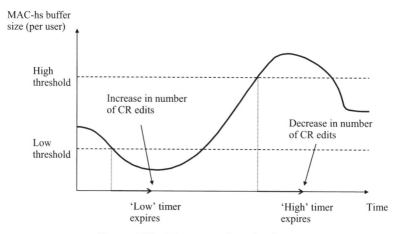

Figure 4.21 Flow control mechanism.

RNC knows how many PDUs belonging to that user it is allowed to forward to the BS in every HS-DSCH interval. The flow control mechanism is depicted in Figure 4.21.

Hence, the MAC-hs FC algorithm basically adjusts the HSDPA user bit rate over Iub as a function of the experienced HSDPA user air interface bit rate. This implies that the PDU buffering time in the BS is relatively constant, independent of the air interface bit rate.

4.3.8 MAC-hs packet scheduler

The MAC-hs packet scheduler is a key element that determines the overall behaviour of the system [29]–[40]. For each TTI, it determines which UE (priority queue), or UEs (code-multiplexing), the HS-DSCH should be allocated to and, in collaboration with the link adaptation mechanism, at what data rate. A significant increase in capacity can be obtained if, instead of allocating radio resources sequentially (round-robin scheduling), the scheduler employs channel-dependent scheduling – that is, the scheduler prioritises transmissions to UEs with favourable current channel conditions. Also, as described in Section 5.3, traffic priorities can be taken into account – for example, to prioritise streaming services ahead of background services.

When comparing scheduling algorithms, we need to distinguish between two kinds of variations in service quality:

- rapid variations in service quality; and
- long-term variations in service quality.

Rapid variations in service quality are due, for example, to multipath fading and variations in the interference level. For many packet data applications, relatively large short-term variations in service quality are acceptable or go unnoticed. Long-term variations in service quality are due, for example, to the distance between the UE and BS. These variations must often be restricted.

A practical scheduling strategy exploits short-term variations while maintaining some degree of long-term fairness between users. In principle, system throughput decreases the more fairness is enforced. Therefore, a trade-off must be reached. Typically, the higher the system load, the greater the discrepancies between different scheduling strategies.

Channel-dependent schedulers must estimate the current radio conditions of the UE. Therefore, each UE that uses high-speed services transmits regular channel quality reports to the BS via the HS-DPCCH. The scheduler might also use other information available in the BS to assess UE radio conditions. An extensive study on HSPDA packet scheduling for both NRT and RT services is provided in [37].

4.4 Introduction to high-speed uplink packet access (HSUPA)

High-speed uplink packet access (HSUPA) or enhanced uplink packet access (EUPA) is a part of the technology improvements in 3GPP R6 for uplink packet data transfer [28]. HSUPA allows users to take advantage of faster uplinks with lower latency when sending large amounts of data; it also improves the efficiency of the radio link, increasing effective throughput, without changing uplink modulation. Some examples of HSUPA-available peak data rates are reported in Table 4.6.

Table 4.6 Example of HSUPA available peak data rates.

Effective coding rate	User data rate with 1 code (kb/s)	User data rate with 2 codes (Mb/s)	User data rate with 4 codes (Mb/s)	User data rate with 6 codes (Mb/s)
2/3	640	1.28	2.56	3.84
3/4	720	1.44	2.88	4.32
4/4	960	1.92	3.84	5.76

The main characteristics of HSUPA are:

- Node B controlled uplink scheduling.
- HARQ protocol between the UE and Node B.
- Possibility of shorter TTI (2 ms).

The following sections introduce the main features of HSUPA. More information on the covered topics can be found in [19] and [41].

4.4.1 Physical layer models for HSUPA

The physical layer E-DCH model with the DCH and HS-DSCH is depicted in Figure 4.22. In the uplink, the information carried on the E-DPCCH consists of the E-TFCI, RSN and 'happy' bit (see Section 4.4.4). The E-DPCCH is sent with a *power offset* relative to the DPCCH. The power offset is provided to the terminal by the SRNC through RRC signalling.

In the example of Figure 4.22, the DPCH active set contains four cells: Cell d_1–Cell d_4. The E-DCH active set can be identical or a subset of the DCH active set, as in this example. HARQ ACK/NACKs are transmitted by each cell of the E-DCH active set on the E-HICH. The E-DCH absolute grant is only sent by the serving E-DCH cell (Cell e_s in Figure 4.22) on the E-AGCH. E-DCH relative grants can be transmitted by each cell of the E-DCH active set on the E-RGCH. Only the E-RGCH of the cells belonging to the serving RLS have the same content and, thus, can be combined by the UE. The serving E-DCH cell and the HS-DSCH serving cell must be identical (Cell H_s = Cell e_s).

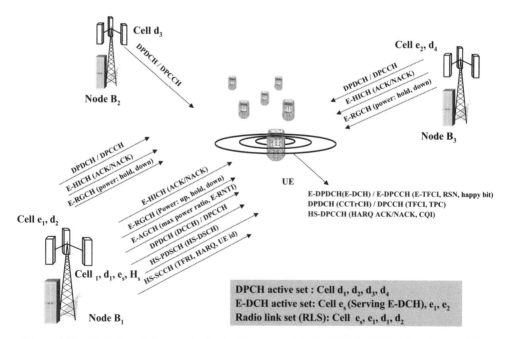

Figure 4.22 Uplink and downlink physical layer models for HSPA (HSUPA and HSDPA).

4.4.2 Protocol architecture

The protocol stack for HSUPA is illustrated in Figure 4.6. As explained thoroughly in the following sections, at the UE end the new MAC-es/-e entity handles HARQ retransmissions, scheduling and MAC-e multiplexing, and E-DCH TFC selection. In the Node B, the new MAC-e entity deals with HARQ retransmissions, scheduling and MAC-e demultiplexing. In the SRNC, MAC-es provides in-sequence delivery (reordering) of MAC-e PDUs and combines data coming from different nodes in case of soft handover. The support of E-DCH implies no change to the MAC-c and MAC-hs entities.

4.4.2.1 MAC architecture at UE end

The MAC architecture at the UE end is depicted in Figure 4.23. As shown in the figure, MAC-d C/T multiplexing is bypassed. In the MAC-e header, the data description indicator (DDI) identifies the carried logical channel, MAC-d flow and MAC-d PDU size. TSN stands for 'transmission sequence number' on the E-DCH.

The MAC-es/e handles E-DCH specific functions, such as HARQ, multiplexing and TSN setting, and E-TFC selection. In particular:

- The *HARQ entity* is responsible for storing and retransmitting MAC-e payloads; it also provides the E-TFC, retransmission sequence number (RSN) and power offset to L1. The L1 derives the redundancy version (RV) of the HARQ transmission from the RSN, CFN and, in case of a 2-ms TTI, from the subframe number. The HARQ entity is configured with RRC signalling through the MAC control SAP.
- The *multiplexing and TSN setting* entity concatenates multiple MAC-d PDUs into MAC-es PDUs, and multiplexes one or more MAC-es PDUs into a single MAC-e

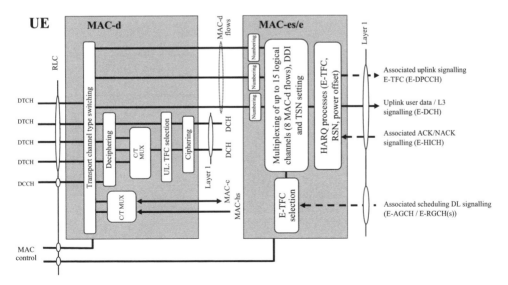

Figure 4.23 UE-end MAC architecture [19].

PDU, to be transmitted in the next TTI. The entity is also responsible for managing and setting the TSN in the MAC-es PDU for each logical channel.

• The *E-TFC selection* entity controls the multiplexing function and selects the E-DCH TFC according to the scheduling information (relative and absolute grants) it receives from the UTRAN.

4.4.2.2   MAC architecture at UTRAN end

The overall UTRAN MAC architecture is shown in Figure 4.24.

In the SRNC, there is one MAC-es entity for each UE that uses the E-DCH. The MAC-es comprises queue distribution, reordering, macrodiversity selection and disassembly. In particular:

• The *reordering queue distribution* function routes the MAC-es PDUs to the correct reordering buffer.
• The *reordering* entity sorts the received MAC-es PDUs according to the received TSN and Node B tagging (CFN and subframe number in the case of a 2-ms TTI). There is one reordering queue per logical channel.
• *Macrodiversity selection* is performed in the case of soft handover. The reordering function receives all MAC-es PDUs from each Node B in the E-DCH active set.
• The *disassembly* entity removes MAC-es PDU headers and delivers MAC-d PDUs to the MAC-d.

In the Node B, there is one MAC-e entity for each UE that uses E-DCH and one E-DCH

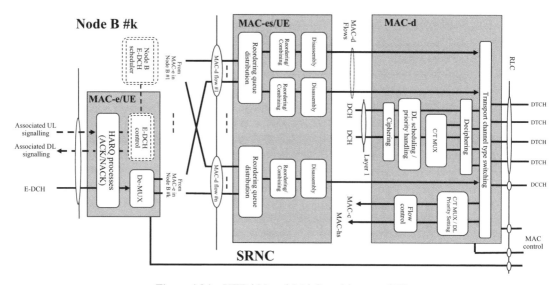

Figure 4.24   UTRAN-end MAC architecture [19].

scheduler function. In particular:

- The *E-DCH scheduling* function allocates radio resources for E-DCH reception between mobile terminals. Scheduling grants are determined and transmitted based on scheduling requests (E-DCH scheduling principles are described in Section 4.4.4).
- The *E-DCH control* function transmits scheduling grants based on the received scheduling requests.
- The *demultiplexing* function demultiplexes MAC-e PDUs, which are then forwarded to the associated MAC-d flow.
- The *HARQ* entity supports multiple processes of the Stop And Wait HARQ protocol. Each process is responsible for generating an ACK or NACK indicating the delivery status of E-DCH transmissions. The HARQ entity handles all tasks that are required by the HARQ protocol.

There is one Iub transport bearer per MAC-d flow (MAC-es PDUs carrying MAC-d PDUs of the same MAC-d flow).

4.4.3 HARQ protocol

HARQ protocol is based on synchronous downlink ACK/NACK transmission and uplink retransmissions. The number of HARQ processes depends on the TTI value (eight for 2 ms and four for 10 ms). The number of retransmissions is limited by the SRNC. The E-DCH with HARQ supports intra-Node B (softer handover) and inter-Node B (soft handover) macrodiversity. Incremental redundancy is implemented with chase combining. As instructed by the SRNC, the terminal either uses the same incremental redundancy version for all transmissions, or sets it as a function of E-TFC, RSN and transmission timing.

HARQ protocol related signalling comprises transmission of the TSN (in-band in the MAC-es header for reordering purposes) and RSN (in the E-DPCCH) in the uplink, and a report to indicate either ACK or NACK in the downlink (in the E-HICH).

In the Node B, HARQ protocol enables lower BLER requirements for the first UE transmission. As a consequence, the effective required uplink transmit power is reduced.

4.4.4 Node B controlled scheduling

The Node B issues scheduling grants to indicate to the UE the maximum amount of uplink resources it may use. Scheduling grants are generated based on scheduling requests and QoS-related information the Node B receives from the UE and SRNC, respectively (see Figure 4.22 and Section 5.4 for more details). Scheduling grants have the following characteristics:

- They are only used for E-DCH TFC selection algorithm in the UE.
- They control the maximum allowed E-DPDCH/DPCCH power ratio of active HARQ processes. For inactive processes, the power ratio is zero and the UE is not allowed to transmit scheduled data.
- They can be sent once per TTI or at a slower rate.

- Absolute grants provide an absolute limitation to the maximum amount of uplink resources the UE may use. They are valid for one UE, a group or for all terminals in the cell. An absolute grant contains the E-RNTI of the UE or group of mobiles for which the grant is intended, the maximum E-DPDCH/DPCCH power ratio the UE is allowed to use, and, in case of a 2-ms TTI, an HARQ process activation flag indicating whether the primary absolute grant activates or deactivates one or all HARQ processes. (If two identities are allocated to a UE, this flag is also used to switch the UE from its primary E-RNTI to its secondary E-RNTI for both the 2-ms and 10-ms TTI.)
- Relative grants increase or decrease the resource limitation compared with the previously used value. Updates of relative grants may be sent by the serving and non-Serving BSs as a complement to absolute grants. As illustrated in Figure 4.22, a relative grant from the serving E-DCH RLS can take one of the three values: 'up', 'hold' or 'down'; a relative grant from the non-serving E-DCH RL can be either 'hold' or 'down'. The 'hold' command is sent as a DTX. The 'down' command corresponds to an 'overload indicator'.

In the UE, the maximum allowed E-DPDCH/DPCCH power ratio for the transmission of scheduled data in active HARQ processes, called the *serving grant* (SG), is used for E-TFC selection algorithm as the maximum allowed power ratio for the transmission of scheduled data in active HARQ processes. Each absolute grant and relative grant is associated with a specific uplink E-DCH TTI (i.e., a HARQ process). This association is implicitly based on the timing of the E-AGCH and E-RGCH. The SG is updated according to the following algorithm [19]:

- Primary absolute grants always affect the SG, whereas secondary absolute grants only affect the SG if the last primary absolute grant was set to 'inactive' and, in the case of a 2-ms TTI, the process activation flag was set to 'All', or if the latest absolute grant that affected the SG was the secondary one. When transition to the secondary E-RNTI is triggered, the UE updates the SG with the latest received absolute grant on the secondary E-RNTI.
- If no 'absolute grant' is received in a TTI, the UE follows the 'relative grant' of the serving E-DCH RLS, which is interpreted relative to the UE power ratio in the previous TTI for the same HARQ process. The UE calculates its new SG by applying a *Delta* compared with its last used power ratio (see [17] for details). When the UE receives a 'hold' (i.e., DTX) from the serving E-DCH RLS, the SG remains unchanged.
- When the UE receives a 'down' command from at least one non-serving E-DCH RL, it is interpreted relative to the UE power ratio in the previous TTI for the same HARQ process as the transmission which the relative grant will affect (see [17] for the new SG calculation). In this case, the UE must ensure that its SG is not increased during one HARQ cycle by any E-AGCH or E-RGCH signalling.
- When the UE receives a scheduling grant from the serving E-DCH RLS and a 'down' command from at least one non-Serving E-DCH RL, the new SG is set to the minimum between the resulting SG from the non-serving E-DCH RL and the resulting SG from the serving RLS.

The UE requests resources from BSs in the form of *scheduling information* and *happy bit*.

The UE is always 'happy', except when it has power available to send data at higher rates, and the total buffer content would require more than X ms to be transmitted with the current SG times the ratio of active processes to the total number of processes (1 for a 10-ms TTI), where X is an RRC-configurable parameter. Scheduling information relates to the logical channels (related scheduled MAC-d flows), where the SRNC requires such reporting. Scheduling information comprises:

- *Logical channel ID* of the highest priority channel with data in its buffer.
- *UE buffer occupancy*: that is, the status of the highest priority logical channel with data in its buffer, as a fraction of the total reported buffer and total buffer status (in bytes).
- *UE power headroom* (UPH): that is, the ratio of the maximum UE transmission power and the corresponding DPCCH code power.

When the UE has scheduled data to send, the scheduling information is sent to the serving E-DCH RLS in a MAC-e PDU. This is also valid when the UE is not allowed to transmit scheduled data, because it has no serving grant available or it has received an absolute grant preventing it from transmitting in any process.

4.4.5 Non-scheduled transmissions

In order to minimise signalling overhead and scheduling delays, the SRNC may configure the UE for non-scheduled transmission. In this case, the UE may send data at any time using the E-DCH, without receiving any scheduling command from the Node B.

Typical examples of data that may use non-scheduled transmission are the signalling radio bearers (SRBs) and guaranteed bit rate (GB) services.

Non-scheduled transmissions are defined per MAC-d flow. The resource for non-scheduled transmission, denoted as *non-scheduled grant* is provided by the SRNC in terms of the maximum number of bits that can be included in a MAC-e PDU. Multiple non-scheduled MAC-d flows may be configured in parallel by the SRNC. Logical channels mapped onto a non-scheduled MAC-d flow can only transmit up to the non-scheduled grant configured for that particular MAC-d flow and cannot transmit data using a scheduling grant. Scheduled logical channels cannot use a non-scheduled grant. Scheduled grants are always considered on top of non-scheduled transmissions. The logical channels are served in the order of their priorities until the non-scheduled grant and scheduled grants are exhausted, or the maximum transmit power is reached.

References

[1] 3GPP, R99, TS 23.060, GPRS – Service Description – Stage 2, v. 3.16.0.
[2] 3GPP, R99, TS 04.65, Mobile Station (MS) – Serving GPRS Support Node (SGSN); Subnetwork Dependent Convergence Protocol (SNDCP), v. 8.2.0.
[3] 3GPP, R99, TS 04.64, Logical Link Control (LLC) Layer Specification, v. 8.7.0.
[4] 3GPP, R99, TS 04.60, Radio Link Control/Medium Access Control (RLC/MAC), v. 8.27.0.
[5] 3GPP, R99, TS 08.18, BSS GPRS Protocol (BSSGP), v. 8.12.0.
[6] 3GPP, R99, 08.58, BSC-BTS, Layer 3 Specification, v. 8.6.0.

[7] 3GPP, R6, TS 45.005, Radio Transmission and Reception, v. 6.11.0.

[8] 3GPP, R99, TS 03.64, Overall Description of the GPRS Radio Interface; Stage 2, v. 8.12.0.

[9] 3GPP, R99, TS 05.01, Physical Layer on Radio Path – General Description, v. 8.9.0.

[10] 3GPP, R99, TS 05.02, Multiplexing and Multiple Access on Radio Path, v. 8.11.0.

[11] 3GPP, R99, TS 05.04 Digital Cellular Telecommunications System; Modulation, v. 8.4.0.

[12] 3GPP, R6, TS 25.301, Radio Interface Protocol Architecture, v. 6.4.0.

[13] 3GPP, R6, TS 23.060, General Packet Radio Service (GPRS); Service Description; Stage 2, v. 6.10.0.

[14] 3GPP, R6, TS 25.323, PDCP Protocol Specification, v. 6.3.0.

[15] 3GPP, R6, TS 25.324, BMC Protocol Specification, v. 6.4.0.

[16] 3GPP, R6, TS 25.322, RLC Protocol Specification, v. 6.4.0.

[17] 3GPP, R6, TS 25.321, MAC Protocol Specification, v. 6.8.0.

[18] 3GPP, R6, TS 25.308, UTRA High Speed Downlink Packet Access (HSDPA); Overall Description; Stage 2, v. 6.3.0.

[19] 3GPP, R6, TS 25.309, FDD Enhanced Uplink; Overall Description; Stage 2, v. 6.6.0.

[20] 3GPP, R6, TS 23.107, Quality of Service (QoS) Concept and Architecture, v. 6.3.0.

[21] H. Kaaranen, A. Ahtiainen, L. Laitinen, S. Naghian and V. Niemi, *UMTS Networks: Architecture, Mobility and Services*, John Wiley & Sons, 2nd Edition, 2005, 406 pp.

[22] 3GPP, R6, TS 25.331, RRC Protocol Specification, v. 6.7.0.

[23] 3GPP, R6, TS 25.302, Services Provided by the Physical Layer, v. 6.5.0.

[24] 3GPP, R6, TS 25.212, Multiplexing and Channel Coding (FDD), v. 6.6.0.

[25] 3GPP, R6, TS 25.211, Physical Channels and Mapping of Transport Channels onto Physical Channels (FDD), v. 6.6.0.

[26] 3GPP, R6, TS 25.213, Spreading and Modulation (FDD), v. 6.4.0.

[27] K. Helmersson *et al.*, Performance of downlink shared channels in WCDMA radio networks, *IEEE, VTC, Spring, 2001*, pp. 2690–2694, v. 4.

[28] 3GPP, R5, TS 25.308, High Speed Downlink Packet Access (HSDPA); Overall Description; Stage 2, v. 5.7.0.

[29] 3GPP, R5, TS 25.321, Medium Access Control (MAC) Protocol Specification, v. 5.12.0.

[30] 3GPP, R5, TS 25.306, UE Radio Access Capabilities, v. 5.12.0.

[31] 3GPP, R5, TS 25.213, Spreading and Modulation (FDD), v. 5.5.0.

[32] 3GPP, R5, TS 25.214, Physical Layer Procedures (FDD), v. 5.11.0.

[33] 3GPP, R5, TS 25.331, Radio Resource Control (RRC) Protocol Specification, v. 5.14.0.

[34] 3GPP, R5, TS 25.433, UTRAN Iub Interface NBAP Signalling, v. 5.13.0.

[35] 3GPP, R5, TR 25.858, UTRA High Speed Downlink Packet Access: Physical Layer Aspects, v. 5.0.0.

[36] H. Holma and A. Toskala (eds), *WCDMA for UMTS*, John Wiley & Sons, 3rd Edition, April 2004, 450 pp.

[37] P. J. Ameigeiras Gutiérrez, Packet scheduling and quality of service in HSDPA, PhD thesis, Ålborg University, Denmark, October 2003.

[38] J. Proakis, *Digital Communications*, McGraw-Hill, 2nd Edition, 1989.

[39] 3GPP, R5, TS 25.877, High Speed Downlink Packet Access (HSDPA) – Iub/Iur Protocol Aspects, v. 5.1.0.

[40] Ericsson, WCDMA Evolved: High-speed Packet-data Services, Review No. 2, 2003, pp. 56–65.

[41] 3GPP, R6, TR 25.808, FDD Enhanced Uplink; Physical Layer Aspects, v. 6.0.0.

5

QoS Functions in Access Networks

David Soldani, Paolo Zanier, Uwe Schwarz, Jaroslav Uher,
Svetlana Chemiakina, Sandro Grech, Massimo Barazzetta and
Mariagrazia Squeo

The QoS management functions in access networks are responsible for efficient utilisation of radio interface and transport resources. In this chapter, we focus on radio resource management (RRM) algorithms, which are needed to guarantee QoS, maintain the planned coverage area and offer high spectral efficiency. RRM functions can be divided into power control (PC), handover control (HC), admission control (AC), load control (LC) and packet scheduling (PS).

Power control is a connection-based function needed to keep interference levels at minimum. Handovers are needed in cellular systems to handle the mobility of the terminals across cell boundaries. Handover control is also a connection-based function. The other RRM algorithms – that is, admission control, load control and packet scheduling – are required in order to guarantee the QoS and to maximise the cell throughput for a mix of different bit rates, service applications and quality requirements. AC, PS and LC are cell-based QoS management functions.

The RRM algorithms with QoS differentiation are based on a particular subset of the QoS attributes associated with the radio access bearers established for carrying distinct service applications. The differentiated treatment of users/services across the access network and between multiple radio access technologies is provided by handling the available resources according to service-specific QoS requirements, such as traffic class, priority and bit rates (maximum and guaranteed). Network parameters are also differentiated based on the subset of QoS attributes and consistently set for 'optimal' QoS provisioning.

Typical locations of the QoS management functions in UTRA/GERA cellular networks are illustrated in Figure 5.1. In the following sections, we present how the corresponding algorithms may be designed with QoS differentiation. Besides this, some insight into the mobility management functions (handoffs, a.k.a. handovers) for optimal

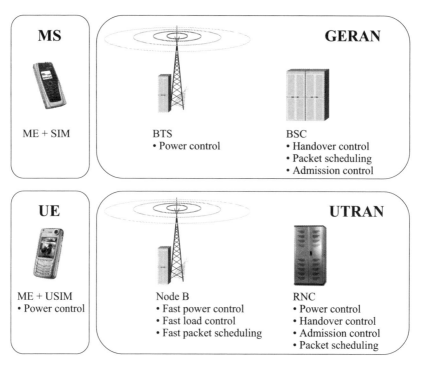

Figure 5.1 Typical locations of QoS management functions in UTRA and GERA networks.

connection performance (i.e., QoE, in terms of service retainability) in 2G, 3G and multiradio networks are also given. The main QoS and QoE aspects in 3GPP–WLAN inter-working scenarios are also pointed out at the end of this chapter. Means and methods for the optimal settings of the introduced parameters are discussed in Chapter 10.

5.1   QoS management functions in GERA networks

This section describes two examples of QoS management functions for GERA networks. The first algorithm is proposed for R97/98 (E)GPRS, where there is no guaranteed bit rate for packet-switched (PS) traffic. The second resource allocation scheme is valid for R99 (or later) EGPRS networks, where packet-switched streaming, conversational, interactive and background QoS classes are supported. More information on EGPRS radio interface can be found in Chapter 4.

5.1.1   Radio interface

In GSM, one time slot is allocated to a certain circuit-switched (CS) connection. The TS is dedicated to the MS for the whole call duration. In (E)GPRS, unlike GSM, one time slot can be shared among at most eight PS connections called 'temporary block flows'

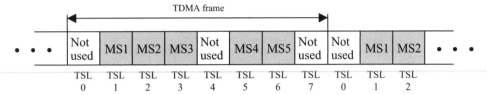

Figure 5.2 Example of time slot allocation for circuit-switched connections.

(TBFs). Although access to and allocation of the time slots is different in (E)GPRS, the systems can coexist with the GSM because the use of physical resources is the very same. Anyway, once a time slot is allocated for a CS connection, it cannot be used for any PS one. The same applies for the opposite case – that is, a time slot allocated for a PS temporary block flow cannot be allocated for a CS connection. Figure 5.2 shows the consecutive time slots in the time domain used by mobile stations. If a mobile station is in active state (e.g., performing a CS call), it uses one time slot within the TDMA frame.

The connection allocated on a time slot for CS traffic can reach a throughput of at most 14.5 kb/s. This is sufficient for high-quality digital voice transmission. An allocated time slot for PS traffic can transmit a radio block once every 20 ms ($\sim 5 \times 4$ ms). Every radio block is encoded using a certain coding scheme. The coding scheme is selected according the radio conditions experienced by the mobile station. The GPRS uses four coding schemes (CS-1 to CS-4) and EGPRS nine modulation coding schemes (MCS-1 to MCS-9).

5.1.1.1 Territory setting

The process of assigning the radio interface resources (time slots) to the CS and PS connections is denoted as 'territory setting'. There are only two ways an operator can specify the distribution of PS and CS channels within BTS transceivers (TRX). The methods are depicted in Figures 5.3 and 5.4.

Both figures show a time slot organisation within a BTS with four transceivers. The first two time slots on the first TRX are allocated for the BCCH and SDCCH. The rest is either used for CS or PS traffic. The CS-X sign, where X is a number from 1 to n, means Mobile Station X, which is performing a circuit-switched call. Every time slot marked by CS-X means there is a circuit-switched channel allocated for the call performed by MS X. The PS-Y sign, where Y is a number from 1 to n, means Mobile Station Y, which is performing a packet-switched call. Every time slot marked by PS-X equals an allocated TBF for the call performed by MS Y.

Figure 5.3 shows the allocation of two separate groups of contiguous time slots to PS and CS users. CS calls cannot be allocated into the PS pool and *vice versa*. The sizes of the pools can be dynamically changed, though, according to the traffic distribution in the cell. This is possible because all the *free time slots* (or *guard time slots*) are located at the border of both pools. The main advantage of this strategy is the high utilisation of available packet-switched resources (time slots) since they are consecutively located next to each other. A channel allocation algorithm can be optimised for each pool separately, since each traffic type has different requirements for the management of free time slots.

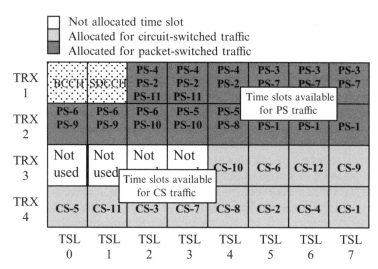

Figure 5.3 Resource allocation example in the (E)GPRS system: net separation between time slots allocated for PS and CS connections.

PS connections are usually shorter than circuit-switched ones. This is because in most cases the CS traffic is continuous and a time slot allocated at the beginning of the CS session must last until the end. The disadvantage of this mechanism lies in the free time slot rescheduling at the border of the two pools. The mechanism must ensure that all free time slots are always rescheduled towards the borders of the two pools.

Figure 5.4 shows no strict traffic-type separation in the set of available time slots. The channels are simply allocated in the first free time slots for the CS traffic and first less utilised time slots in the PS traffic. The rescheduling of free time slots is no longer a

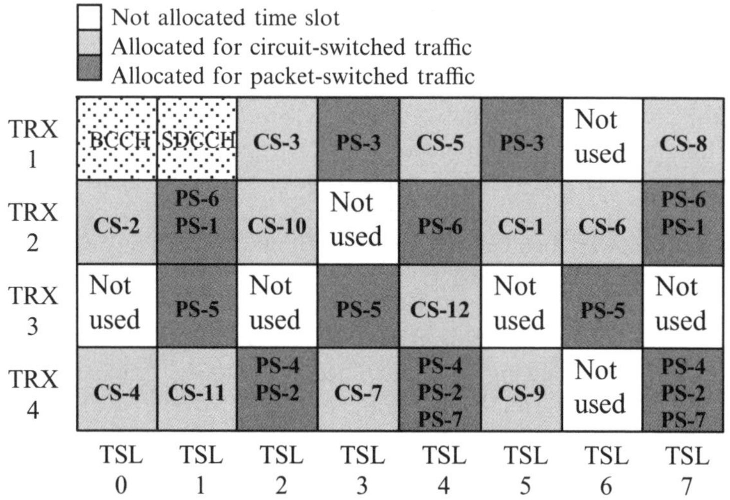

☐ Not allocated time slot
▨ Allocated for circuit-switched traffic
■ Allocated for packet-switched traffic

	TSL 0	TSL 1	TSL 2	TSL 3	TSL 4	TSL 5	TSL 6	TSL 7
TRX 1	BCCH	SDCCH	CS-3	PS-3	CS-5	PS-3	Not used	CS-8
TRX 2	CS-2	PS-6 PS-1	CS-10	Not used	PS-6	CS-1	CS-6	PS-6 PS-1
TRX 3	Not used	PS-5	Not used	PS-5	CS-12	Not used	PS-5	Not used
TRX 4	CS-4	CS-11	PS-4 PS-2	CS-7	PS-4 PS-2 PS-7	CS-9	Not used	PS-4 PS-2 PS-7

Figure 5.4 Resource allocation example in the (E)GPRS system: no PS–CS traffic separation.

problem and the amount of available resources is higher. However, the possibility of optimised channel allocation for the two traffic types is not possible here. When free channels are not consecutive, multislot connections cannot use them efficiently.

5.1.1.2   Scheduling function

In (E)GPRS networks, scheduling is a mechanism located in the BSC that fairly distributes radio resources to different, allocated temporary block flows. It can also prioritise one or more temporary block flows above others. As an example, the packet-switched multislot connections PS-2, PS-4 and PS-11, depicted in Figure 5.3, have to share the Time Slots 2, 3 and 4 in the first TRX. The BSC must allow these connections to have their own transmission and reception time.

The scheduling of the user and signalling data is performed at (E)GPRS Medium Access Control (MAC) level. In the downlink, it is the scheduling mechanism that decides whose data should be transmitted at a certain time. As an example, a weighted round robin (WRR) scheduler or deficit round robin (DRR) scheduler, described in Sections 5.1.2 and 5.1.3, respectively, can be used for this purpose. In the uplink, transmission turn multiplexing is controlled using a channel allocation mechanism [1].

5.1.2   QoS differentiation in R97/98 EGPRS radio access networks

In this section, we describe a simple packet scheduler implementation solution that does not offer bit rate guarantees to PS traffic, but is able to differentiate throughput (i.e., a portion of received transmission turns) according to the TBF priority. Differentiation is based on the allocation retention priority (ARP) attribute, which has a value ranging from 1 to 3, 1 being the highest priority. Note that to ensure compatibility of different cellular network releases there is a one-to-one mapping between ARP and R97/98 precedence class values (see Chapter 3). The ARP is listed in the QoS profile of the PDP context [1].

As described in Chapter 3, the BSC knows the ARP value associated with each TBF. The BSC also has an internal one-to-one mapping from each of the three ARP values to a parameter called *scheduling step size* (SSS). The SSS value associated with a TBF is used by the packet scheduler to allocate the turn of that particular TBF in a weighted round robin (WRR) fashion. The inverse of the SSS value yields the scheduling weight for each TBF. Thus, the lower the SSS value, the higher the scheduling priority of the TBF in question. *Transmission turn value* (TTV) is an auxiliary variable introduced in the WRR scheduler to control the transmission turns for each TBF. Every time the TBF is transmitting, the SSS value is added to the TTV. The TBF with the lowest TTV will transmit the next transmission turn. In our case, the SSS ranges from 1 to 12. A TBF associated with an SSS value of 3 would transmit four radio blocks during the period when a TBF with an SSS value of 12 transmits only one radio block.

An example of how the above WRR scheduling works is illustrated in Figure 5.5, where four TBFs are competing for one TSL transmission window. During five scheduling steps, TBF_1, which has the lowest SSS, gets the transmission turn most often. The initial state of the TTV for each TBF is visible in the first scheduling turn. There,

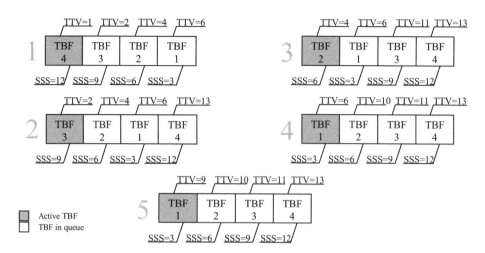

Figure 5.5 Five scheduling turns describing the functionality of the weighted round robin function.

although TBF$_4$ has the highest SSS value, it also has the lowest TTV. Thus, it will get to transmit before the others. After 20 ms, when the next scheduling step is performed for the following radio block, TBF$_4$ is placed at the end of the queue, since its TTV is increased by an SSS value of 12, thus becoming the highest in the queue. Next to transmit is the one with currently the lowest TTV (i.e., TBF$_3$). After SSS and TTV summation, TBF$_3$ is placed before TBF$_4$. The same scheme is applied for all other scheduling steps. In the last two steps, TBF$_1$, after four scheduling turns, is still the one with the lowest TTV and, thus, gets again the time slot to transmit.

The throughput R_j experienced by TBF$_j$ in one time slot (TSL) can be estimated by:

$$R_j = \frac{\dfrac{1}{SSS_j}}{\displaystyle\sum_{i=1}^{N} \dfrac{1}{SSS_i}} R_{TSL} \qquad (5.1)$$

where SSS_i is the SSS value associated with TBF$_i$, N is the total number of temporary block flows sharing the same TSL and R_{TSL} is the average bit rate the TSL in question can offer during the measurement period.

Equation (5.1) is only an approximation, since it ignores the facts that, due to different radio conditions, some users need more retransmissions than others, and that radio blocks may contain different amounts of user data depending on the modulation and coding scheme used. In terms of scheduling, the PS considers each time slot independently. Thus, a TBF may get to transmit more often on one slot than on the other due to the different number of users in both slots. Equation (5.1) is thus not sufficient for the network administrator to assess the quality experienced by the users of the services and to ensure that the contracted QoS is sustained. Hence, in Chapter 9, we propose a more adequate tool for monitoring the throughput of the diverse packet flows in the Base Station Subsystem (BSS).

The spectral efficiency gains provided by service differentiation according to the packet scheduling algorithm described above were investigated in [2]. The scenario parameters are reported in Table 5.1. Three different service types were offered in the network – that is, video streaming, web browsing and MMS messaging. For each service, a user satisfaction criterion based on throughput was defined. Streaming, being a real time service, was deployed with the highest possible priority – that is, the SSS value for the bearer carrying this service was set to 1. MMS traffic, being a typical best-effort application without any particular QoS requirement, was handled with the lowest scheduling priority – that is, the MMS SSS was set to 12. Conversely, the SSS value related to web browsing varied from 4 to 9. Several combinations of traffic mixes and a number of dedicated GPRS time slots were analysed by means of dynamic simulations. It was found that a network with QoS differentiation activated and the SSS set to 1, 4 and 12 – for video streaming, web browsing and MMS, respectively – at a given QoE (percentage of satisfied users), could carry roughly double the load compared with the non-differentiated case, where all bearer services were associated with the same SSS value. The attained capacity gains are depicted in Figure 5.6.

Table 5.1   Simulation parameters [2].

Element	Element description
Scenario	The network consists of 25 sites placed on a regular hexagonal grid with 1.5-km inter-site distance. Each site contains three sectors, leading to a total of 75 cells. Non-synchronised BTSs in 900-MHz band
Antennas	Directional (65-deg radiation pattern)
GPRS territory	Only 1 TRX per BTS. 3, 5 or 7 TSLs available for GPRS in each BTS
GPRS connections	RLC layer works in AM. Link adaptation with coding schemes CS-1 and CS-2. Release of downlink TBFs is delayed 1 second
Multislot class	Channel allocator assigns 3 TSLs in the downlink and 1 TSL in the uplink
Power control	Uplink power control activated
Cell reselection	1-s delay + new TBF establishment delay is considered in cell reselection
Reuse	4/12 in BCCH layer
TCP configuration	MSS = 1460 kB; AWND = 64 kB; initial CWND = 2MSS; SGSN buffer size = 64 kB
Simulated time	TDM frame resolution (4.6 ms). Real time simulated from 2.3 h (high-load cases) to 4.6 h (low-load cases)
Offered services	Streaming on Priority Class 1, web browsing on Priority Class 2, MMS on Priority Class 3
Traffic load	Based on different traffic mix. 0.5 to 7 Erlangs.

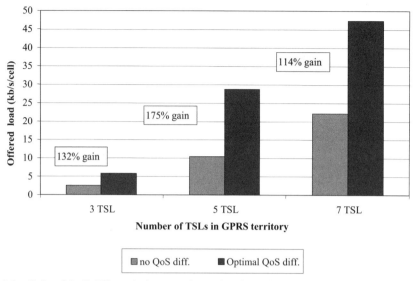

5.1.3   QoS differentiation in R99 or later EGPRS radio access networks

In R99 (or later) GERA networks, QoS attributes are the same as the QoS attributes defined for UTRAN [3]. Four priority classes with different requirements for data transmission are specified. Traffic of conversational and streaming QoS class is offered with guaranteed bit rate. Interactive and background QoS classes are carried on the best effort.

The resource allocation schemes in the BSC depend on the traffic characteristics: guaranteed bit rate (GB) bearer services undergo admission control with radio resource reservation, non-guaranteed bit rate (NGB) bearers are scheduled based on the associated priority attributes, but no target delay or throughput performances are guaranteed.

5.1.3.1   Admission control function

Admission control (AC) decides whether a call can be admitted to the network or should be rejected because of lack of radio resources (TBFs). For bearer services of conversational or streaming classes, admission control is performed as part of the PDP context activation procedure.

5.1.3.2   Deficit round robin scheduler

In R99 and later releases, one of the alternatives to the WRR scheduler is the deficit round robin (DRR) scheduler. The DRR is a fair scheduling algorithm that makes packets of larger size wait longer in a queue than smaller packets.

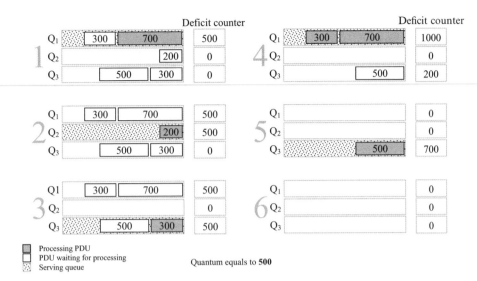

Figure 5.7   Six scheduling turns describing the functionality of the deficit round robin algorithm.

The DRR scheduler uses two attributes. The first is the *deficit counter*, which represents the maximum number of bits transmitted in a single scheduling turn. The second attribute is the *quantum*, a number that is added to the deficit counter of the active queue every scheduling step. Since the DRR scheduler schedules packets and not bits, the number of packets transmitted in a single scheduling turn depends on the size of the packets. If the size in bits of, say, two consecutive packets in the active queue is smaller or equal to the deficit counter, both packets are transmitted in the same turn. The number representing the total size of transmitted packets is subtracted from the deficit counter. If the size in bits of the first packet in the queue is bigger than the deficit counter, no packet is transmitted and no change of the deficit counter happens. If there are no packets after serving the active queue, the deficit counter of that particular queue is reset to 0. The behaviour of the DRR scheduler is depicted in Figure 5.7.

The example depicted in Figure 5.7 shows how the DDR algorithm manages three queues within six steps. The quantum is set to 500. Initially, the first and third queues (Q_1 and Q_3) contain two packet data units (PDUs) waiting for processing – Q_2 has only one. The DRR algorithm processes the queues in round robin fashion, which means that it serves a queue on each turn, going repeatedly through the queues in order Q_1, Q_2 and Q_3. No queue prioritisation is presented in this example. In Step 1, Q_1 is the queue being processed. As we can see from the figure, the size of the first packet (700) is larger than the sum of the deficit counter and the quantum (0 plus 500). Thus, the packet is not scheduled and its size is not subtracted from the deficit counter. In Step 2, the packet is scheduled instead, since its size (200) is smaller than the sum of the deficit counter and the quantum (0 plus 500). Even though the deficit counter after processing the PDU is 300 (500 minus 200), it is cleared because no other packet data unit is waiting in the queue. In Step 3, following the same logic, only the first PDU is processed (300 is smaller than 0 plus 500). This time, since there is another packet in the queue, the deficit counter is not cleared and it gets the value 200 (500 minus 300). At this point, two more steps (4 and 5) are enough to empty all the queues.

Every queue can be associated with a priority value. This value may be calculated based on the QoS traffic class, allocation retention priority and, in the case of the interactive class, also on the traffic handling priority. The queues with higher priority can be processed more often than those with the lower priority. Connections requiring higher resources for their services can then be placed in queues with higher priority. In this way, utilisation of the whole network resources can also be higher than the case without prioritisation.

The quantum can differ for each QoS service profile, thus also providing differentiation at the service profile level. As an option, the quantum can be calculated as a function of the guaranteed bit rate specified for each service profile. There is a possibility that one queue can represent one connection. In this case the quantum can be calculated not only as a function of the guaranteed bit rate but also as a function of packet size. If the transmitted packet size is involved, the quantum should be recalculated at a given number of steps because user traffic can change and the network can adapt to a connection's resource requirements.

When the DRR algorithm is combined with a mechanism that takes into account R99 traffic classes, it has the advantage over the WRR algorithm of providing a set of parameters, which can be fine-tuned to obtain the optimal performance of the network. Figure 5.8 shows the overall spectral efficiency gains of the combination of the DRR

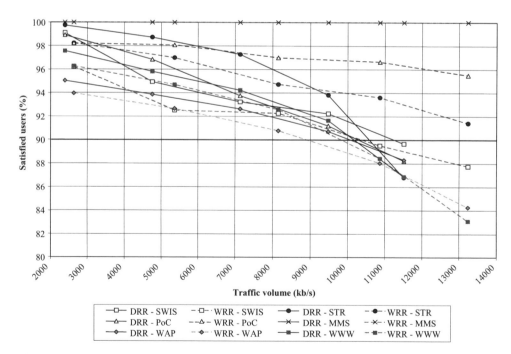

Figure 5.8 Gains in terms of spectral efficiency (network throughput at a given QoE): DRR vs. WRR scheduling algorithm.

Table 5.2   EGPRS simulation parameters.

Scenario type	Macro-cells
No. of BTSs	75
EGPRS MS capability	2 TSLs in downlink
Scheduler	Based on DRR
Service	Network traffic proportion (%)
WAP	64.72
MMS	5.65
PoC	10.6
STR	2.1
SWIS	2.26
WWW	14.67

algorithm and the R99 QoS mechanism compared with the WRR algorithm in the EGPRS network. The main simulation parameters are reported in Table 5.2. Figure 5.8 shows the performance of six services: SWIS, PoC, WAP, STR, MMS and WWW. The satisfaction criterion is at least 90% of satisfied users of each service at the same time. The satisfied user of a particular service is one who fulfils the service criteria defined in Table 10.4. Figure 5.8 shows that the service performing the worst in the WRR case is WAP. The curve representing WAP performance is the first to go below the 90% satisfaction border line. WAP users start to be unsatisfied when the network throughput reaches 9000 kb/s. The weakest service of the DRR case is also WAP, but as a result of fine-tuned parameters the satisfaction of the users continues until after reaching 10 000 kb/s of the network throughput, which yields approximately a spectral efficiency gain of 12%.

5.1.4   Handovers and cell reselection in 2G networks

In 3GPP specifications for 2G networks, several features are specified to reduce the outage generated during the transition of a connection from one cell to another. Such outages may decrease the quality perceived by the end-user, or may generate bad speech quality if the handover is not triggered properly. This section describes these aspects, treating briefly handover for speech calls and cell reselection for GPRS and EDGE radio access networks.

5.1.4.1   Handover procedure

The handover procedure has the scope to select the best available radio link from the point of view of signal level, quality, interference or traffic balancing, in order to guarantee to the user a seamless transition from one cell to another.

In 2G, the MS, BSS and MSC are involved in this process. Handovers are mainly triggered by threshold comparison or by periodic checks. The MS measures the downlink level and quality from its serving and neighbouring cells and reports these to the BSS on the slow associated control channel (SACCH). The BTS monitors the uplink signal level and quality of the served MS, and also measures the level of interference on the idle traffic channels. The BSC processes uplink and downlink measurements and decides whether a handover is necessary or not and what type of handover should be performed. Then, the BSC evaluates the target cell among the neighbouring cells reported by the MS. Usually, the MSC is simply notified by the BSC that the procedure took place, but sometimes it takes an active part in the handover process – for example, if the handover is inter-BSC. As opposed to 3G networks, the handoff in the BSC is always hard. In fact, the MS is connected to only one cell at a time.

5.1.4.2 Handover types

There are three basic groups of handoffs: radio resource handovers, imperative handovers and traffic reason handovers. Each handover type is assigned a priority level, so that the BSC algorithm can also make decisions in cases where measurement conditions satisfy the criteria of more than one handover type.

Radio resource handover
Radio resource handovers are triggered by checking the following measurements:

- uplink/downlink signal level (Rx_Level);
- uplink/downlink quality (Rx_Qual);
- uplink/downlink interference (I_Level);
- power budget.

Level- and quality-triggered handovers take place when the received uplink or downlink signal strength or quality falls below a pre-defined threshold.

Interference is detected on a time slot basis; if it becomes too high the BSC may perform either an inter-cell or intra-cell handover by selecting an interference-free connection. Indeed, selection of a different channel may improve end-user perception of speech quality and avoid call drops. Interference is detected in the case of bad quality when the received level (uplink or downlink) is above a certain threshold.

Finally, the power budget handover ensures that the mobile station is always allocated to the cell with the minimum path loss and is triggered when the path loss of a neighbour cell is estimated to be lower than that of the serving cell.

Level, quality and interference handovers are triggered by threshold crossings, while the power budget cause is checked periodically. For optimal QoE, handover using radio criteria should be handled taking into account the following priority order: interference, followed by quality, signal level and, finally, power budget.

Imperative handover
Imperative handovers are performed if one of the following conditions is satisfied:

- the distance between the MS and BTS exceeds a maximum threshold;
- an O&M command to empty the cell is issued;
- rapid field drop is detected; and
- the serving cell is congested and a directed retry is needed.

In the first case, handover is triggered because the distance between the MS and BTS is such that synchronisation can no longer be guaranteed. The second case occurs when a cell needs to be emptied in order to perform a maintenance procedure. Ongoing calls are handed over to adjacent cells. An imperative handover procedure is started in order to face a very fast reduction in uplink signal level (rapid field drop) – due to shadowing, for example. Finally, in case the serving cell is congested, directed retry handovers are used during the call setup phase to redirect it to a cell with lower traffic.

Imperative handovers are faster than other handover types, because there are fewer conditions to be fulfilled. Indeed, they are designed to react immediately to a critical situation, which would otherwise cause the call to be dropped.

Traffic reason handover
Handovers may also occur for reasons other than radio link control – for example, to control the traffic distribution between cells, or between layers. These are the so-called 'traffic reason handovers'. Their purpose is to smooth out the load over the network, avoiding the congestion of crowded areas and allowing in this way a better call setup success rate.

5.1.4.3   Target cell evaluation process

Target cell evaluation is done in the BSC in order to find the best target to counterbalance the cause for the handover. The BTS receives reports about the best candidate target cells measured by the mobiles in the downlink, while uplink measurements are executed directly by the BTS. Uplink and downlink measurements are then sent to the BSC, which chooses the best target cell based on Rx level and quality requirements, as well as the actual load of the candidate cells.

These measurements are averaged to avoid ping-pong effects. So, depending on the size of the averaging window, handover may be fast or slow. Fast handovers may generate excessive signalling load; on the other hand, when handovers are slow, the user might experience bad quality before the connection is successfully handed over to the target cell. This happens, for example, in the case of rapid field drops, or when the mobile is moving very fast. So, the averaging window influences end-user perception a lot, and its best value is always a trade-off between the need to minimise delays and avoid the user experiencing bad quality, and the need to avoid ping-pong effects that may generate bad quality and excessive signalling load as well.

5.1.4.4   Cell reselection procedures

The handover process in GPRS standby or ready state is called a 'cell reselection procedure'.

As in the handover case, there is only one connection between the MS and BSC at a time; moreover, in a packet-switched data call, there is some latency between the release of the radio connection in the old cell and its re-establishment in the new cell. This gap is due to the fact that the MS usually performs the cell change autonomously. As a result of the network not being involved in this process, it takes some time for the BSC to forward the data to the new cell. Cell reselection adversely affects the QoE when the user moves from one cell to another. Hence, it is essential for optimal service performance to keep this gap as short as possible. In order to do this, 3GPP specifications introduced a new procedure called the 'network-controlled cell reselection' (NCCR) [6]. The type of supported cell reselection procedure is communicated to the MS using a new information element in broadcast messages called the 'network control' (NC) order. For cell reselection purposes, only two values of this parameter are relevant (for more information see [6]):

- *Network Control Order 0 (NC0) – normal MS control.* The MS performs cell reselection autonomously based on field strength measurement criteria.
- *Network Control Order 2 (NC2) – network control.* The MS sends measure reports to the network, and the BSC commands cell reselection. NC2 applies only to an MS in GPRS ready state having a data transfer ongoing or immediately after it.

In the following sections, some details about NC0 and NC2 are reported.

Normal MS control

In this case, the cells to be monitored are defined in the neighbour cell list broadcast on the packet broadcast control channel (PBCCH). If the PBCCH does not exist, the normal GSM neighbour list is used. For the sake of simplicity, let us consider the case when only the broadcast channel (BCCH) is supported. In idle mode, when camped in the cell, the MS measures the received signal of the neighbour cells broadcast on the BCCH. Since there are no packet common control channels (PCCCHs) in place and no extra features are available, cell update is executed autonomously by the MS. Once selected the new cell and, if in packet transfer mode, the MS establishes a new link at the logical link control (LLC) layer with the SGSN, which informs the BSC of successful cell change (flushing procedure). So, there is a gap during which the BSC is not aware of cell change, where it simply sees the mobile disappear from the old cell. During this time, data are buffered in the BSC, which executes flow control towards the SGSN, in order to avoid buffers overflowing. All these aspects contribute in increasing the gap in data transmission so that the MS has enough time to decode the system information in the new cell to start the data transfer again. A typical gap in cell reselection is shown in Figures 5.9 and 5.10, where streaming of a video clip has been interrupted by the cell reselection procedure. The gap is measured at the application layer based on any unacceptable values of round trip time. As the gap is normally quite long (e.g., higher than 30 s in Figure 5.10,), in the following sections we will describe briefly how new features may reduce this data interruption.

Network-controlled cell reselection

Network-controlled cell reselection (NCCR) is a feature that speeds up the cell reselection process commanding the MS to send measurement reports to the BSC, and giving the network full control of the procedure. In the case of non-stationary handsets that cross cell borders, this provides better service performance. The mode of operation is

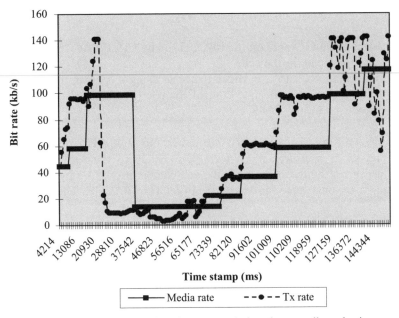

Figure 5.9 Gap in streaming data transmission due to cell reselection.

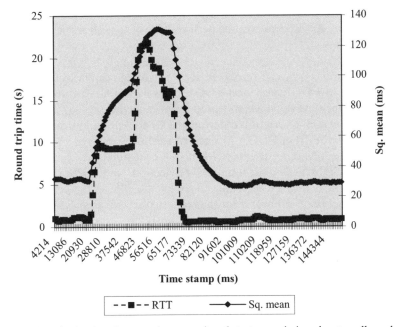

Figure 5.10 Round trip time increase in streaming data transmission due to cell reselection.

communicated to the MS in two possible ways:

- Broadcasting the NC2 value on the PBCCH (or BCCH, if PBCCH is unavailable) to all MSs in ready state in the cell.
- Commanding the NC2 to a particular mobile using a 'packet measurement order' message on the packet associated control channel (PACCH), if the MS is in packet transfer mode, or on the PCCCH.

After receiving this message, the MS continuously monitors all carriers in the neighbour cell list, and sends measurement reports periodically to the network. Based on the received measures, the network starts the cell change procedure by sending a 'packet cell change order' message to the MS on the PCCCH or PACCH. The message contains:

- the BSIC and BCCH frequency of the new cell;
- the NC mode to be initially applied in the new cell.

The network regards the procedure as successfully completed when it realises that communication has been established with that MS through the new cell. On the other hand, the MS regards the procedure as successfully completed when it has received a response to a 'packet channel request' message on the new cell. The procedures for the MS in idle or data transfer mode are shown in Figure 5.11.

Service outage can thus be measured in the BSC as the difference between the instant it orders the cell change using the 'packet cell change order' message and the instant it receives the 'packet channel request' message from the terminal. Normally, service outage with the NCCR is just a few seconds, providing a relevant benefit to the quality perceived by the user when he/she is on the move between cells.

In practice, the NCCR improves the performance of the cell reselection procedure for several reasons:

- *Efficient allocation of EDGE resources.* By controlling cell reselection, the network is able to separate EDGE and non-EDGE capable mobiles in order to avoid them sharing the same time slots, which would result in inefficient usage of radio resources. The autonomous cell reselection criterion (NC0) does not allow distinguishing GPRS MS cell reselection from EDGE MS cell reselection.
- *Efficient inter-system cell reselection.* Multi-RAT terminals will be able to camp either in GSM or WCDMA cells. Although cell reselection can be performed autonomously by the terminal, it may be desirable to direct the MS to the cell in the RAT that provides the most efficient bearer for the requested service application. That would not be possible if the cell was selected autonomously.
- *Packet-switched traffic totally controlled by the network.* This means that traffic balancing may also be done for the packet-switched domain, while autonomous cell reselection would not allow load balancing among different traffic classes.
- *Faster cell reselection and thus shorter gaps in data transmission.* This is possible because the network is aware of the cell change process. The gap is of course dependent on the service application that is running. For example, with file transfer protocol (FTP) the gap is normally between 2 and 5 s.

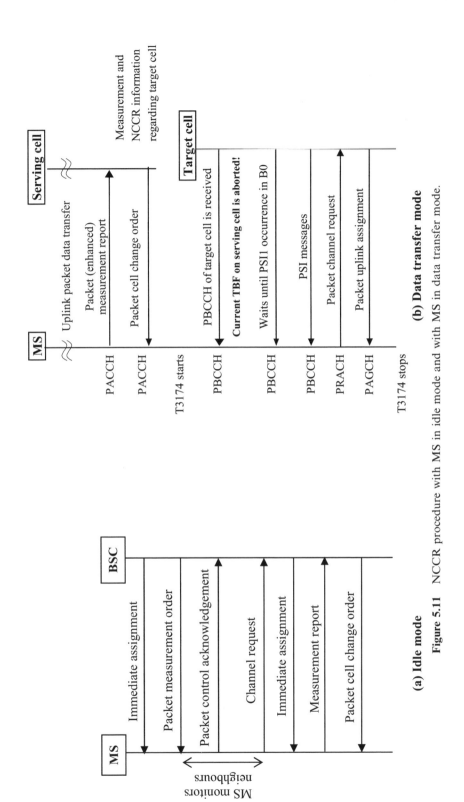

Figure 5.11 NCCR procedure with MS in idle mode and with MS in data transfer mode.

(a) Idle mode

(b) Data transfer mode

Anyway, using the NCCR, the MS still needs to decode all the system information when accessing the new cell. Hence, outage can be further reduced if the process of system information decoding is speeded up. This is possible if another feature, called 'network-assisted cell change (NACC) is implemented together with NCCR.

Network-assisted cell change
NACC aims at reducing service outage time by giving a means to the network to assist the MS before and during cell change. It may be supported together with the NCCR. The feature reduces cell reselection delay by:

- Providing the MS in data transfer mode – via the actual serving cell – with a part of the system information (SI), or packet system information (PSI) for a faster resumption of data transfer on the new cell (the terminal would need to receive such information from the target cell).
- Supporting the 'packet PSI status' procedure, where the MS indicates to the network which PSI related to the target cell has already been acquired on the serving cell, so that the network may provide the terminal with just the missing data [7]. Utilisation of the 'packet PSI status' message requires that the PCCCH is implemented in the serving cell. In 3GPP GERAN R4, a similar message called 'packet SI status' has been introduced to support this feature on the CCCH as well [8].

When the NCCR and NACC are used together, a typical break in the ongoing service decreases from seconds to a few hundreds of milliseconds, thus improving end-user perception quality and allowing mobile users a seamless transition from one cell to another in data transfer mode.

5.2 QoS management functions in UTRA networks

This section describes examples of the radio resource management (RRM) functions with QoS differentiation for UTRA FDD networks that enable operators to offer new services in a cost-effective way [9]–[14]. More details and implementation aspects of the proposed QoS functions may be found in [15]. Enhancements of the described radio resource allocation schemes for high-speed downlink and uplink packet access are described in Section 5.3 and 5.4, respectively.

5.2.1 Admission control

Admission control (AC) decides whether a radio bearer (RB) service can be established or modified in the radio access network or not. The algorithm is run for each cell, separately for uplink and downlink directions, during the radio bearer setup or recon-figuration procedure [16]. When a connection is set up or modified, AC with QoS differentiation assigns a *resource request priority* (RRP) to the bearer service in question based on the QoS profile it receives from the core network (CN) [3]. The RRP value is a parameter for the operator to set. The lower the RRP value the higher the assigned priority. New radio link requests are arranged into priority queues and served following the strict priority principle (branch additions have top priority) and, at a given priority, based on their arrival times (FIFO). Resource requests (RRs) are rejected if either the *maximum queuing time* or the corresponding *maximum allowed queue length* is exceeded. These parameters may be set differently depending on bearer service characteristics, such as QoS class, allocation retention priority (ARP) and traffic handling priority (THP), in the case of interactive class [3]. Except for the overload situation, defined by:

$$P_{Total} = P_{NGB} + P_{GB} > P_{Target} + Offset \tag{5.2}$$

where P_{Total} is the total current transmission (or received) power in the cell, exploited by the served traffic with guaranteed (GB) and non-guaranteed bit rate (NGB), and $P_{Target} + Offset$ is the overload threshold, which can be set differently for uplink and downlink directions, NGB bearer services are always admitted, whereas GB traffic is not admitted if either (5.2) or the following inequality is satisfied:

$$P_{GB} + \Delta P_{GB} > P_{Target} \tag{5.3}$$

where ΔP_{GB} is the estimated power increase in the cell if the bearer in question is admitted. ΔP_{GB} may be calculated for the downlink using the following equation, which is a modified version of the formula used in [17] to calculate the initial radio link power:

$$\Delta P_{GB,DL} = \frac{\rho R}{W} \left(\frac{p_{tx,CPICH}}{\rho_c} + (1 - \alpha) P_{TxTarget} - P_{TxTotal} \right) \tag{5.4}$$

where ρ and R are the required E_b/N_0 and maximum bit rate of the bearer in question, $p_{tx,CPICH}$ is the power of the common pilot channel in the cell, ρ_c is the energy per chip per noise spectral density (CPICH E_c/N_0) received by the mobile, W is the chip rate (3.84 Mchip/s) and α is the code orthogonality factor ($\alpha = 1$ means 'perfect orthogonality'). In the uplink direction, a weighted sum of the integral and derivative formulas presented in [17] may be utilised, namely:

$$\Delta P_{GB,UL} \approx \beta \frac{\Delta L}{1 - \eta} P_{RxTotal} + (1 - \beta) \frac{\Delta L}{1 - \eta - \Delta L} P_{RxTotal} \tag{5.5}$$

where β is the weight for the operator to set and η the uplink load factor, which is given by:

$$\eta = \frac{I_{own} + I_{oth}}{P_{RxTotal}} \tag{5.6}$$

where I_{own} is the received power from users in the own cell and I_{oth} comes from users of the surrounding cells. The fractional load ΔL may be simply estimated as:

$$\Delta L = \frac{1}{1 + \frac{W}{\rho R}} \tag{5.7}$$

During SHO, diversity branches are not set up if the following condition (valid for the downlink only) is satisfied:

$$P_{GB} + \Delta P_{GB} > P_{Target} + Offset \tag{5.8}$$

where P_{GB} is actual uncontrollable power in the target cell and ΔP_{GB} is the estimated power increase in the same cell due to the radio link in question.

5.2.2 Packet (bit rate) scheduler

The bit rates of the admitted NGB bearer services are scheduled based on the actual radio resource priority values and, at a given priority, based on the arrival times of the corresponding capacity requests (CRs), following the FIFO principle. The bit rate allocation method is based on the *minimum* and *maximum allowed bit rates*, which define the lowest and highest limits of the transport format set (TFS) that can be allocated to the requesting bearer service in the uplink or downlink directions. These parameters may be differentiated depending on the bearer service characteristics.

In order to increase the degree of fairness (reduce variance in allocated bit rates) among peers, the radio resource priority value of a persisting CR in the PS queue may be modified taking into account the actual active session throughput of that particular connection in the downlink direction, being the most critical. The new radio resource priority value of the bearer in question may be calculated as follows:

$$RRP_i(t) = RRP_i - x\frac{R^i_{Target} - R_i(t)}{\max\{TFS_i\}} \tag{5.9}$$

where RRP_i is the initial priority assigned by AC (see Section 5.2.1) to the bearer in question i; x is the *priority step parameter*, which can be set differently for different cells; R^i_{Target} is the target bit rate of that particular connection; $R_i(t)$ is the measured average active session throughput at time t during packet data transmission; and $\max\{TFS_i\}$ is the maximum bit rate of the TFS of the dedicated channel used to transport the corresponding user data.

During uplink or downlink bit rate allocation, the PS follows the best-effort model and relies upon the power budget left by GB and NGB active (a) and inactive (i) connections for that particular direction; that is:

$$P^{NGB}_{Allowed} = P_{Target} - (P_a + P_{i,GB} + kP_{i,DTX}) \tag{5.10}$$

The power of inactive GB traffic $P_{i,GB}$ takes into account the required power of the bearer services just admitted, but not yet on air, whereas $P_{i,DTX}$ is the power that needs to be reserved for the bearer services in discontinuous transmission (DTX) during their idle or reading periods. Since it is unlikely that all inactive connections get active at the same time, we introduce k as a management parameter to restrict reserved power to a lower value: k ranges from 0 to 1, where 1 is the most conservative number that yields the lowest effectiveness, since in this case all required power would be reserved. In the case of SHO, the allocated bit rate is the minimum of the bit rates scheduled for each of the links of the radio link set.

Capacity requests are rejected if they stay longer in the PS queue than the value specified by the *CR maximum queuing time*. Allocated bit rates may be rescheduled if the ongoing communication has lasted longer than the corresponding *granted minimum DCH allocation time*. Both parameters may be set differently depending on the distinct bearer service characteristics.

Dedicated channels (DCHs) are used for high-rate packet data transmission. Bearer services that have been longer in DTX than the corresponding *inactivity timer* are moved to Cell_FACH (forward access channel) state and the corresponding allocated resources are released. When new data exceed a certain threshold in the buffer of the radio network controller (RNC) or UE, a new CR is sent to the PS and subsequently another DCH is allocated to the bearer in question based on the selected scheduling algorithm, as explained in the following sections. The inactivity timer may be a differentiated parameter.

5.2.2.1 Fair throughput scheduling

In the case of fair throughput (FT) scheduling, the unexploited power in (5.10) is shared fairly in terms of throughput among the different bearer services every scheduling period. This is achieved by matching the bit rates and thus the estimated transmit powers of the bearer services to sum up to (5.10). Power estimates in the bit rate increase algorithm are based on (5.4) or (5.5) depending on bit rate scheduling direction.

An example of bit rate allocation for up to five capacity requests in a PS queue using the FR algorithm is illustrated in Figure 5.12. The chronological order of the capacity requests (CRs) is written in brackets and the priorities of the bearers in question represented by different tints. In the figure, the possible bit rates in the TFS are: 128, 144, 256 and 384 kb/s, where 128 kb/s and 384 kb/s are the dedicated channel (DCH) *minimum* and *maximum allowed bit rate*, respectively. When five capacity requests are in the queue, only four of them are allocated 128 kb/s.

5.2.2.2 Fair resources scheduling

In the case of fair resource (FR) scheduling, all users get the same power and the actual bit rates depend on the received signal-to-interference ratio (SIR). This means that users

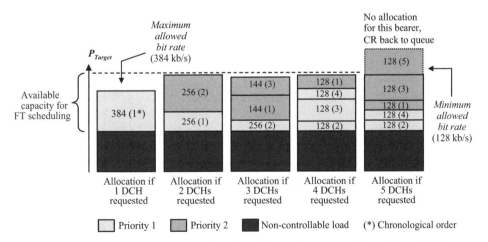

Figure 5.12 Example of bit rate allocation for 1–5 capacity requests in packet scheduler queue when FT scheduling is used [15].

close to the base station will be allocated higher bit rates than the users at cell borders. In the case of QoS differentiation, different factors (*weights*, denoted by w_i) could be used for the power, depending on the used service (TC) and subscription profile. The weights can have values between 0 and 1. The algorithm we present is an enhancement of FT scheduling described in the previous section and, for the downlink direction, consists of the following steps:

1. $P_{NGB\,Allowed}$ is calculated as shown in (5.10).
2. $P_{NGB\,Allowed}$ is evenly divided among all radio links prior to scheduling (or rescheduling) – that is, $p_{tx,link} = P_{NGB\,Allowed}/N$, where N is the number of calls under scheduling.
3. $p_{tx,link}$ is used in the formula below to calculate the 'virtual' bit rate (R_i) of that particular session (i):

$$R_i = \frac{W}{\rho} \frac{\dfrac{w_i}{\sum_j w_j} p_{tx,link}}{P_{tx,target}(1 - \alpha) + \dfrac{P_{tx,cpich}}{\rho_c} - P_{tx,tot}} \qquad (5.11)$$

where w_i are the differentiated power weights introduced above, and all other terms are as defined in (5.4), from which (5.11) was derived. This will now result in different bit rates depending on the corresponding path loss and interference situation.
4. For each session, the virtual bit rate is rounded upwards to the next bit rate in the TFS. For those cases where R_i exceeds the maximum bit rate in the TFS, it is rounded down to the maximum bit rate in the TFS (see Figure 5.13).
5. A temporary TFS for each session is constructed. In this TFS the highest bit rate is the one found in Step 4 for each session separately and the lowest ones are replaced by 0.
6. FT scheduling is performed as presented in Section 5.2.2.1, yet uses the temporary TFS generated in the previous step. Sessions with 0 bit rates are not scheduled.

Figure 5.14 shows an example of bit rate allocation using the algorithm presented above for the five cases presented in Section 5.2.2.1. Capacity requests are the ones illustrated in Figure 5.12 for FR scheduling. In this case, the possible bit rates in the TFS are: 64, 128,

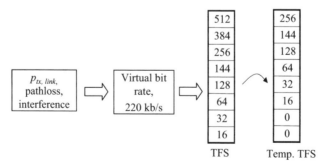

Figure 5.13 Example of transport format set (TFS) construction when the FR scheduling algorithm is used [15].

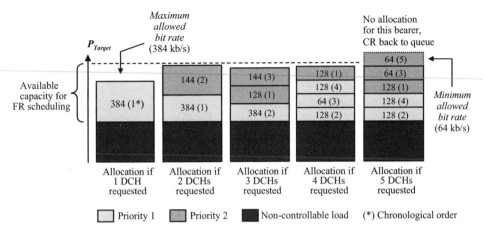

Figure 5.14 Example of bit rate allocation for 1–5 capacity requests in packet scheduler queue when FR scheduling is used [15].

144, 256 and 384 kb/s, where 64 kb/s and 384 kb/s are the dedicated channel (DCH) *minimum* and *maximum allowed bit rate*, respectively. Although the minimum allowed bit rate for the DCH has been reduced to 64 kb/s, when five capacity requests are in the queue only four of them get a possibility to transmit.

In order to compensate for slow-fading fluctuations, propagation path losses and interference conditions, E_c/N_0 and $P_{tx,tot}$ values in (5.11), should be continuously monitored for each mobile terminal in the cell and the allocated bit rate rescheduled whenever needed, regardless of the value of the *granted DCH minimum allocation time* parameter.

5.2.3 Load control

One important task of radio resource management functionality is to ensure that the system is not overloaded and remains stable. If the system is properly planned, and admission control and the packet scheduler work sufficiently well, overload situations should be exceptional. If overload is encountered, however, the load control functionality returns the system back to the targeted load, denote by P_{Target} in Section 5.2.1. Possible load control actions are: reduce the throughput of packet data traffic, hand over to another WCDMA carrier or to GSM, decrease the bit rates of real time users (e.g., AMR speech codec) and drop calls in a controlled and differentiated fashion [16].

For instance, a simple solution with QoS differentiation was proposed in [15]. In that work, LC only supports reduction of the bit rates of NGB bearer services when (5.2) is satisfied. The DCH bit rate may be downgraded only when the allocation time of the carried service lasts longer than the corresponding *DCH granted (overload) minimum allocation time*. Bit rates are reduced, starting with bearers with the lowest or given priority, based on their arrival times (FIFO), but none of the sessions is released.

5.2.4 Power control

The group of functions that control the level of transmitted power in order to minimise interference and keep the quality of the communication are: uplink/downlink outer-loop power control, uplink/downlink inner-loop power control (also presented as fast closed-loop PC in the literature), uplink/downlink open-loop power control and the slow power control applied to downlink common channels. A comprehensive description of the above algorithms and a detailed performance analysis thereof can be found, for instance, in [16] and [17].

The possibility for the network operator to provide differentiated power control to bearer services with diverse performance requirements is rather limited, since the control of power occurs mostly at the physical layer without any discrimination of the carried information. For one signalling connection (or UE) and multiple bearer services, there might be several transport channels (TrCHs) multiplexed onto one code composite transport channel (CCTrCH) carried on the same physical channel that PC takes place. However, different E_b/N_0, BLER/BER values may be set depending on the bearer service characteristics for outer-loop PC and power estimates, such as power increase/decrease and initial power calculations [17].

5.2.5 Handover control

Handover control (HC) is a connection-based function of the radio access network that supports three categories of handover: hard handover (HHO), soft and softer handover (SHO). HHO and SHO are, respectively, network (NEHO) and mobile (MEHO) evaluated handoffs. Hard handover means that all the old radio links in the UE are removed before the new radio links are established. Hard handovers are seamless (not perceptible to the user) only for NRT communications. Handovers that require a change of carrier frequency (inter-frequency handover) or radio access technology (RAT) are always performed as hard handover. Soft handover means that radio links are added and removed in such a way that the UE always keeps at least one radio link to the UTRAN. Soft handover is performed by means of macrodiversity, which refers to the condition that several radio links are active at the same time. Soft handover can be used when cells operated on the same frequency are changed. Softer handover is a special case of soft handover where the radio links that are added and removed belong to the same Node B.

A comprehensive description of handover procedures and relevant control parameters can be found in [19] and [20]. In the following sections, we focus on the main aspects of intra-frequency, inter-frequency and inter-system (from WCDMA to GSM) handovers for optimal service retainability. The differentiation aspects in handling radio links carrying service applications with different quality requirements are pointed out. In the following sections, a description of the fundamental measurement reporting criteria in UTRAN is also included. Service performance during inter-system handovers is described in Section 5.5.

5.2.5.1   Reporting of measurements

Measurement reporting from mobile terminals to the radio network controller encompasses the definition of neighbouring cell lists, measurement reporting criteria and reporting of measurement results [19].

Neighbouring cells are defined on a cell-by-cell basis in the radio network configuration database. Each cell in the radio access network has an individual set of neighbouring cells. A neighbouring cell may be located in the radio network subsystem (RNS) or in the base station subsystem (BSS) of any public land mobile network (PLMN). A neighbouring cell in the WCDMA network is identified by:

- UTRAN cell identifier:
  - PLMN identifier (MCC and MNC);
  - RNC identifier;
  - Cell identifier.
- Location area code (LAC).
- Routing area code (RAC).
- UTRA RF channel number.
- Primary CPICH scrambling code.

A neighbouring cell in the GSM/EGPRS network is defined by:

- Cell global identification (CGI):
  - mobile country code (MCC);
  - mobile network code (MNC);
  - location area code (LAC);
  - cell identifier (CI).
- BCCH frequency (GSM 900, GSM 1800 or GSM 1900).
- Base station identity code (BSIC):
  - base station colour code (BCC);
  - network colour code (NCC).

Terminals receive the neighbouring cell list information through the system information blocks (SIBs) on the broadcast control channel (BCCH) or, while moving in Cell_DCH of connected mode, via RRC measurement control messages on the dedicated control channel (DCCH) [19]. In the neighbouring cell information, the network provides to the UE a *cell-individual offset* (CIO) parameter, which is added to the measured quantity before the UE compares the measured value with the related reporting criterion.

Measurement reporting criteria depend on the handover type. Each measurement type (e.g., intra-, inter-frequency and inter-RAT) normally has separate measurement reporting criteria (set of measurement parameters). The measurement reporting criteria are defined on a cell-by-cell basis and transmitted to the UE together with the above identifiers using the BCCH and measurement control messages on the DCCH. Each measurement is controlled and reported independently from each other. During the Cell_DCH of connected mode, by using measurement control messages handover control may set up, modify or release any measurement type. When the measurement reporting criteria are fulfilled, the mobile station reports the measurement results to the RNC.

Intra-frequency measurement reporting criteria may be defined as follows:

- Reporting mode: event-triggered.
- Measurement quantity: CPICH E_c/N_0.
- Reporting quantities: primary scrambling code (SC), CPICH RSCP, CPICH E_c/N_0 and cell synchronisation information for both active (cells participating in soft handover), monitored (neighbouring cells) and detected set cells.
- Main events for the reporting criteria [19]:
  - 1A: a primary CPICH enters the reporting range;
  - 1B: a primary CPICH leaves the reporting range;
  - 1C: an inactive primary CPICH becomes better than an active one;
  - 1D: change of best cell;
  - 1E: a primary CPICH becomes better than an absolute threshold;
  - 1F: a primary CPICH becomes worse than an absolute threshold;
  - 6A: the UE transmission power is higher than an absolute threshold;
  - 6B: the UE transmission power is lower than an absolute threshold;
  - 6F: the UE Rx–Tx time difference for a radio link included in the active set is higher than an absolute threshold
  - 6G: the UE Rx–Tx time difference for a radio link included in the active set is lower than an absolute threshold

Reporting Events 1E, 1F, 6A and 6B can be used, among others (see Section 5.2.5.6 and 5.5) for inter-frequency and/or inter-RAT (GSM/EGPRS) handovers. Event 1D may be employed for handling the intra-frequency mobility with HSDPA. Events 6F and 6G may be used by the HC to remove the interfering radio link from the active set.

Inter-frequency measurement reporting criteria may be defined as follows:

- Reporting mode: periodic.
- Measurement quantity: CPICH E_c/N_0.
- Reporting quantities: SC, CPICH E_c/N_0 and CPICH RSCP on an unused frequency.
- Maximum number of reported cells: 6 (six) monitored set cells per reported unused frequency.
- Amount of reporting: RNP parameter.
- Reporting interval: RNP parameter.
- DPCH compressed mode status info: compressed mode parameters to be used by the UE in order to perform inter-frequency measurement.

Inter-RAT (GSM/EGPRS) measurement-reporting criteria may be defined as follows:

- Reporting mode: periodic.
- Measurement quantity: GSM carrier RSSI. BSIC verification is required only for the target GSM carrier before the execution of inter-RAT handover.
- Reporting quantity: GSM carrier RSSI.
- Reported cells: cells of other RATs within the active set.

- Maximum number of reported cells: 6 (six) monitored set cells.
- Amount of reporting: RNP parameter.
- Reporting interval: RNP parameter.
- DPCH compressed mode status info: compressed mode parameters to be used by the UE in order to perform either GSM carrier RSSI measurement or GSM initial BSIC identification.

A mobile station with a single receiver is unable to receive more frequencies or RATs at the same time. In this case, compressed mode in needed while the mobile station makes the required inter-frequency or inter-RAT measurements. Compressed mode for uplink and downlink directions is described in [19]–[20].

5.2.5.2 Soft handover

The SHO final decision is taken in the RNC. Based on the event-triggered measurement reports received from the terminal, HC orders the mobile station to add, replace or remove cells from its active set. The soft handover functionality is illustrated in Figure 5.15. The maximum active set size (maximum number of cells participating in SHO) is a parameter for the operator to set. In the example of Figure 5.15 the UE may be simultaneously connected to two cells at maximum (the maximum active set size is

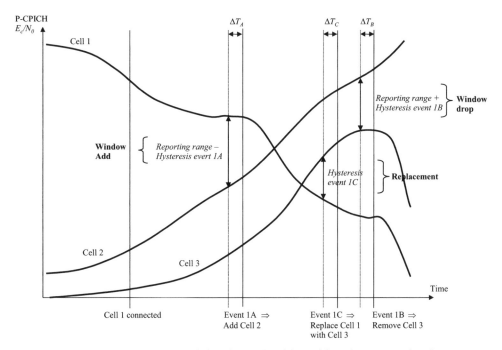

Figure 5.15 WCDMA soft handover algorithm with active set equal to 2.

assumed to be 2). The algorithm works as follows [20]:

- If *Pilot* $E_c/N_0 >$ *Best pilot* $E_c/N_0 -$ *Reporting range* $+$ *Hysteresis Event 1A*   for a period of ΔT_A and the active set is not full, the cell is added to the active set. This event is called *Event 1A* or radio link addition.
- If *Pilot* $E_c/N_0 <$ *Best pilot* $E_c/N_0 -$ *Reporting range* $-$ *Hysteresis Event 1B*   for a period of ΔT_B, then the cell is removed from the active set. This event is called *Event 1B* or radio link removal.
- If the active set is full and *Best candidate pilot* $E_c/N_0 >$ *Worst old pilot* $E_c/N_0 +$ *Hysteresis Event 1C* for a period of ΔT_C, then the weakest cell in the active set is replaced by the strongest candidate cell (i.e.. strongest cell in the monitored set). This event is called *Event 1C* or combined radio link addition and removal.

Reporting range is the threshold for soft handover, *Hysteresis Event 1A* is the addition hysteresis, *Hysteresis Event 1B* is the removal hysteresis, *Hysteresis Event 1C* is the replacement hysteresis and ΔT is the time to trigger for that particular event. These parameters may be differentiated based on QoS attributes, such as traffic classes, and thus set differently depending on the characteristics of upper layer protocols. If the radio link consists of multiple transport channels (CCTrCH) carrying different bearer services, the parameter settings for the most demanding traffic should be used. *Best pilot* E_c/N_0 is the strongest measured cell in the active set, *Worst old pilot* E_c/N_0 is the weakest measured cell in the active set, *Best candidate pilot* E_c/N_0 is the strongest measured cell in the monitored set and *Pilot* E_c/N_0 is the measured and filtered quantity.

5.2.5.3 Intra-frequency hard handover

Intra-frequency hard handovers are required in situations when inter-RNC soft handovers are not possible (e.g., due to Iur congestion). Hard handover is based on the event-triggered periodic intra-frequency measurement results the UE reports when the RNC is unable to increase the active set following an Event 1A or 1C. The handover decision may be based on the reported E_c/N_0 of the best cell of the active set, the measured E_c/N_0 of the target cell and a parameter (threshold) to prevent repetitive hard handovers between cells. Different thresholds (margins) may be defined for radio links carrying traffic with different QoS requirements, such as traffic class. In the case of CCTrCHs, the threshold set for the most demanding bearer services may be used.

5.2.5.4 Intra-frequency mobility with HSDPA

The mobility procedures for HSDPA users are affected by the fact that transmission of the HS-PDSCH and HS-SCCH to a user occurs only through one of the cells participating in SHO, denoted by the HS-DSCH cell. The serving HS-DSCH cell is determined by the RNC. The connectivity on HSDPA is achieved by means of a synchronised change of the serving HS-DSCH cell, which means that during the cell change procedure the transmission and reception of HS channels is done at a certain time dictated by UTRAN. More information on the mobility aspects of using HSDPA can be found in [16], [19] and [23].

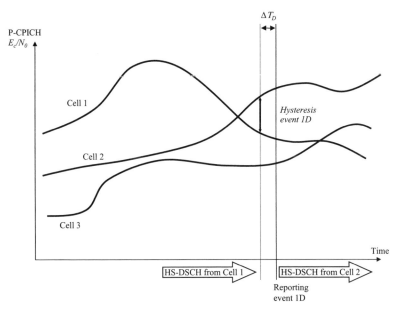

Figure 5.16   Best serving HS-DSCH cell measurements.

As discussed in Section 5.2.5.1, for HSDPA, Measurement Event 1D has been defined, which is called the measurement event for the best serving HS-DSCH cell. The UE reports the best serving HS-DSCH cell to the serving RNC based on the measured P-CPICH E_c/N_0 or RSCP of potential candidate cells. The procedure is illustrated in Figure 5.16, where H_{1D} is the hysteresis margin of Event 1D and ΔT_D is the time to trigger parameter. It is possible to configure this measurement event so that all cells in the user's candidate set are taken into account, or to restrict the measurement event so that only the current cells in the user's active set for dedicated channels are considered. The hysteresis margin may be used to avoid fast change of the serving HS-DSCH cell for this particular measurement event. Also in this case, it is possible to specify the *cell-individual offset* (CIO) to favour certain cells – for instance, to extend their HSDPA coverage area.

Another possibility is to use periodic reporting for downlink intra-frequency CPICH E_c/N_0 and uplink SIR_{error} measures [24], when serving HS-DSCH cell change and SHO for the associated DPCH are allowed. In this case, the RNC orders the UE to report Events 1A, 1C, 1B, 6F and 6G. In the measurement reporting criteria, the parameter settings for HSDPA may be defined differently with respect to ordinary cells. When the first HSDPA-capable cell is added to the active set, SIR_{error} reporting in the cells participating in the SHO and E_c/N_0 measurements in the UE may be started simultaneously, if the UE supports HSDPA and has a PS radio access bearer established in Cell_DCH with an active set size greater than 1. Possible triggers for serving HS-DSCH cell change are:

- *Periodical CPICH E_c/N_0 measures*: cell change is initiated if the CPICH E_c/N_0 of the current serving HS-DSCH cell is lower than the actual CPICH E_c/N_0 value of the best cell in the active set minus a threshold for the operator to set.

- *Periodical UL SIR$_{error}$ measures*: the *SIR$_{error}$* of the actual serving HS-DSCH cell is below a certain threshold defined by the network administrator.
- *Event 1B*: the serving HS-DSCH cell is removed from the active set.
- *Event 1C*: the serving HS-DSCH cell is replaced by another cell.
- *Failures in serving HS-DSCH radio link*: loss of synchronisation or drifting in Rx–Tx time difference.
- *Serving HS-DSCH radio link is handed over to DRNC*: if there are other HSDPA-capable cells in the active set under the SRNC, serving HS-DSCH cell change is executed towards a candidate cell in the SRNC.

The target HSDPA-capable cell may be selected based on its allocated HSDPA power status, measured uplink *SIR$_{error}$* and/or downlink CPICH E_c/N_0.

Intra-Node B HS-DSCH to HS-DSCH handover

Once the serving RNC decides to make an intra-Node B handover from a source HS-DSCH cell to a new target HS-DSCH cell under the same Node B, as illustrated in Figure 5.17, the serving RNC sends a synchronised radio link reconfiguration prepare message to the Node B, as well as a radio resource control (RRC) radio bearer reconfiguration message to the user. At a specified time index where handover from the source cell to the new target cell is carried out, the source cell stops transmitting to the user, and the MAC-hs packet scheduler in the target HSDPA cell is thereafter allowed to control transmission to the user. Similarly, the terminal starts listening to the HS-SCCH (or several HS-SCCHs depending on the MAC-hs configuration) from the new target cell – that is, the new serving HS-DSCH cell. This also implies that the CQI reports from the user are measured from the channel quality corresponding to the new target cell. The PDUs buffered in the source cell, which have never been transmitted and/or are pending retransmission, assuming the Node B supports MAC-hs preservation, are moved to the

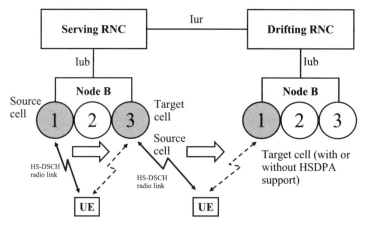

The user is moving to another cell

Figure 5.17   Sketch of an HSDPA user making an intra and inter Node-B change in serving HS-DSCH cell.

target cell during HS-DSCH handover. If this functionality is not supported, then handling of an incomplete PDU is the same as in the inter-Node B handover case presented in the next section.

During intra-Node B HS-DSCH to HS-DSCH handover, it is likely that the associated DPCH is in two-way softer handover. Under such conditions, the uplink HS-DPCCH may also be regarded as being in two-way softer handover, so Rake fingers for demodulation of the HS-DPCCH are allocated to both cells in the user's active set. This implies that uplink coverage of the HS-DPCCH is improved for users in softer handover and no power control problems are expected.

Inter-Node B HS-DSCH to HS-DSCH handover
In inter-Node B HS-DSCH to HS-DSCH handover the serving HS-DSCH source cell is under one Node B, while the new target cell is under another Node B, and potentially also under another RNC, as illustrated in Figure 5.17. Once the serving RNC decides to initiate such a handover, a synchronised radio link reconfiguration prepare message is sent to the drifting RNC and the Node B that controls the target cell, and a radio resource control (RRC) radio bearer reconfiguration message is sent to the terminal. In the source cell, all buffered PDUs for the user are deleted, including PDUs pending retransmission. At the same time, the flow control unit in the MAC-hs of the target cell starts requesting PDUs from the MAC-d in the serving RNC, so that it can start transmitting data on the HS-DSCH to the user. Lost data are recovered through upper layer retransmissions. For instance, when RLC protocol realises that the PDUs it has originally forwarded to the source cell are not acknowledged, it will initiate retransmissions, which basically implies forwarding the same PDUs to the new target cell that were deleted in the source cell. For applications running over unreliable protocols, such as UDP and unacknowledged mode RLC, the PDUs that are deleted in the source cell prior the handover are lost.

HS-DSCH to DCH handover
Handover from an HS-DSCH to a DCH may be potentially needed for HSDPA users that are moving from a cell with HSDPA to a cell without HSDPA (R99 compliant only cell), as illustrated in Figure 5.17. Triggers for an HS-DSCH fallback to a DCH may be: IFHO/ISHO measurements for other bearer services multiplexed in the same radio link, low uplink SIR_{error} value, removal of the HS-DSCH radio link due to Event 1B/1C/1D, radio link failure or Rx–Tx time difference measurement, and relocation of the HS-DSCH radio link. Once the serving RNC decides to initiate such a handover, a synchronised radio link reconfiguration prepare message is sent to the involved Node Bs, as well as a radio resource control (RRC) radio bearer reconfiguration message to the user. Similarly to inter-Node B HS-DSCH to HS-DSCH handover, HS-DSCH to DCH handover results in resetting of the PDUs in MAC-hs in the source cell. This subsequently requires recovery via higher layer retransmissions.

3GPP R5 specifications support the implementation of handover from a DCH to an HS-DSCH. This handover type, for instance, may be employed if a user is moving from a non-HSDPA-capable cell to an HSDPA-capable cell, or to optimise the load balance between HSDPA and DCH use in a cell.

Handover delay is estimated to be below 500 ms. In practice, the actual handover delay will depend on RNC implementation and the size of the RRC message that is sent to the user during the handover phase and the data rate on the Layer 3 signalling channel on the associated DPCH.

5.2.5.5 Intra-frequency mobility with HSUPA

The change of serving cell and/or serving radio link set (RLS) for E-DCH scheduling is supported via RRC signalling. In particular [37]:

- For the serving RLS, the radio access network may select the RLS with the highest data throughput and, for the serving cell, the cell that provides the best downlink quality.
- When an E-DCH serving cell change is triggered, the network updates the serving grant in the UE, and all L2-deactivated processes become active. Processes can be enabled/disabled via RRC.

5.2.5.6 Inter-frequency and inter-system hard handovers

Inter-frequency handovers are possible when the operator has at its disposal two or more FDD frequencies. Several frequencies on the same site may be used for high-capacity cells or macro- and microlayers using different frequencies. Handover between distinct carriers may take place due to coverage, quality or loading reasons, or simply because the network administrator wishes to offer certain services on dedicated carriers, such as the ones supporting HSDPA. Also, fast-moving mobiles can be handed over from micro- to macrofrequencies. High mobility is detected based on the frequency of active set updates.

Figure 5.18 shows how coverage and capacity can be extended by utilising two frequencies arranged in three different layers.

Inter-RAT handover is required to complement WCDMA coverage areas with GSM radio access technologies. The mobile station must support both WCDMA and GSM radio access technologies before inter-RAT handover is possible. For circuit-switched (CS) services both handover from WCDMA to GSM and handover from GSM to WCDMA are possible. Inter-RAT handovers of packet-switched (PS) services between WCDMA and GSM/EGPRS are based on cell reselection procedures. HC in UTRAN supports network-initiated cell reselection from WCDMA to GSM/EGPRS in the Cell_DCH or Cell_FACH states of connected mode [19]. The latter is

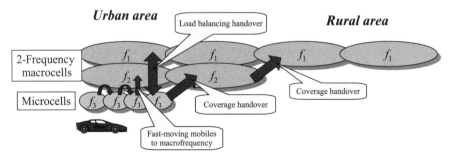

Figure 5.18 Examples of WCDMA inter-frequency handovers.

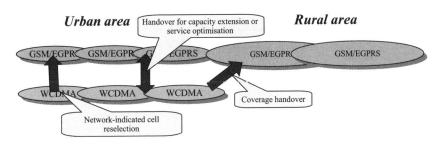

Figure 5.19 Examples of inter-RAT handovers: coverage extension with GSM and load sharing between GSM and WCDMA. Hierarchical cells (macro, micro) can be utilised with idle mode control by using HCS parameters.

possible only for NGB services. In the Cell_PCH and URA_PCH states of connected mode, the mobile station handles the reselection of cells autonomously. The RNC views cell reselection from GSM/EGPRS to WCDMA as RRC connection establishment, and mobile-initiated cell reselection from WCDMA to GSM/EGPRS as Iu connection release. Possible inter-RAT handover (or cell reselection) causes are the ones listed above for inter-frequency handover. After the hard handover decision from WCDMA to GSM, the RNC initiates an inter-RAT relocation procedure in order to allocate radio resources from the GSM BSS. If resource allocation is successful in the GSM BSS, the RNC orders the mobile station to make an inter-RAT handover from UTRAN to GSM. In the case of network-initiated cell reselection from WCDMA to GSM/GPRS, the RNC sends a cell change command to the UE, and the UE is responsible for continuing the already existing PS connection via the GSM/GPRS radio access network. The decision algorithm for inter-RAT handover from GSM to WCDMA is located in the GSM base station controller (BSC). When a radio access bearer is handed over from one radio access technology to another, the core network is responsible for adapting the quality of service (QoS) parameters of the radio access bearer according to the new (GSM/GPRS or WCDMA) radio access network. Figure 5.19 shows how coverage and capacity can be extended to moving terminals from one system to another.

Coverage and quality reasons handovers
The main handover (uplink and downlink) triggers for detecting coverage edges are:

- *UE Tx power*: the UE sends a measurement report to the RNC when its transmission power exceeds a certain reporting power threshold (Event 6A) for a pre-defined time interval (hysteresis). Upon receiving the measurement report from the UE, HC starts inter-frequency/RAT measurements.
- *Uplink DCH quality*: HC starts inter-frequency/RAT measurements when uplink quality is worse than the BER/BLER target for a pre-defined time interval (hysteresis).
- *Downlink DPCH power*: HC starts inter-frequency/RAT measurements when the transmitted radio link power exceeds a certain power threshold for a pre-defined time interval (hysteresis).
- *CPICH RSCP* or E_c/N_0: HC starts inter-frequency/RAT measurements when the measured CPICH RSCP or E_c/N_0 value of all active set cells has become worse than the reporting threshold (Event 1F).

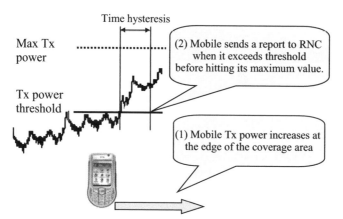

Figure 5.20 Event-triggered 6E: UE transmission power approaches its maximum power capability.

Figure 5.20 illustrates an example of how coverage edges can be detected when UE transmission power approaches its maximum value. All the thresholds of the above triggers are defined on a cell-by-cell basis by attaching a specified measurement control parameter set to a specific cell. The parameter may be differentiated based on the QoS classes of the carried bearer services. In the case of CCTrCH, the parameter settings of the most demanding class should be used. If two or more handover causes are reported simultaneously, a trigger-for-coverage reason should have higher priority with respect to a load- or service-based HO (see next section). The priority between inter-frequency and inter-RAT handovers depends on the value assigned to the service handover information element (IE) received from the CN [23]. If included, the service handover IE determines whether the RAB should or should not be handed over to GSM, or shall not under any circumstances be handed over to GSM – that is, UTRAN must not initiate handover to GSM for the UE unless the RABs with this indication have first been released with the normal release procedures. If the RNC does not receive such a parameter either in the 'RAB assignment request' or 'relocation request' message, inter-frequency handover normally has higher priority.

In the case of inter-frequency handovers, if there is more than one target frequency to be measured, the measurement order may be random or set by the network administrator. The handover path (target cell) and handover decision may be based on the measurement results of the best neighbouring cell, such as CPICH RSCP and E_c/N_0 values, and relevant RNP parameters. These parameters (margins and thresholds) can be differentiated for radio links carrying traffic with diverse QoS requirements (e.g., traffic class). In the case of the CCTrCH, the parameter settings of the most demanding bearer service may be used.

In the case of inter-RAT handovers, the handover path (target cell) and handover decision are based on the measured GSM carrier RSSI values, and relevant RNP parameters. These parameters (thresholds) can be differentiated for radio links carrying traffic with diverse QoS requirements (e.g., traffic class). In the case of the CCTrCH, the parameter settings of the most demanding bearer service may be used. If several neighbour cells meet the required radio link properties at the same time, the potential target cells may be ranked according to a pre-assigned priority level, and the cell with the highest priority may be selected. In the case of CS voice/data services, the RNC should always verify the BSIC of the target cell before execution of inter-RAT handover to GSM so that the mobile station can synchronise with the GSM cell before handover execution, and to verify the identification if two or more neighbour cells have the same *BCCH frequency*. In the case of PS (GB and NGB) services, verification should be ordered only in the case of *BCCH frequency* collision between neighbour cells.

Load- and service-based handovers

Load-based handovers are used for balancing the traffic (load) between different WCDMA frequency layers or between WCDMA and GSM/GPRS cellular networks. Service-based handovers are used for directing UEs to preferred RAT or hierarchical WCDMA layer.

Load-based handover triggers may be checked periodically – for instance, every time new interference measurements are received from the base station. In particular, in a WCDMA cell the following reasons may trigger a load-based handover procedure:

- *Too high uplink or downlink interference*: that is, total transmission or received wideband power in the cell exceeds a certain threshold.
- *The capacity request rejection ratio (CRRR) for NGB traffic is higher than a threshold.*
- *The cell is running out of downlink spreading codes.*
- *The cell is hardware- or logical resources limited*: for example, the BS is congested, or there is no Iub AAL2 transmission capacity left.

Load-based handovers may be initiated when one of the above-measured quantities is above the corresponding threshold for a pre-defined time interval (hysteresis). Handovers for that particular reason may be stopped only when the measured metric returns to below the threshold and remains smaller than that for a pre-defined time interval (delay), as shown in Figure 5.21. Each threshold and time to trigger is a parameter for the operator to tune.

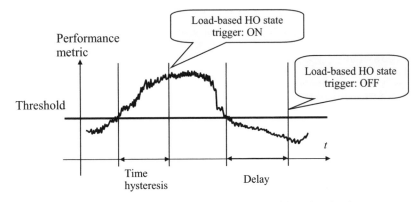

Figure 5.21   Measurement procedure for load-based HO triggers.

Table 5.3 Cell-based service priority handover profile.

Service type		Target RAT or layer
CS conversational	Speech	e.g. 'GSM'
	Transparent data	e.g. 'GSM'
PS conversational	Speech	e.g. 'WCDMA'
	Real time data	e.g. 'WCDMA'
CS streaming	Non-transparent data	e.g. 'WCDMA macro'
PS streaming	Real time data	e.g. 'WCDMA macro'
Interactive	THP1–NGB data	e.g. 'WCDMA micro' or 'HSDPA'
	THP2–NGB data	e.g. 'WCDMA micro' or 'HSDPA'
	THP3–NGB data	e.g. 'WCDMA micro' or 'HSDPA'
Background	NGB data	e.g. 'WCDMA micro' or 'HSDPA'

Service-based handover actions may be taken periodically, and the *checking period* is usually a parameter for the operator to set.

HC orders from the network elements the needed measurements as a function of the selected handover type. The type of handover depends on the service handover IE value (received from the CN) and relevant control parameters. For that purpose, two mapping tables may be employed for load- and service-based handovers. Table 5.3 maps the service type the UE is using onto the target cell. Table 5.4 combines this information with the priority values the HC receives from the CN. In the tables, the target RAT or layer, as well as the combined service priority lists, are entries for the operator to set. Target layers may be handled according to rules for the operator to set, which may be based on *hierarchical cell structure* (HCS) priorities. The HCS is a cell-based parameter, which needs to be set for the serving cell and all cells of the neighbouring list. When a mobile is not in the preferred RAT or WCDMA layer, the idea is to hand off that particular communication type to the available target cell with the highest priority. Service- or load-based handovers are not performed for multiservice connections where the combined service priority lists are not compatible.

Table 5.4 Combined priority table for load and service based handovers.

Handover service priority IE (from CN)	Cell-based service priority information (from Table 5.3)	Combined service priority list (example)
'RAB should be handed over to GSM'	Target RAT or layer	Target RAT or layer
'RAB should not be handed over to GSM'	Target RAT or layer	Target RAT or layer
'RAB shall not be handed over to GSM'	Target RAT or layer	Target RAT or layer
Service Priority IE not available	Target RAT or layer	Target RAT or layer

The connections that need to be handed over for either service or load reasons may be selected based on SRNC location, UE measurement activities, possibility of repeating the handover, load caused in the cell, etc. When a terminal is selected, HC investigates the availability of the target cell from the neighbouring cell list, which is the same as the one used for coverage-or-quality-reason handovers. The handover procedure is stopped if the target cell for that particular terminal is not available. Except for GSM/GPRS, the load of the target cell should also be checked before a service- or load-based handover may take place. Handovers or network-controlled cell reselections (in the case of GPRS) are performed only for those connections that are in Cell_DCH state. Service- and load-based handover measurements, inter-frequency and inter-RAT handover measurements as well as the criteria adopted for the selection of the target layer are the ones described for coverage and quality reasons.

5.2.6   Capacity gains of service differentiation in UTRAN

The feasibility and added values of the QoS management functions presented in the previous sections for WCDMA were analysed by means of dynamic simulations in [15]. Performance results showed enough flexibility for the operator to handle distinct services in a cost-effective manner and to sustain the required QoS of real time services – that is, PoC, audio and video streaming – without any admission control or *a priori* physical radio channel reservation.

Customised user satisfaction criteria were defined for the offered services and spectral efficiency was computed as the system load (mean cell throughput normalised with respect to the chip rate, 3.84 Mchip/s) at which 90% of users of the worst performing service were satisfied.

Within this operation constraint, using the fair throughput scheduling algorithm of Section 5.2.2.1, the spectral efficiency gain provided just by prioritisation was 30% and the additional capacity gain when the non-guaranteed bit rate services were also provisioned with differentiated parameters was about 20%. Performance results in terms of percentage of satisfied users and average cell throughput (taken over all simulated cells) normalised with respect to the chip rate are shown in Figures 5.22 and 5.23, respectively. In the non-prioritised case with cell-based parameters using Service Mapping 1 (denoted as NPM1), all resource requests had the same priority and none of the relevant parameter values were differentiated. (In Mapping 1, speech was served as CS conversational class and SWIS as PS streaming class. These were the only services offered with a guaranteed bit rate; all other applications were treated on a best-effort basis (BE or non-guaranteed bit rate, NGB). Furthermore, PoC, video and audio streaming, and WAP/MMS were mapped onto PS interactive class. WAP/MMS were supposed to be placed by the operator behind the same APN. Dialup connections – which comprised, for example, FTP, HTTP and email traffic – were carried on PS background class.) In the prioritised case with cell-based parameters using Service Mapping 1 (denoted by PM1), services were mapped as in NPM1 and used the same parameter values, but in this case they were offered with different priorities. Speech calls had top priority, followed by SWIS, PoC and all other service applications. Within interactive class, using different traffic-handling priorities, PoC was handled first, followed by audio and video streaming, and then WAP/ MMS. Dialup, on the background class, was the service with the lowest priority, and

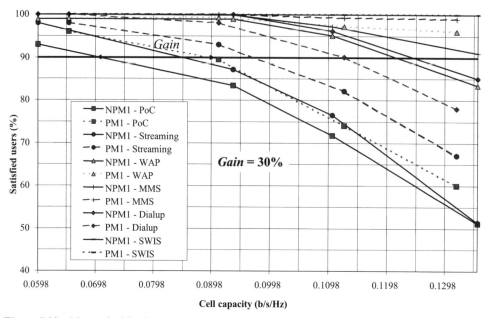

Figure 5.22   Non-prioritised (NPM1) vs. prioritised scenario with cell-based parameter settings (PM1), using Mapping 1 [12].

Reproduced by permission of © IEEE 2005.

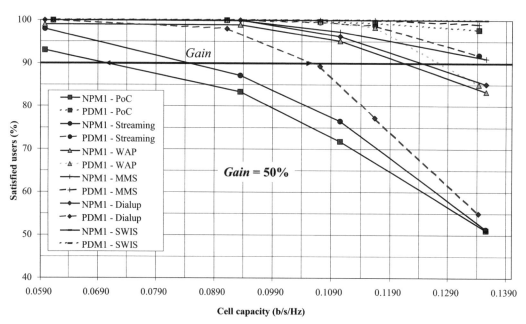

Figure 5.23   Non-prioritised scenario (NPM1) vs. prioritised case with differentiated parameters (PDM1), using Mapping 1 [12].

Reproduced by permission of © IEEE 2005.

hence served last. In the prioritised case with differentiated parameters using Service Mapping 1 (denoted as PDM1), services were mapped as in NPM1, but offered with different priorities and parameter values. To further improve PoC and streaming performance, the *minimum allowed bit rate* for WAP/MMS and Dialup connections was reduced, and the *DCH granted minimum allocation time* for DCHs carrying PoC and streaming increased, whereas for Dialup this was reduced (see [15] for more information).

Ultimately, the extra-spectral efficiency gain provided by the fair resources (FR) scheduling algorithm described in Section 5.2.2.1, with respect to the fair throughput (FT) scheduling presented in Section 5.2.2.2, depended heavily on the offered bearer services and traffic mix characteristics: the more that traffic is offered with a wide range of bit rates, the more spectral-efficient FR scheduling is.

Simulation results did not take into account any limitation entailed by Iub (interface between the base station and radio network controller) bottlenecks, hardware capabilities and/or availability of orthogonal codes. Yet, from average cell throughput and the probability of a user (connection) being inactive, we conclude that code tree limitation was not an issue in this study.

5.3   HSDPA with QoS differentiation

It is generally acknowledged that data services will have a huge rate of growth over the coming years and will likely become the dominating source of traffic load in 3G mobile cellular networks. Example data applications to supplement speech services include multiplayer games, instant messaging, online shopping, face-to-face videoconferences, movies, music, as well as personal/public database access (see Chapter 2 for more information on service applications). As more sophisticated services are introduced, a major challenge facing cellular system design is to facilitate a mixture of diverse services with very different QoS requirements and achieve simultaneously high spectral efficiency. One mean of simplifying the handling of such different services is to map them onto a limited number of traffic classes, which, depending on their characteristics, are then mapped onto different transport channels. This is the cornerstone of QoS differentiation in R99 of UTRAN specifications. The QoS mechanisms for this release were extensively covered in Section 5.2.

In order to further enhance the QoE of data users (in terms of higher throughput and shorter delays) and improve spectral efficiency, R5–R6 specifications introduced new transport channels (HS-DSCH in the downlink and E-DCH in the uplink, respectively). This section presents QoS differentiation for the downlink shared channel. Uplink-related issues are covered in Section 5.4.

What are the weaknesses of R99 transport channels (DCH, FACH) that the HS-DSCH aims to correct? The forward access channel (FACH) is a common channel offering low latency. However, as it does not apply fast closed-loop power control it exhibits limited spectral efficiency and is in practice limited to carrying only small amounts of data. The dedicated channel (DCH) is the 'basic' transport channel in UTRAN and supports all traffic classes. The data rate is updated by means of variable spreading factors (VSFs) while the block error rate (BLER) is controlled by inner and outer loop power control mechanisms. However, the power and hardware efficiency of the DCH is limited for burst and high data rate services, since channel reconfiguration is a rather slow process (in the range of 500 ms). Hence, for certain Internet services, although the system provides a high maximum bit rate allocation, DCH channel utilisation can be rather low. In order to enhance trunking efficiency, R99 specifications proposed the downlink shared channel (DSCH). This provides the possibility to time-multiplex different users (as opposed to code-multiplexing them). The benefit of the DSCH over DCH is its fast channel reconfiguration time and packet scheduling procedure (in the order of 10-ms intervals). In [18], simulations suggested that the efficiency of the DSCH can be significantly higher than for the DCH for bursty high data rate traffic.

The HSDPA concept can be seen as an evolution of the DSCH, and the transport channel is thus denoted the high-speed DSCH (HS-DSCH). HSDPA introduces several adaptation and control mechanisms in order to enhance peak data rates, spectral efficiency, as well as QoS control for bursty and downlink-asymmetrical packet data.

The following subsections focus on the implications of supporting QoS in HSDPA. For a detailed description of HSDPA functionality, please refer to Section 3.4.3 and [19]–[36].

5.3.1   Radio access bearer attributes

The RAB attributes that can be used in the RNC for channel-type selection (DCH or HS-DSCH) and parameter setting were listed in Table 4.4. Notice that guaranteed bit rate and transfer delay are only available for conversational and streaming traffic classes. Due to the nature of the carried traffic, in the following we assume that conversational class is not going to be mapped onto the HS-DSCH.

5.3.2   QoS information provided to MAC-hs

When a new HSDPA user is admitted, the RNC provides the Node B with information about the new MAC-d flow. The information elements for HS-DSCH MAC-d flows are reported in Table 5.5.

Comparing Table 5.5 to Table 4.4, we notice that, although an ARP is sent to the Node B, some other traditional QoS attributes that could be useful to the MAC-hs packet scheduler, such as TC and THP, are not. Alternatively, other HSDPA-specific QoS parameters are available to the Node B [22]:

- *Scheduling priority indicator* (SPI): indicates the relative priority of each priority queue (HS-DSCH data frame). The SPI ranges from 0 to 15, where 0 corresponds to the lowest priority and 15 to the highest. The Node B can use it when scheduling the HS-DSCH. Note that in the case of multiplexing of logical channels there may be more than one priority queue per HSDPA user (MAC-d flow).
- *Discard timer* (DT): defines the time to live for a MAC-hs SDU starting from the instant it arrives in an HSDPA priority queue. The Node B uses this information to discard out-of-date MAC-hs SDUs. The discard timer can have values between 20 and 7500 ms.

Table 5.5 HS-DSCH MAC-d flows information elements [22].

IE/Group name	Presence	Range	Semantics description
HS-DSCH MAC-d flow-specific information		1 ... < maxnoofMACdFlows >	
> HS-DSCH MAC-d flow ID	M		
> Allocation/Retention priority	M		
> Binding ID	O		Shall be ignored if bearer establishment with ALCAP.
> Transport layer address	O		Shall be ignored if bearer establishment with ALCAP.
Priority queue information		1 ... < maxnoofPrioQueues >	
> Priority queue ID	M		The HS-DSCH MAC-d flow ID shall be one of the flow IDs defined in the HS-DSCH MAC-d flow-specific information of this IE. Multiple priority queues can be associated with the same HS-DSCH MAC-d flow ID.
> Associated HS-DSCH MAC-d flow	M		
> Scheduling priority indicator	M		
> T1	M		
> Discard timer	O		
> MAC-hs window size	M		
> MAC-hs guaranteed bit rate	O		
> MAC-d PDU size index		1 ... < maxnoofMACdPDUindexes >	
≫ SID	M		
≫ MAC-d PDU size	M		
> RLC mode	M		

- *MAC-hs guaranteed bit rate* (MAC-hs GBR): this indicates the guaranteed number of bits per second the Node B should deliver through the radio interface under normal operating conditions (provided that there are data to transmit). Although in Table 4.4 GBR is specified only for streaming and conversational users, MAC-hs GBR may also be defined for other traffic classes.

Similarly, new HSDPA-specific measurements made at the Node B are available to the RNC [22]:

- *Non-HSDPA power measurement*: this is an average measurement of the total BS transmit power used by all channels except the HS-PDSCH and HS-SCCH. This measurement is defined in [24], where it is named 'transmitted carrier power of all codes not used for HS-PDSCH or HS-SCCH transmission'.
- *HS-DSCH-provided bit rate*: this is defined, for each priority class (SPI), as the total number of MAC-d PDU bits whose transmission over the radio interface has been considered successful by MAC-hs during the last measurement period, divided by the duration of the measurement period (the minimum averaging period is 100 ms). This measurement is defined in [22] and [27].
- *HS-DSCH-required power*: this is the sum, over all active HS-DSCH connections belonging to a given priority class (one SPI), of the minimum powers necessary to meet the MAC-hs GBR. (*Note*: power is reported only for those classes where MAC-hs GBR is different from 0.) Although the measurement is defined in [22], 3GPP specifications do not indicate how the Node B should measure this.

Figure 5.24 summarises information exchange between RNC and Node B.

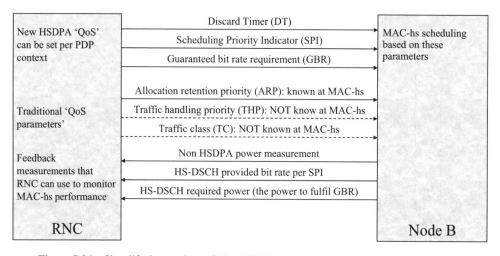

Figure 5.24   Simplified overview of the HSDPA interface between the RNC and BS.

5.3.3 Setting the HSDPA QoS parameters

When the HS-DSCH is selected for a new user, the admission control algorithm in the RNC produces the values for SPI, discard timer and MAC-hs GBR. This may be achieved through the static mapping proposed in Table 5.6, where it is assumed that conversational class is always carried on the DCH.

Notice here that MAC-hs GBR (target bit rate BR_{target}) for some of the traffic classes and/or THP may be set to 0. Also, the discard timer could be set to 'void', meaning that no discard timer value is set. As further discussed in Section 5.3.9, setting the discard timer and buffer thresholds for MAC-hs flow control should be carefully co-ordinated.

This simple approach allows the MAC-hs packet scheduler, described in detail in Section 5.3.10, to discriminate between five distinct bearer services over HSDPA. Notice that 3GPP does not define strict rules on how the packet scheduler should prioritise different SPI flows. Nevertheless, at the time of writing, we do not foresee any need for setting the ARP parameter for HSDPA users and, thus, mapping the SPI to other attributes of the QoS profiles than the parameters proposed in Table 5.6.

5.3.4 HSDPA power allocation

The 3GPP specification [22] allows two different methods for allocating the HSDPA power in the Node B. Power can be allocated dynamically or statically by the CRNC, or the Node B can exploit any unused power for HSDPA transmission. Since the latter option cannot be easily integrated with the wideband power-based RRM algorithms presented in Section 5.2, in the following we assume that power is dynamically allocated by the CRNC [32]. The downlink power budget from the RNC point of view is illustrated in Figure 5.25.

The stack on the left-hand side of Figure 5.25 does not show how the power resource is shared between different bearer services and physical channels. QoS differentiation requires discriminating between services with GBR – denoted by GB (i.e., bearer services of conversational or streaming QoS class for DCH, and bearer services with non-null MAC-hs GBR for HS-DSCH, see Table 5.6 – and those with no bit rate guarantees – denoted as NGB (i.e., bearer services of interactive or background QoS class for DCH, and bearer services with null MAC-hs GBR for HS-DSCH, see Table 5.6). This distinction is made on the right-hand side of Figure 5.25.

Table 5.6 Simple mapping of traffic class and THP to SPI, GBR and discard timer.

Traffic class		SPI	MAC-hs GBR	Discard timer (DT)
Streaming		4 (high)	$BR_{target\_4} =$ GBR (RAB attribute)	Defined as a function of the RAB attribute transfer delay
Interactive	THP1	3	BR_{target_3}	$DT_{\_3}$
	THP2	2	BR_{target_2}	$DT_{\_2}$
	THP3	1	BR_{target_1}	$DT_{\_1}$
Background		0 (low)	BR_{target_0}	$DT_{\_0}$

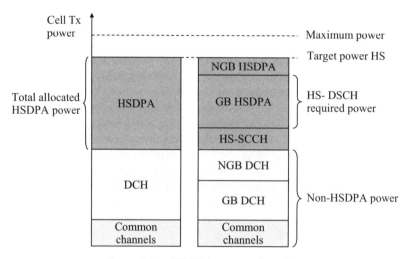

Figure 5.25   HSDPA power allocation.

The sum of common channel power and DCH traffic (GB and NGB) is the quantity that the Node B reports with the non-HSDPA power measurement (see Section 5.3.2). The quantity is highly time-variant due to the fast closed-loop control mechanism that is applied to every DPCH, while HSDPA power is fairly constant as the MAC-hs packet scheduler tries to use the same HS-PDSCH power in every TTI (see Section 4.3). However, since the allocated power includes both the HS-PDSCH and the HS-SCCH (which latter is power-controlled), there might be some power fluctuations. Another consequence of the HSDPA being fairly constant is the fact that the needed power control headroom for the DCH is smaller (see Section 7.2), thus the target power threshold (P_{Target_HS}) can be set closer to the maximum cell power.

GB HSDPA in this context refers to the fraction of HSDPA power that is used to fulfil the target bit rate requirement for all HSDPA users (SPIs) with non-null MAC-hs GBR (HS-DSCH required power in Section 5.3.2). As already mentioned, MAC-hs GBR may also be provided for interactive and background traffic classes. NGB HSDPA is the power share used for the remaining HSDPA users (SPIs).

The power used for NGB is categorised as 'available' power, which means that the CRNC can adjust it in order to make sure there is enough space for GB services, as explained in the following sections. Finally, it is important to point out that, although the CRNC allocates HSDPA power, the MAC-hs packet scheduler decides the power split between GB and NGB services, as presented in Section 5.3.10.

5.3.5   Channel-type selection and admission control

When it comes to selecting whether the DCH or HS-DSCH should be used for a specific RAB, admission control should first take into account what type of service is requested.

Table 5.7 shows how the TC and THP could be used to select whether a service can be mapped onto the HS-DSCH or not. In the example, conversational is the only traffic

Table 5.7 Traffic classes that can be mapped onto the HS-DSCH.

Traffic class		Allowed to use HS-DSCH
Conversational		No
Streaming		Yes
Interactive	THP1	Yes
	THP2	Yes
	THP3	Yes
Background		Yes

class that is not allowed to use the HS-DSCH [34]. In practice, the operator should be able to modify the table so that – for instance, in the case of a low bit rate service such as PoC mapped onto interactive (THP2) – it is possible to set the corresponding entry in Table 5.7 to 'no'.

Assume the requested traffic class can be mapped onto the HS-DSCH. A quality-based HSDPA algorithm, which aims at only admitting new users if it is estimated that it can be served with their target bit rate while still being able to satisfactorily serve all the allocated users with the same or higher priority (according to SPI), may be formulated as follows [36]. A new GB HSDPA user of priority k is admitted if:

$$P_{HSDPA} \geq P_{new} + \sum_{x \geq k} P_k + P_{HS\text{-}SCCH} + P_0 \tag{5.12}$$

where P_{HSDPA} is the allocated HSDPA transmission power, P_{new} is the estimated power required for the new user, $P_{HS\text{-}SCCH}$ is the estimated power required for transmitting the HS-SCCH (a typical average value for the macrocellular environment is 0.4–0.6 W) and P_0 is a configuration parameter which represents a safety power offset [36].

Note that the second term of (5.12) corresponds to the estimated required HS-DSCH transmission power to serve all allocated users that have the same or higher priority than the new user. Hence, the algorithm automatically results in a higher admission probability of users with high priority (SPI value) compared with low-priority users. The power offset P_0 is introduced to compensate for potential estimation errors of P_{new} and HS-DSCH-required power (P_k) from the Node B, as well as to provide an instrument for controlling trade-offs between admitting new users that afterwards would experience poor quality (low value of P_0) compared with rejecting new users that would have been successfully served on HSDPA (high value of P_0). The decision rule in (5.12) can be applied to both new users entering the system and users entering a cell via a handover from a neighbouring cell – that is, via a synchronised HS-DSCH cell change procedure (see Section 5.2.5.4).

Power increase P_{new} denotes the power needed to fulfil the QoS requirements of the new HSDPA user (i.e., the MAC-hs GBR, BR_{target}). A simple power increase estimator can be based on HS-DSCH-required power P_k and aggregate HS-DSCH bit rate measures BR_k provided for the SPI class to which the user belongs. Thus:

$$P_{new} = \frac{BR_{target}}{BR_k} P_k \qquad (5.13)$$

The target bit rate requested by the new user (BR_{target}) is equal to the MAC-hs GBR for the SPI the user belongs to (see Table 5.6). The target bit rate is always different from 0, since admission control is performed only for HSDPA GB users. A more refined way than (5.13), which takes into account the SINR at the user as a function of the expected average bit rate, can be found in [36].

In deriving (5.13), it is assumed that $BR_k \propto P_k$, which is true for cases where HS-DSCH transmission power is kept constant, and the number of TTIs where the user is scheduled is proportional to BR_k. This not strictly true when radio channel aware packet schedulers are used in combination with adaptive modulation and coding [38].

A special case arises when the new HSDPA user is the first in the cell. Then, the power needed by equal or higher priority users, second term in (5.12), is 0 and P_{new} cannot be estimated using (5.13), yet it may be set according to a parameter (with a specific value for each SPI class) or simply derived from the path loss and average SINR (configurable parameter) as presented in Section 7.2.

Section 5.3.6 deals with the problem of releasing the connection of HSDPA users which are inactive – that is, those that have insufficient or no data to transmit.

5.3.6   HS-DSCH release for inactivity

HS-DSCH inactivity may be detected based on a combination of MAC-d throughput and RLC buffer status. MAC-d flow is released if low utilisation is detected for a certain time period (*inactivity timer*). To avoid ping-pong effects, it is useful to define a parameter (*penalty timer*) during which the user cannot use the HS-DSCH.

In order to properly support QoS differentiation, it is possible to enhance the basic approach described above specifying different inactivity and penalty timers for different SPIs. The advantage of using short-inactivity timers is that we quickly free unused HSDPA resources and associated DPCHs, while the disadvantage is that we need to set up a new transport channel if more data arrive after expiration of the *inactivity timer*. Similarly, the advantage of using long-inactivity timers is that it is more likely that no additional data arrive when the HS-DSCH is released, while the disadvantage is that HSDPA resources including associated DPCH are reserved for a longer period.

5.3.7   Overload control with DCH and HS-DSCH users

In case both the DCH and HS-DSCH are used in a cell and the total transmitted power becomes too high – that is, (5.2) is satisfied – the RNC should take action to bring the cell back to normal operating status – that is, P_{Total} below P_{Target_HS}. The typical action, which can be applied to HSDPA as well, is reduction of the allocated bit rates or number of users by intra- or inter-frequency handovers (see Sections 5.2.5.4 and 5.2.5.6) or drop connections in a controlled manner. The selection of HSDPA users can be done, for instance, based on SPI. Note that the RNC has another degree of freedom since it may start from either HSDPA or DCH users. This can be decided using a configurable parameter. Anyway, overload actions will always affect NGB users (DCH and HSDPA) first.

5.3.8 HSDPA handover algorithm with QoS differentiation

As explained in Section 5.2.5.4, the intra-frequency HSDPA handover algorithm supports the following options:

- HS-DSCH to DCH handover (also used for congestion control, as pointed out in Sections 5.3.7 and 5.3.10).
- Intra- and inter-BS HS-DSCH to HS-DSCH handover.

If the QoS requirements of HSDPA users (e.g., MAC-hs GBR) have also to be guaranteed in the case of handovers to the target cell, the algorithm has to take these constraints into account, before deciding to make a handover from one cell to another. This is important, because not only the number of active HSDPA users (and their GBR, SPI, discard timer, etc.), but also the resources allocated to HSDPA (HS-PDSCH codes and HSDPA power) may be very different in the target cell. Therefore, it is necessary to control the available resources in the target cell as well. In principle, the task could be done using the same load estimation algorithm that is applied for initial channel-type selection to determine whether a user should be allowed to use HSDPA (see Section 5.3.5). However, if a handover request from an HSDPA user is rejected due to, say, high downlink load in the target cell, uplink HS-DPCCH coverage might not be sufficient to maintain the quality of the current connection. Hence, in the target cell, HSDPA handover candidate users should have priority over new HSDPA users.

5.3.9 Flow control algorithm in Node B and RNC handling of Iub congestion

The basic principle behind the MAC-hs flow control algorithm, as described in Section 4.3, is that the credits (CRs) allocated to a particular HSDPA user are decreased if the MAC-hs related buffer exceeds a *high threshold* over a certain time period, while they are increased if the MAC-hs buffer becomes lower than a *low threshold* for a certain time period.

Since different users may operate at different bit rates – with different discard timer settings, etc. – it is necessary to have independent settings of the high/low buffer thresholds for different SPI classes.

3GPP R5 specifications do not provide any explicit signalling procedure over the Iub interface for the RNC to set the MAC-hs high/low buffer thresholds. Hence, the Node B should be able to handle these parameters for different users based on the information it receives from the RNC and local knowledge of the performance for different active HSDPA users. In that case a simple two-step approach may be used:

- The initial setting of high/low thresholds is set according to the GBR, discard timer, SPI, etc., which are provided by the RNC.

- Afterwards, the high/low thresholds are dynamically updated according to the actual bit rate provided to the user so that no PDUs are deleted in the Node B due to expiration of the discard timer.

Given this starting point, high/low buffer thresholds could be set based on the following considerations.

If the MAC-hs buffer size is close to the high threshold and the user is operating at a bit rate BR that avoids the deletion of PDUs due to expiration of the discard timer (DT_x, for SPI x, the class to which the user belongs) the *high threshold* should be set according to:

$$High\_threshold \leq DT_x BR \qquad (5.14)$$

Notice that the Node B can measure the BR as the average bit rate provided to the user during the recent past (averaging period equal to N times the number of HS-PDSCH TTIs). Similarly, if the MAC-hs buffer size equals the low threshold, then the maximum bit rate BR_{max} the user can be provided during the next averaging period $T_{HS\text{-}DSCH}$ equals:

$$BR_{max} = \frac{Low\_threshold}{T_{HS\text{-}DSCH}} \qquad (5.15)$$

Hence, the low threshold should be set according to:

$$Low\_threshold \geq T_{HS\text{-}DSCH} BR_{max} \qquad (5.16)$$

with

$$High\_threshold \gg Low\_threshold \qquad (5.17)$$

Note that BR_{max} (maximum bit rate supported in the next TTI) is known to MAC-hs in the Node B, since it basically depends on the CQI reported by the UE (see Section 4.3).

At the other end of the Iub, the RNC is responsible for sending PDUs to the BS according to the received credits (CRs) from the MAC-hs flow controller(s) and the limited Iub HSDPA capacity. During times with no Iub congestion, the behaviour of the RNC is fairly simple, as it just forwards PDUs to the BS according to the allocated CRs. During congestion (i.e., allocated CRs exceed the available HSDPA Iub capacity) the RNC needs to do some prioritisation among the users (e.g., according to SPI). The algorithm should follow a consistent strategy for prioritisation (sharing of excess HSDPA capacity), like the one used in the MAC-hs packet scheduler (see Section 5.3.10).

5.3.10   Packet scheduler

The total allocated HSDPA power is time-variant, since it depends on the concurrent DCH load (see Section 5.3.4). The MAC-hs packet scheduler is responsible for accommodating, within the limited and variable resources, HSDPA users with different QoS requirements and those experiencing different air–interface conditions. There can be several solutions to such a scheduling problem (e.g., see [34] and [35]). In the following, we describe the requirements for a MAC-hs packet scheduler based on the packet-scheduling method proposed for WCDMA in Section 5.2.2.

Figure 5.26 shows the packet scheduler environment and interactions with other blocks in MAC-hs.

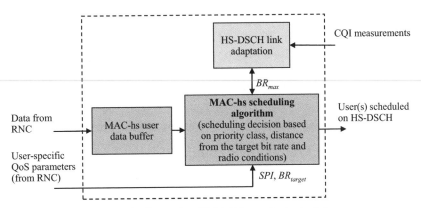

Figure 5.26   Example of MAC-hs packet scheduler.

Assume there are several HSDPA users (belonging to different SPI classes) admitted in a cell supporting HSDPA. During the scheduling period, HSDPA users that have either no data to transmit in the buffer or their target bit rate requirement is satisfied (i.e., the actual offered bit rate is above the target) are not included in the candidate scheduling list. All the remaining users (including those with a zero target bit rate) should be ranked according to the following metrics:

- Scheduling priority (SPI).
- The difference between the target bit rate (BR_{target}, whose values are specified in Table 5.6) and the offered bit rate (BR, average bit rate provided to the user during the last N TTIs).
- Maximum bit rate (BR_{max}), which can be provided by the link adaptation manager to the user in the next TTI.

If the system can serve only one user during each TTI, this ranking is enough to decide who will be served in the next TTI. If the access network supports 10 up to 15 HS-PDSCH codes, the MAC-hs packet scheduler should be able to support code-multiplexing of multiple HSDPA users in one TTI (maximum of four users, [26]), since many of the users are expected to use a maximum of five HS-PDSCH codes anyway. In this case, the packet scheduler will select the K highest ranking users (where K is the number of users that are code-multiplexed in one TTI). The value of K may be derived from the following factors:

- Number of HS-SCCH codes allocated in the cell (see Section 3.4.3).
- Number of HS-PDSCHs supported by the UE.
- Power needed to satisfy each user target bit rate (e.g., estimated using (5.13)).
- Total allocated HSDPA power (see Figure 5.25).

A simple and robust method for splitting the available HSDPA resources among the K code-multiplexed users is to evenly divide the available codes and power among them.

In the case of traffic congestion in the cell, in order to free allocated resources, the RNC may terminate the connections of the HSDPA users that have been on stall (admitted in the cell but never scheduled) for a long time. Note that this mechanism does not depend on the overload control actions taken when the total transmitted power exceeds the maximum planned power in (5.2), as described in Section 5.3.7.

5.4 HSUPA with QoS differentiation

QoS handling for HSUPA differs from the HSDPA case presented earlier, because uplink transmissions are not under the direct control of the Node B. As a consequence the QoS differentiation described in Section 5.3 cannot be applied directly to HSUPA and new mechanisms need to be developed.

During service setup the SRNC performs admission control and defines the priority of each bearer service. Afterwards, it performs channel-type selection. Typically, the services with guaranteed bit rate (GB) – for example, bearers of conversational or streaming class or signalling radio bearers (SRBs) – are mapped onto DCHs, while NGB services are mapped onto the E-DCH. However, in case the operator decides not to use any DCH at all, even GB bearer services and SRBs may be mapped onto the E-DCH, as illustrated in Figure 5.27. Such services are supported through non-scheduled transmissions.

In the case of GB services on the E-DCH, certain resources are assigned to the UE by the SRNC. This is communicated to the Node B and UE during initial signalling. The UE can transmit the data on the E-DCH with the assigned bit rate without requiring resources from the Node B. These resources are controlled by the SRNC as in the DCH case (see Figure 5.27). The Node B implements the other E-DCH functionalities (e.g., HARQ).

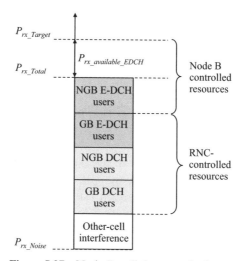

Figure 5.27 Node B uplink power budget.

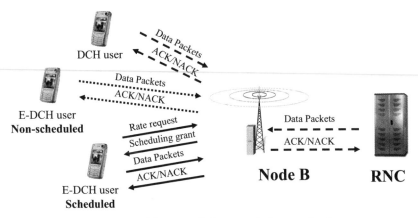

Figure 5.28 Different uplink transmission mechanisms.

For NGB services the UE sends bit rate requests to the Node B when it has uplink data to transmit. The Node B estimates the amount of resources available to the E-DCH and distributes them among E-DCH users based on rate requests, priorities, buffer status, etc. The Node B sends scheduling grants to the UEs. The UE selects the appropriate E-TFC based on the received grant, power remaining after DCH transmission and amount of data to transmit. More details and background information on HSUPA may be found in Section 4.3.

Figure 5.28 illustrates the difference in uplink transmission for DCH, non-scheduled E-DCH and scheduled E-DCH users. These mechanisms with QoS differentiation are described in more detail in the following sections.

5.4.1 QoS control

The RAB attributes for HSUPA QoS control are available in the SRNC according to 3GPP R99 principles. For QoS-based E-TFC selection, multiplexing of logical channels in MAC-e PDUs and HARQ operation, the SRNC provides the UE with the following information [37]:

- Logical channel priority for each logical channel (as in 3GPP R5).
- Mapping between logical channel(s) and MAC-d flow(s) (as in 3GPP R5).
- Allowed MAC-d flow combinations in one MAC-e PDU.
- Power offset for reference E-TFC(s). The power offsets for the other E-TFCs are then calculated so that quality (protection of a MAC-e PDU) when using any of the E-TFCs is identical to that of the reference E-TFC(s).
- The E-DPCCH power offset. This is used to set the protection level for E-DPCCH transmissions.
- HARQ profile (power offset and maximum number of transmissions) per MAC-d flow. The power offset is used in E-TFC selection to regulate the BLER operating point for transmission. The 'maximum number of transmissions' attribute is used in HARQ operation to regulate the maximal latency and residual BLER of MAC-d flows.

- A non-scheduled grant (valid only for MAC-d flows that are configured for non-scheduled transmission).
- Maximum number of E-DPDCH channelisation codes along with minimum spreading factor.
- Periods for sending scheduling information applicable when the UE is or is not allowed to transmit scheduled data.

(Using RRC signalling, the UE is in addition informed about: the E-HICH, E-RGCH, E-AGCH, E-DPCCH and E-DPDCH configuration; E-DCH scheduling information parameters; grant information; mapping between logical channels and MAC-d flows; and for each MAC-d flow, the MAC-d flow specific power offset, the maximum number of transmissions and the multiplexing list.)

For scheduling and resource reservation, the SRNC provides the following QoS-related parameters to base stations in the E-DCH active set [37]:

- Power offsets for reference E-TFC(s). The Node B then calculates the power offsets for the other E-TFCs.
- E-DPCCH power offset.
- HARQ profile (power offset and maximum number of transmissions) per MAC-d flow. The power offset is used whenever the Node B needs to convert between rate and power in its resource allocation operation.
- Guaranteed bit rate for logical channels that carry guaranteed bit rate services. This is used for allocating grants to mobile terminals.
- Non-scheduled grant for non-scheduled transmission MAC-d flows. This is used in the Node B to reserve sufficient radio resources.
- Maximum uplink UE power (minimum between the UE maximum transmit power and maximum allowed uplink Tx power configured by the SRNC).
- Scheduling priority per logical channel of logical channels mapped onto the E-DCH and the corresponding mapping between logical channel identifier and DDI value. This information enables the Node B to consider the QoS-related information of the logical channels for efficient scheduling.

Logical channels mapped onto dedicated channels are always prioritised over those mapped onto the E-DCH. To determine E-TFC states (blocked or supported), the UE uses the power offsets for the reference E-TFC(s), the signalled power offset attributes for its MAC-d flows, the required E-TFC dependent back-off and UE remaining power. When calculating remaining power, HS-DPCCH, DPCCH, DPDCH and E-DPCCH powers are taken into account. E-TFC selection is performed in the UE based on logical channel priorities as in 3GPP R99, maximising the transmission of higher priority data.

5.4.2 HSUPA dynamic resource handling

In HSUPA, the resource utilisation decision is distributed among the SRNC, UE and Node B. The SRNC assigns the *target value of received uplink power* to the Node B (P_{rx_Target} in Figure 5.27). The UE sends the scheduling request to the Node B, indicating its buffer occupancy and power headroom, as well as its happiness status (on the E-

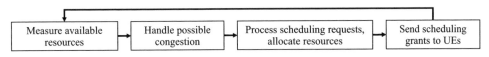

Figure 5.29 Node B packet-scheduling algorithm.

DPCCH). The first scheduling request is sent in a MAC-e PDU as a non-scheduled transmission. If the UE has already assigned E-DCH resources, the scheduling request is sent along with the data in a MAC-e PDU. Then, every scheduling period the Node B measures the received total wideband power in the cell (denoted as P_{rx_Total} in Figure 5.27, or by RTWP in [37]) and calculates the available resources (power budget) for E-DCH scheduling, as a difference between the target power value, defined by the SRNC, and the actual total received power. The packet scheduler can use this available power budget (denoted by $P_{rx\_available\_EDCH}$ in Figure 5.27) for allocating E-DCH users. The Node B processes the scheduling requests from UEs and sends scheduling grants based on their priority and its own scheduling requests. The Node B can also monitor the resource utilisation of already existing E-DCH users and redistribute radio resources among them. The flow chart for Node B packet scheduling is shown in Figure 5.29. Moreover, to issue relative grants the CRNC may also send to the Node B a *target value for the non-serving E-DCH to total E-DCH power ratio (PR)* per cell. The Node B should then issue non-serving radio link relative grant 'down' commands when the following conditions are met:

$$\begin{cases} P_{rx\_Total} > P_{rx\_Target} \\ PR > PR_{Target} \end{cases} \tag{5.18}$$

The *PR* is the ratio of power from UEs for which this cell is a non-serving RL and the total E-DCH power. Received non-serving E-DCH power and total E-DCH power may be estimated from the E-TFC information on the E-DPCCH and a *reference power offset*. The *reference power offset* is defined per UE using the same value range as the MAC-d flow specific HARQ offset and signalled from SRNC to Node B for this calculation. When using the E-TFI for computing the E-DCH power received in a cell which is a part of a multi-cell RLS, the Node B allocates the computed power equally divided among all cells in the RLS, regardless of whether the RLS contains the E-DCH serving cell or not [37].

After receiving the scheduling grant, the UE should select the E-TFC that maximises the transmission rate of the highest priority data. This needs to be done taking into account the following constraints:

- Logical channel (MAC-d flows) priorities.
- Allowed combinations of MAC-d flows in one MAC-e PDU.
- Power offset (ratio between E-DPDCH and DPCCH power) with respect to the highest priority MAC-d flow.
- Amount of data to transmit.
- Estimate of the power remaining after DCH TFC selection.

As already discussed, GB services (i.e., bearer services with strict quality requirements) mapped onto the E-DCH, or DCH, are subject to admission control. In order to support this function and enable noise floor estimations (see Figure 5.27), the Node B signals to the CRNC the following measures/estimates:

- Total **RTWP** (as in 3GPP R5).
- Provided bit rate per logical channel priority and per cell, taking into account only the logical channels mapped onto the E-DCH.

Ultimately, the CRNC may handle resources between cells based on a *load excess indicator* it receives from the Node B when the frequency of 'down' commands for which this cell is a non-serving radio link exceeds a pre-defined level.

5.4.3   Simulation results

Figure 5.30 presents the gain that can be achieved moving the scheduling functionality for HSUPA from the RNC to the Node B [37].

In particular, the first graph shows the PDF of the noise rise distribution for RNC and Node B based schedulers. It can be observed that Node B packet scheduling allows the system to more efficiently control the total received uplink power than is the case with RNC scheduling, and the required power headroom to prevent the system from entering overload status is consequently reduced. Hence, the average uplink load to meet the specified noise rise outage constraint can be increased, and the average cell throughput consequently improved.

This is confirmed by the right-hand part of Figure 5.30, which illustrates average cell throughput as a function of average number of users per cell. Node B scheduling provides a cell throughput improvement compared with RNC scheduling of between 6 and 9%, almost independently of the average number of users per cell.

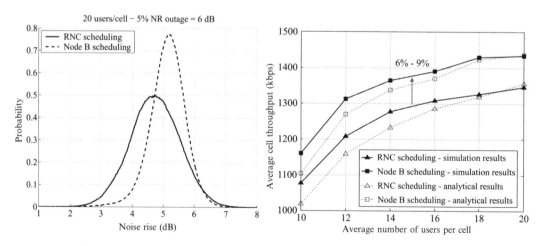

Figure 5.30   Simulation results: RNC vs. Node B uplink packet scheduling.

5.5   Service performance in UTRA-GERA networks

Service control and performance in the multiradio environment is essential with the step-by-step roll-out of a new radio access technology like WCDMA on top of GSM. The new WCDMA network infrastructure adds new capacity and service capabilities while the existing GSM network serves as a coverage extension where WCDMA coverage has not yet been built. During the co-existence of multiple radio access technologies, dynamic load balancing between the different access networks enables efficient usage of invested infrastructure and spectrum. Furthermore, with the gradual penetration of dual-mode GSM/WCDMA handsets and subsequent replacement of single-mode GSM handsets in the market, load balancing becomes an economical necessity for deployment of multi-radio networks. According to the forecast, half of the mobiles in Western Europe will be dual-mode capable by 2008 (see Figure 5.31).

In moving users between different access technologies, in order to provide the service application performance levels necessary for satisfactory QoE, the network must take into account QoS constraints. Thus, in the following we will further elaborate service control and performance aspects in GSM/WCDMA networks. It is clear that the following discussion affects only dual-mode terminals. Single-mode GSM terminals are agnostic to the newer WCDMA network.

5.5.1   Service control

Control mechanisms to steer mobiles to a certain access exist in idle mode and in connected mode. However, in idle mode all mobiles listen to the same broadcast cell reselection parameters – that is, it is not possible to steer specific mobiles to a preferred access system. In connected mode, mobiles can be pushed individually to the preferred

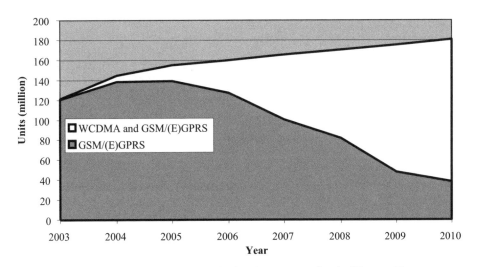

Figure 5.31   GSM/WCDMA mobile handset penetration in Western Europe.
Source: Strategy Analytics 2004.

access during the call setup phase – for example, with inter-system directed retry – or during an ongoing call with inter-system handover. The trigger to initiate the procedure can come from the radio access network, more specifically the radio controller (RNC in WCDMA, BSC in GSM), or from the core network. For CS services, it is the MSC that gives a handover recommendation to the radio controller over the A or Iu interfaces, and for PS services the SGSN gives the cell reselection recommendation over the Gb or Iu interfaces [25]. But, it is always the radio controller who has the final decision about the handover/cell reselection command because it is aware of coverage and cell load. The triggers for a system change can be (see also Section 5.2.5):

- Coverage (only initiated in the radio access network).
- Cell load (radio access network).
- QoS attributes (either by core network or radio access network).
- Service (only via core network – the radio access network is unaware of the service but is aware of its QoS attributes).
- IMSI (either by core network or even radio access network), etc.

Concerning the idle mode camping strategy, it is in principle possible to let mobiles camp in GSM and then at call setup or during the call activate service control. However, dual-mode mobiles before 3GPP R6 do not forward CS video requests in GSM so that the network cannot even initiate a directed retry to WCDMA (see [41], [42]). Thus, if a CS video service is launched, in practice dual-mode mobiles will have to camp by default in WCDMA as long as there are still a significant amount of pre-R6 dual-mode terminals in the market.

5.5.2 QoS renegotiation

The QoS attributes of a service are agreed between the terminal and the core network during service initiation. If the terminal happens to be in a network that cannot provide the requested QoS, negotiations between mobile and core network take place and will lead to either downgraded QoS parameters or to a rejection of the service request – for example, when the service application in the mobile runs only with the originally requested QoS attributes. (Note that limitations for downgrading QoS can also come from the radio access network – for example, if EGPRS is unsupported or the serving cell is congested.) When QoS has been downgraded and the terminal enters later a network that supports the originally requested QoS (e.g., during a system change from GSM/GPRS to WCDMA), the new SGSN can initiate a QoS upgrade back to the higher QoS. The new SGSN receives originally requested QoS parameters through the old SGSN. *Vice versa*, QoS downgrading can also occur in a system change from WCDMA to GSM/GPRS.

5.5.3 Handover/Cell reselection performance for PS services

In this section, we first analyse the interruption gaps for PS services in a UTRAN to GERAN system change. Performance in a GERAN to UTRAN cell change are discussed at the end of the section.

5.5.3.1 From UTRAN to GERAN

Many commercial WCDMA networks have at roll-out already experienced network-initiated inter-system cell reselection – in contrast to the pure mobile-based inter-system cell reselection of idle mode.

During cell reselection, a number of delay components contribute significantly to the overall interruption gap: after the mobile has received the last data packet in WCDMA, it needs to switch to the GSM frequency and there acquire for the first time the broadcast system information (SI), which takes on average 2.4 s. From the GSM cell information, the UE may initiate uplink TBF establishment followed by location area update (LAU) and routing area update (RAU) procedures. If performed separately, the delay introduced by the two procedures is about 7.5 s, which is the longest part of the overall delay. Only after that is downlink TBF established, upon which TCP delays may add again significantly to the delay budget. The 2 s indicated can easily be longer if packets are dropped or lost during the system change. As TCP is unaware of, and today uncoordinated with, lower protocol layers, TCP delays vary strongly from case to case depending on how TCP retransmission time-outs (RTO) kick in. Lost IP packets can easily add tens of seconds to the interruption gap. In the case of unacknowledged streaming traffic (e.g., UDP), the interruption gap is shorter by an amount equal to the TCP delays. In total, this leads to a gap for network-initiated inter-system cell change of about 13 s on average for acknowledged traffic (TCP) and 11 s for unacknowledged traffic (UDP) – see Figure 5.32.

With the support of the Gs interface between the SGSN and MSC/VLR, LAU and RAU can be combined and delays are further decreased by 4–5 s [43]. Usage of the packet BCCH in GSM/GPRS can further reduce the time for acquiring system information by around 1 s on average; and if inter-system network-assisted cell change (NACC) is used, the UE is given the needed system information (SI) to access the target cell even before the system change from UTRAN to GERAN, which completely saves the time spent

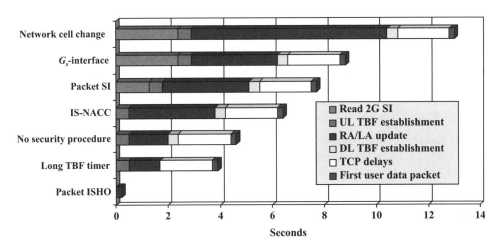

Figure 5.32 Service breaks for TCP traffic in UTRAN to GERAN system change (UDP when omitting TCP delays).

reading the SI during the gap [44], [45]. RA/LA updates can be made faster when optionally omitting security procedures during the service gap, and the establishment of a downlink TBF can be made redundant if TBF timers are just long enough. With these tricks the break in service is less than 4 s, as shown in Figure 5.32.

The ultimate improvement for PS interruption times however came with the handover for PS services in 3GPP R6, where the same principles as for CS services are applied – 'make before break' [46]:

- Resources in the target cell are allocated for the mobile while it is still in the source cell.
- Target cell system information required for the mobile to access the new cell is sent via the source cell while the rest is acquired in the target cell.
- To address the assigned resources in the target cell P-TMSI is allocated in the target cell while the mobile is still in the source cell.

Normally, the TCP congestion control mechanism assumes that, once a connection is established, the end-to-end path it traverses is relatively stable. However, the TCP node point of attachment may change due to cellular mobility events like inter-system cell reselection. If the TCP sender is unaware of the change, it will continue to send packets to the network at the same rate as before the point of attachment changed. This can clearly lead to suboptimal TCP behaviour. Further, if the mobility event causes enough packets to be lost, TCP may be forced to wait for RTO to expire before continuing to transmit data (the timer value can be large in cellular networks, in the order of many seconds).

A way to reduce TCP outage times could be the lightweight mobility detection and response (LMDR) algorithm [47]. This describes a network-layer independent mechanism by which mobile hosts can propagate path-change notifications to the server. Notification could be done via some unused bits in the TCP header or, more formally, using a new TCP option. Once the sender receives the notification, it can adjust its behaviour for better performance. In particular, it can first reassess the congestion situation on the new path (by reverting back to slow start following reception of the notification) and, second, immediately resend packets lost on the old path (instead of waiting for RTO expiry). While – in the light of PS handover – TCP retransmission timeouts (RTOs) are less likely to occur, LMDR still might be beneficial for adjusting to a different congestion situation after handover.

5.5.3.2 From GERAN to UTRAN

During cell reselection from GERAN to UTRAN, quite similar procedures have to be performed: acquiring 3G system information, setting up the signalling connection (RRC connection), RAU/LAU and setting up the radio bearer (user plane). The resulting delay components sum up to an overall interruption break of about 5 s with some variation depending on whether combined or separated RAU/LAU is performed and which bit rate is used for the SRB signalling bearer in WCDMA (3.4 kb/s or 13.6 kb/s).

If TCP traffic is used, again TCP delays must be taken into account, which can vary from less than a second to tens of seconds (e.g., in case buffered packets in the source systems are not properly transferred to the target system).

Also, in 3GPP R6, PS handover will be the ultimate feature to get interruption times down to the subsecond area [46].

5.5.4 Handover performance for CS services

For conversational CS services there is only one way of changing the access system – that is, by handover (in contrast to non-real-time packet data services).

The *reliability* of inter-system voice handover has proven to be well above 90% in time-critical WCDMA to GSM handovers due to cessation of WCDMA network coverage. Call drops in the WCDMA to GSM direction occur predominantly during the handover preparation phase – that is, when GSM neighbour signal strength and identity (BSIC) are measured. For other reasons – such as load, service, IMSI, etc. – the time to execute handover is not as critical and, thus, handover success rates should be even higher. The most critical factors for reliable WCDMA to GSM handovers are:

- RNC triggers and neighbour measurement algorithm;
- GSM network planning (frequency reuse, interference, etc.);
- neighbour cell list management (updated and optimal) [48].

In the GSM to WCDMA direction, it is BSC implementation that counts.

Improvements in reliability and execution speed can be expected with future enhancements to HO preparation algorithms, parameter optimisation, automatic configuration update via NMS (e.g., neighbour cell lists), UE measurement performance, etc.

Inter-system handover is a hard handover. Thus, there is a small service break as in intra-GSM handover. Interruption breaks have been measured as follows:

- In the WCDMA to GSM direction, the downlink gap is about 120 ms and the uplink ~ 200 ms. Typically, only one of the two gaps is relevant because the user is either talking or listening during inter-system handover.
- In the GSM to WCDMA direction, the gap is slightly larger, ~ 250 ms in the downlink and ~ 350 ms in the uplink. The longer handover gap is related to the SFN decoding that has to be done during the service gap.

Service gaps in inter-system voice handovers are thus somewhat longer than those in intra-GSM hard handovers (60–80 ms have been measured), but from a mobile user perspective the service interruption is not noticeable especially in the WCDMA to GSM direction.

5.5.5 Service performance and terminal capabilities

If a user has CS and PS services ongoing in parallel (e.g., voice and web browsing), he/she uses a multi-RAB connection in WCDMA, while in GSM/(E)GPRS he/she uses dual-transfer mode (DTM). The latter requires the feature to be supported in terminals and in the network. DTM support, however, did not come right away with the first commercial WCDMA-capable terminals (R99). This means in practice that such a non-DTM dual-mode terminal, when leaving WCDMA coverage and using multi-RAB, cannot continue

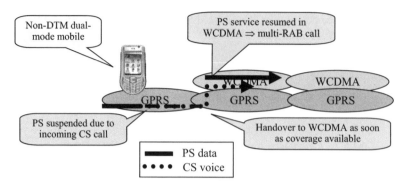

Figure 5.33 A suspended PS connection will be resumed when non-DTM terminals change to WCDMA.

the PS service. The CS service has priority and will be continued in GSM with the normal inter-system handover procedure. The PS service will be suspended and only resumed when the CS call terminates or the mobile goes back to WCDMA (see Figure 5.33).

If the UE also supports DTM, the PS service can be continued in GSM/(E)GPRS. The procedure again involves inter-system handover for CS connection; and when CS handover is completed the PS connection will be reinitiated. While the CS service will not see any user-perceivable degradation PS connection has an interruption gap similar to the one registered in the cell reselection procedure described in Section 5.5.3. This is also subject to improvements (currently discussed in 3GPP).

Another aspect of terminal capabilities in multiradio environments is the support of EDGE or EGPRS. EGPRS is not necessarily supported in every cell in 2G networks. For instance, of the overlaying cells one might support EGPRS while the others support only GPRS. As already pointed out in Section 5.1.4, it is very beneficial to direct an EGPRS-capable mobile directly to the EGPRS-enabled cell instead of choosing simply a neighbour cell with good enough signal strength, which then might be GPRS-only. In that case, the PS service will first have lower bit rates than the terminal can support and, in addition to that, will undergo a second cell reselection from the GPRS to the EGPRS cell with a certain outage time, which again can degrade the QoE. To avoid this, the radio controller in WCDMA (RNC) must select the EGPRS-capable cell during the handover/network-controlled cell reselection preparation phase, as illustrated in Figure 5.34.

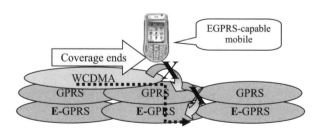

Figure 5.34 Directing an EGPRS-capable mobile directly to an EGPRS-enabled cell improves service performance.

5.5.6   Load balancing between GSM and WCDMA

When trying to efficiently use the existing capacity in GSM and WCDMA networks, ISHOs are needed for dynamic load balancing. Both CS and PS services can be considered to balance load with a few current exceptions, such as CS video services and high bit rate PS streaming services, especially when terminal capabilities limit the bit rate.

Nevertheless, there are three good reasons basic CS voice is the most feasible service to be used for load balancing:

- CS voice will still be the predominant network load even in 3G for many years to come. This allows a sufficient mass to be available for handover;
- voice quality is practically the same in GSM and WCDMA – this means the mobile user will not notice any difference in service quality when moved from one system to the other; and
- as pointed out in Section 5.5.4, the interruption breaks of inter-system hard handover in both directions (GSM ↔ WCDMA) are not noticeable to the end-user. So, system change on its own does not impose any service degradation, which can be claimed for PS services only when PS handover is available (see above).

Having said this, there are still some limitations to this otherwise rather straightforward approach. Dual-mode terminals from the early years do not support DTM in 2G. If they were handed over as part of a voice service to 2G, mobile users could not initiate a parallel PS data service in the same way as they could with multi-RAB in 3G [49]. To avoid this confusing situation, non-DTM terminals should be better kept in 3G even if the service is basic voice.

Another reason for including PS services in load balancing is business – for example, in shared networks and MVNOs. Assuming a network operator owns the GSM network outright and has the WCDMA network shared or leased, he might want to have, say, low bit rate PS services served in his own network with (E)GPRS and allow the more bandwidth hungry services be served in the WCDMA where he pays by usage. This is feasible as well but needs to take into account the trade-off between load balancing and impact on PS service performance, as discussed in Section 5.5.3.

5.6   3GPP–WLAN inter-working

Up until R5, the system architecture was limited to access technologies defined internally within 3GPP. 3GPP first deviated from this trend with the introduction of the 3GPP–WLAN inter-working work items in 3GPP R6. Inter-working with WLAN represents one of the main new features introduced in 3GPP R6, together with HSUPA, IMS Phase 2 (which, among other things, introduces access independence), multimedia broadcast multicast service (MBMS) and enablers for PoC.

By the time that work on R6 was starting, WLAN was gaining significant importance as a wireless access technology and 3GPP could not ignore its significance as a local-area (typically indoor) complement to wide-area cellular packet access. 3GPP thus produced specifications that define how WLAN access can be integrated into the 3GPP system architecture. The resulting Stage 1, Stage 2 and Stage 3 specifications can be found in [50]–[53]. 3GPP specifications do not set 3GPP-specific requirements for WLAN access systems, but rely on the existing functionality available in a typical WLAN access network based on IEEE 802.11 standards. In the meantime, IEEE has also been working on QoS extensions to IEEE 802.11, known as IEEE 802.11e, which are currently in draft specification stage [54].

3GPP TR 22.934 introduces six inter-working scenarios, representing the various levels of integration between WLAN and 3GPP networks [50]. These scenarios are outlined next. QoS and QoE considerations vary depending on the deployed inter-working scenario. These considerations will be discussed in Section 5.6.1. The scenarios are:

- *Scenario 1 – common billing and customer care*: the subscriber receives one bill from the mobile operator for the usage of both 3GPP and WLAN services. This does not pose any new requirements on 3GPP specifications.
- *Scenario 2 – 3GPP system-based access control and charging*: authentication, authorisation and accounting for WLAN access are provided by the 3GPP system using (U)SIM credentials. After successful authentication, the subscriber is authorised to receive direct Internet access from the WLAN hot spot. The resulting architecture is illustrated in Figure 5.35 (note that the UE is not necessarily a multi-mode UE).

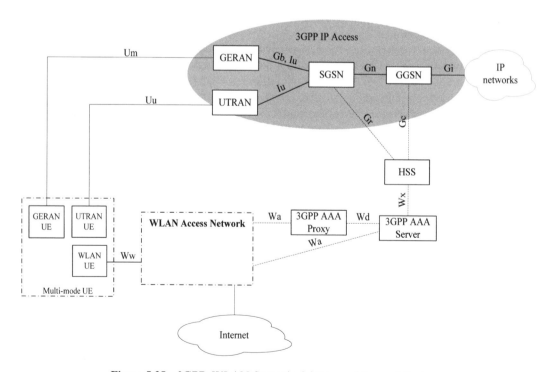

Figure 5.35   3GPP–WLAN Scenario 2 inter-working architecture.

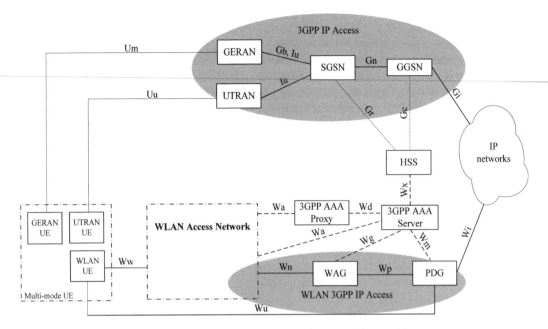

Figure 5.36 3GPP–WLAN Scenario 3 inter-working architecture.

- *Scenario 3 – access to 3GPP system PS-based services*: this inter-working scenario enables the subscriber to access 3GPP PS services through the WLAN access gateway (WAG) and packet data gateway (PDG) – that is, IP Multimedia Subsystem (IMS)-based services, instant messaging, presence-based services, MBMS and operator-hosted corporate access, through WLAN. The resulting architecture is depicted in Figure 5.36.
- *Scenario 4 – service continuity*: this scenario allows the services supported in Scenario 3 to survive a change of access between WLAN and UTRAN/GERAN. The change of access may be noticeable to the end-user, but there will be no need for services to be re-established. Due to the different access network capabilities, there may be a change in service quality after the transition across access technologies.
- *Scenario 5 – seamless services*: this scenario provides seamless service continuity between the access technologies for the services supported in Scenario 3. This is achieved by minimising aspects such as data loss and break time during the switch between access technologies.
- *Scenario 6 – access to 3GPP CS services*: this scenario allows access to services provided by the entities of the 3GPP circuit-switched core network to be accessible over WLAN.

3GPP R6 specifications cover Scenario 2 and 3. Work on further inter-working scenarios will be taken up in R7. The current 3GPP specifications do not include any end-to-end QoS mechanisms for WLAN. The discussion in the following sections will thus be mostly limited to the general QoS and QoE aspects that are relevant to each of the inter-working scenarios.

5.6.1 QoS and QoE aspects in 3GPP–WLAN inter-working

There are various degrees of QoS and QoE considerations associated with the integration of WLAN access networks with 3GPP networks. Each inter-working scenario defined in 3GPP results in a set of QoS and QoE requirements. The present 3GPP–WLAN specifications do not explicitly define any QoS mechanisms associated with the inter-working scenarios. This may eventually change with future releases of the specifications, at least in terms of QoS requirements that cover some of the inter-working scenarios. The discussion that follows is consequently limited to the general QoS and QoE considerations associated with each of the inter-working scenarios.

3GPP–WLAN inter-working Scenario 2 exhibits the same QoE and QoS issues as any other public WLAN deployment. The only differentiator is the delay associated with the authentication and authorisation procedures that may need to take place over a roaming interface, between the AAA proxy on the visited WLAN network and the remote AAA server in the home network. Delays over this interface will translate into latencies incurred when logging onto the WLAN. Traffic associated with access to Internet services over public WLANs is typically not differentiated based on service or subscriber. However, peak rates are typically regulated per subscriber, in order to ensure fair sharing of the available resources.

3GPP–WLAN inter-working Scenario 3 enables access over WLAN to services traditionally available only through GERA and/or UTRA. Depending on the services offered and the load conditions, a QoS solution may be necessary to enhance end-user experience without compromising the efficient use of resources. The type of QoS that can be supported for Scenario 3 access depends on various factors, particularly, amongst other things, the type of transport between the public WLAN hot spot and the 3GPP core network. In some deployment scenarios, a controlled transport solution may allow for consistent end-to-end QoS. Other deployment scenarios may assume plain Internet to connect the WLAN hot spot and the 3GPP core. In this case end-to-end QoS cannot be delivered. Since we are dealing with IP networks, a combination of IETF DiffServ [55] and IEEE 802.11e [54] are the prime candidates for providing prioritised treatment of traffic. Since – in Scenario 3 – traffic is tunnelled between the mobile device and the 3GPP core network using an IPsec tunnel, special care must be taken to ensure that the packet reordering that may occur due to prioritisation procedures will not cause problems with the replay protection mechanisms in IPsec. 3GPP R7 specifications are expected to include QoS procedures for Scenario 3.

3GPP–WLAN inter-working Scenarios 4 and 5 are in the process of being specified in 3GPP R7. The underlying general solution has not yet been widely discussed. One of the candidate enablers for mobility across cellular and WLAN access is mobile IP [56]. In general, the key QoS indicators related to mobility across WLAN and cellular access are:

i. Packet loss during inter-access transition.
ii. Inter-access mobility delay. In the uplink this translates to the time the last packet was sent over the source access to the first packet sent over the target access. In the downlink this is measured by the time the last packet was received over the source access to the first packet received over the target access;
iii. QoS preservation across access technologies.

Since WLAN and UTRA/GERA frequency bands are sufficiently disparate to permit simultaneous operation without causing significant interference in the mobile device, the QoS measured in terms of packet loss (i) and handoff latency (ii) can be optimal, particularly in the case of mobility across WLAN and GERAN. In the case of mobility from WLAN towards UTRAN some delay and packet loss will typically occur while RRC signalling prepares the UTRAN RAB for packet transmission. In the worst cases, there may also be some coverage gaps between cellular and WLAN coverage. This may particularly be the case in those countries that experience limited indoor cellular coverage. Handoff performance will clearly suffer in such scenarios. Priority-based QoS preservation can easily be attainable across access technologies. QoS based on dedicated resources may however impact handoff performance negatively when, say, WLAN coverage is lost abruptly, prior to completion of cellular QoS reservation procedures.

Though not officially stated, the generic access to A/Gb interface specifications 3GPP has imported from the Unlicensed Mobile Access (UMA) Consortium (see [57]–[59]) represents an embodiment of what was envisioned for 3GPP–WLAN inter-working Scenario 6 in [50]. In this scenario, support for voice is the dominant driver. This is enabled by tunnelling GSM protocols over an IPsec tunnel through the residential WLAN towards an IPsec gateway that emulates a BSC. The reference network architecture is illustrated in Figure 5.37. The prime motivation behind this work has been to leverage residential WLAN access to overcome the indoor cellular coverage limitations that are common in some geographical areas. QoS possibilities are thus governed by the QoS capabilities that can be assumed between the subscriber's device and the operator's core network, through the subscriber's residential WLAN. In the event that the mobile operator does not have control over the broadband connection feeding the subscriber's

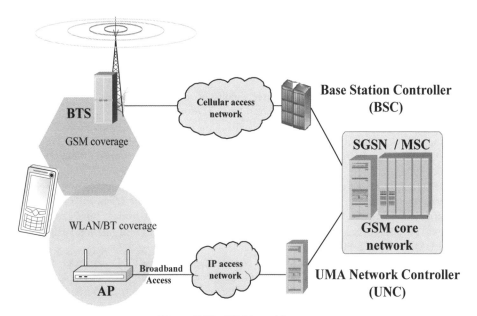

Figure 5.37   UMA architecture.

residential WLAN and the distribution of wireless resources in the subscriber's WLAN, only a best-effort service can be assumed.

An alternative to Scenario 6 for a 3GPP operator to deliver voice services through residential broadband connections, which increasingly include a WLAN link as the last hop, is to deploy a solution based on the 3GPP IMS. Specifications for enabling this are currently being defined in ETSI, under the name of 'next generation networks' (NGNs). It is expected that the first release of specifications will include basic QoS mechanisms such as admission control, whereas end-to-end QoS mechanisms are subject to specification in later releases. Eventually, mobility across WLANs and cellular coverage will become relevant even in the case of NGNs. The same QoS and QoE considerations outlined above for 3GPP–WLAN inter-working Scenarios 4 and 5 will apply.

References

[1] 3GPP R99, TS 03.64, General Packet Radio Service (GPRS); Overall Description of the GPRS Radio Interface; Stage 2, 2004, v. 8.12.0.

[2] A. Kuurne, D. Fernández and R. Sánchez, Service based prioritization in (E)GPRS radio interface, *IEEE, VTC Fall, 26–29 September 2004*, pp. 2625–2629, vol. 4.

[3] 3GPP R99, TS 23.107, QoS Concept and Architecture, v. 3.9.0.

[4] 3GPP R98, TS 05.08, Radio Subsystem Link Control, v. 7.7.0.

[5] 3GPP R98, TS 08.58, BSC-BTS Layer 3 Specification, v. 7.4.1.

[6] 3GPP R98, TS 04.08, Mobile Radio Interface Layer 3 Specification, v. 7.21.0.

[7] 3GPP R98, TS 04.60, Radio Link Control/Medium Access Control (RLC/MAC), v. 7.10.0.

[8] 3GPP R4, TS 44.060, Radio Link Control/Medium Access Control, v. 5.0.0.

[9] D. Soldani and J. Laiho, An enhanced radio resource management with service and user differentiation for UMTS networks, *IEEE, VTC Fall, Orlando, USA, October 2003*, pp. 3473–3477, vol. 5.

[10] D. Soldani and J. Laiho, User perceived performance of interactive and background data in WCDMA networks with QoS differentiation, *WPMC, October 2003, Yokosuka, Japan*, pp. 303–307, vol. 2.

[11] J. Laiho and D. Soldani, A policy-based quality of service management system for UMTS radio access networks, *WPMC, October 2003, Yokosuka, Japan*, pp. 298–302, vol. 2.

[12] D. Soldani, K. Sipilä and A. Wacker, Provisioning radio access networks for effective QoS management: Capacity gains of service differentiation in UTRAN, *IEEE International Symposium on a World of Wireless, Mobile and Multimedia Networks (WoWMoM2005), June, 2005, Italy*.

[13] D. Soldani, A. Wacker and K. Sipilä, An enhanced virtual time simulator for studying QoS provisioning of multimedia services in UTRAN, *Management of Multimedia Networks and Services* (Lecture Notes in Computer Science 3271), Springer-Verlag, pp. 241–254.

[14] ETSI, TR 101112, Selection procedures for the choice of radio transmission technologies of the UMTS, UMTS 30.03, v. 3.2.0.

[15] D. Soldani, QoS management in UMTS terrestrial radio access FDD networks, dissertation for the degree of Doctor of Science in Technology (Doctor of Philosophy), Helsinki University of Technology, October, 2005. See *http://lib.tkk.fi/Diss/2005/isbn9512278340/*

[16] H. Holma and A. Toskala (eds), *WCDMA for UMTS*, John Wiley & Sons, 3rd Edition, 2004, 450 pp.

[17] J. Laiho, A. Wacker, and T. Novosad (eds), *Radio Network Planning and Optimisation for UMTS*, 2nd Edition, John Wiley & Sons, 2006, 630 pp.
[18] K. W. Helmersson and G. Bark, Performance of downlink shared channels in WCDMA radio networks, *Proc. IEEE VTC, Spring 2001*, vol. 4, pp. 2690–2694.
[19] 3GPP R5, TS 25.331, Radio Resource Control Protocol Specification, v. 5.13.0.
[20] 3GPP R6, TS 25.922, Radio Resource Management Strategies, v. 6.1.0.
[21] 3GPP R5, TS 25.214, Physical layer procedures (FDD), v. 5.11.0.
[22] 3GPP R5, TS 25.433, 'UTRAN Iub interface NBAP signalling,' v. 5.13.0.
[23] K. I. Pedersen, A. Toskala and P. E. Mogensen, Mobility management and capacity analysis for high speed downlink packet access in WCDMA, *IEEE, 60th VTC, Fall, September 2004*, pp. 3388–3392, vol. 5.
[24] 3GPP R5, TS 25.215, Physical Layer Measurements FDD, v. 5.7.0.
[25] 3GPP R5, TS 25.413, UTRAN Iu interface RANAP signalling, v. 5.12.0.
[26] 3GPP R5, TS 25.308, High Speed Downlink Packet Access (HSDPA); Overall Description; Stage 2, v. 5.7.0.
[27] 3GPP R5, TS 25.321, Medium Access Control (MAC) Protocol Specification, v. 5.11.0.
[28] 3GPP R5, TS 25.306, UE Radio Access Capabilities, v. 5.11.0.
[29] 3GPP R5, TS 25.213, Spreading and Modulation (FDD), v. 5.6.0.
[30] 3GPP R5, TS 25.435, UTRAN Iub Interface User Plane Protocols for Common Transport Channel Data Stream, v. 5.8.0.
[31] 3GPP R5, TS 23.107, Quality of Service (QoS) Concept and Architecture, v. 5.13.0.
[32] Nokia, *High Speed Packet Access Solution*, white paper.
[33] T. Kolding, K. Pedersen, J. Wigard, F. Frederiksen and P. Mogensen, High-speed downlink packet access: WCDMA evolution, *IEEE VTS News*, No. 1, 2003, pp. 4–10, vol. 50.
[34] P. J. Ameigeiras Gutiérrez, Packet scheduling and quality of service in HSDPA, Ph.D. thesis, University of Ålborg, Ålborg, Denmark, October 2003.
[35] A. Golaup, O. Holland and A. Hamid Aghvami, Concept and optimization of an effective packet scheduling algorithm for multimedia traffic over HSDPA, *IEEE, PIMRC 2005, Berlin, Germany, September 2005*.
[36] K. I. Pedersen, Quality based HSDPA access algorithms, *IEEE VTC 2005 Fall*, accepted.
[37] 3GPP R6, TS 25.309, FDD Enhanced Uplink; Overall Description; Stage 2, v. 6.6.0.
[38] T. E. Kolding, Link and system performance aspects of proportional fair scheduling in WCDMA/HSDPA, *IEEE VTC Fall, October 2003*, pp. 1717–1722, vol. 3.
[39] K. I. Pedersen, T. F. Lootsma, M. Støttrup, F. Frederiksen, T. E. Kolding and P. E. Mogensen, Network performance of mixed traffic on high speed downlink packet access and dedicated channels in WCDMA, *IEEE VTC Fall, September 2004*, pp. 4496–4500, vol. 6.
[40] C. Rosa, Enhanced uplink packet access in WCDMA, Ph.D. thesis, University of Ålborg, Ålborg, Denmark, December 2004.
[41] 3GPP R6, TS 27.001, General on Terminal Adaptation Functions (TAF) for Mobile Stations (MS), v. 6.1.0.
[42] 3GPP R6, TS 48.008, Mobile Switching Centre–Base Station System (MSC–BSS) Interface; Layer 3 Specification, v. 6.10.0.
[43] 3GPP R99, TS 29.018, General Packet Radio Service (GPRS); Serving GPRS Support Node (SGSN) – Visitors Location Register (VLR); Gs Interface Layer 3 Specification, v. 3.11.0.
[44] 3GPP R5, TR 44.901, External Network Assisted Cell Change (NACC), v. 5.1.0.
[45] 3GPP R6, TR 25.901, Network Assisted Cell Change (NACC) from UTRAN to GERAN; Network Side Aspects, v. 6.1.0.
[46] 3GPP R6, TS 43.129, Packet Switched Handover for GERAN A/Gb Mode, v. 6.3.0.
[47] Y. Swami and K. Le, *Lightweight Mobility Detection and Response (LMDR) Algorithm for TCP*, IETF Draft.

[48] R. Guerzoni, D. Soldani, I. Ore and K. Valkealahti, Automatic neighbour cell list optimisation for 3G networks: Theoretical approach and experimental validation, *International Wireless Summit, WPMC05, September 2005, Ålborg, Denmark.*

[49] 3GPP R99, TS 23.060, General Packet Radio Service (GPRS); Service Description,' v. 3.16.0.

[50] 3GPP R6, TR 22.934, Feasibility Study on 3GPP System to Wireless Local Area Network (WLAN) Inter-working, September 2003, v. 6.2.0.

[51] 3GPP R6, TS 23.234, 3GPP System to Wireless Local Area Network (WLAN) Inter-working: System Description, December 2004, v. 6.3.0.

[52] 3GPP R6, TS 24.234, 3GPP System to Wireless Local Area Network (WLAN) Inter-working; User Equipment (UE) to Network Protocols; Stage 3, March 2005, v. 6.2.0.

[53] 3GPP R6, TS 29.234, 3GPP System to Wireless Local Area Network (WLAN) Inter-working; Stage 3, April 2005, v. 6.2.0.

[54] IEEE 802.11 WG, Draft Supplement to Standard for Telecommunications and Information Exchange Between Systems – LAN/MAN Specific Requirements – Part 11: Wireless Medium Access Control (MAC) and Physical Layer (PHY) Specifications: Medium Access Control (MAC) Enhancements for Quality of Service (QoS), IEEE 802.11e/D13, 2005.

[55] IETF, RFC 2475, An Architecture for Differentiated Services.

[56] IETF, RFC 3344, IP Mobility Support for IPv4.

[57] 3GPP R6, TS 43.901, Feasibility Study on Generic Access to A/Gb Interface, v. 6.0.0.

[58] 3GPP R6, TS 44.318, Generic Access to the A/Gb Interface; Mobile Generic Access Interface Layer 3 Specification, v. 6.1.0.

[59] 3GPP R6, TS 43.318, Generic Access to the A/Gb Interface; Stage 2, v. 6.2.0.

6

QoS Functions in Core and Backbone Networks

Renaud Cuny, Heikki Almay, Luis Alberto Peña Sierra and
Jani Lakkakorpi

This chapter describes the QoS and traffic management mechanisms that may be
implemented in the core and backbone parts of the network. Packet core elements
play a key role in QoS management by, for instance, checking mobile station (MS)
requested QoS attributes against the subscriber's profile and by performing admission
control, authorisation (R5) or translation procedures. The UMTS bearer service model
defined in 3GPP is a framework upon which related QoS mechanisms may be implemented. Because QoS is an end-to-end issue, suitable QoS mechanisms are also needed in
the backbone. Note that in case most of the backbone traffic is circuit-switched (CS)
voice, QoS differentiation can only bring minor gains and thus slight overprovisioning is
needed to ensure satisfactory end-user experience. As the portion of packet-switched (PS)
data increases in the backbone, capacity gains brought by QoS differentiation also
increase.

Sections 6.1–6.3 respectively address the CS core, PS core and backbone domains.

6.1 Circuit-switched QoS

This section introduces QoS mechanisms for CS traffic. As explained below, the resources
for CS calls have to be ensured not only in radio and core CS network elements but also in
core transport elements (or backbone).

6.1.1 Architecture of the circuit-switched core network

In 3GPP R4 the split circuit-switched core architecture was defined. In an R4-compliant
mobile service switching centre (MSC) server system, handling of the user plane and
the control plane are separated. The traditional MSC is split into an MSC server that

QoS and QoE Management in UMTS Cellular Systems
Edited by David Soldani, Man Li and Renaud Cuny © 2006 John Wiley & Sons, Ltd

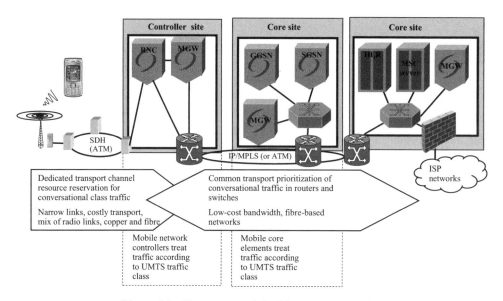

Figure 6.1   Transport and QoS in an R4 network.

takes care of call control and signalling and a media gateway that provides user plane functionality. The key interfaces – Iu-cs, Nb and Mc – can be implemented using IP or asynchronous transfer mode (ATM) transport. The core network is common to both the UMTS terrestrial radio access network (UTRAN) and GSM/EDGE radio access network (GERAN).

Note that the 3GPP architecture totally ignores the distances, physical locations and grouping of network elements. For QoS these are however essential. Core network elements typically reside in a rather small number (typically between 3 and 10) of centralised core sites. Radio network controllers reside either in the core sites or in distributed controller sites. These controller sites often also house GSM base station controllers and media gateways.

In the radio access network (RAN) dedicated transport channels are provided for circuit-switched traffic. Between controller sites and core sites a common packet transport network for all traffic is typically used. In this network, circuit-switched traffic is mixed with packet-switched mobile traffic and conventional data traffic from other sources and the QoS schemes of the used transport technology apply. The transport and QoS in R4 networks is outlined in Figure 6.1.

6.1.2   Circuit-switched services

The most important circuit-switched service is speech. Voice calls dominate in mobile networks both in terms of traffic volume as well as revenue generators. In addition to speech, circuit-switched data services are available.

Speech services are normal voice connections. Call cases include 3G–3G calls as well as calls to the GSM network, public-switched telephone network (PSTN) and IP multimedia. Data services can be divided into transparent and non-transparent, depending on the requirements of the application used. Transparent services are used for delay-sensitive applications requiring a synchronous bearer service. Non-transparent services are used for applications that support an asynchronous bearer service. The benefit of using the non-transparent service is the possibility of retransmissions over an error-prone air interface.

6.1.3 Factors affecting the quality of circuit-switched services

The quality of service perceived by the user (or QoE) is partly determined by general factors such as call setup time and call success rate and partly by the actual connection – that is, speech or video quality, data download time and latency. During the call, handovers may cause interruptions or modifications in the service. These also affect user experience.

For voice services the impact of packet loss, delay and delay variation is dependent on the codecs used. For adaptive multi-rate (AMR) codecs, frame erasure rates of 0.5–1.0% do not seem to cause significant quality degradations [1].

Probably the most important and from the QoS perspective the most challenging circuit-switched data service is video telephony. It requires a constant bit rate, small delay variation and a continuous bit stream. So, a synchronous transparent data bearer is used. In a two-way video conversation the delay caused by the mobile network is an issue.

6.1.4 Circuit-switched core and the 3GPP QoS concept

The 3GPP quality of service concept and architecture specification TS 23.107 [2] does not explicitly specify the QoS mechanisms for the circuit-switched core. Focus in the 3GPP specification is on the TC used in the packet core. These TCs (conversational, streaming, interactive, background) are rather irrelevant as voice and circuit-switched data services are all conversational in nature. The only exception is the short message service (SMS). SMS messages are carried in the signalling network among 3GPP-defined control protocols. The latest major QoS-related changes for the circuit-switched core were first introduced in 3GPP R4. They included the MSC server–media gateway concept and IP transport of core network protocols.

TS 23.107 specifies that the UMTS packet core network shall support different back-bone bearer services for a number of QoS needs. The operator can choose whether IP or ATM QoS capabilities are used. In case of an IP backbone, differentiated services shall be used.

The paragraphs above may seem confusing. 3GPP QoS concept and architecture specifications do not provide any QoS differentiation for the circuit-switched core. The network operator decides the QoS treatment of different logical interfaces. This is rather fundamental as the network operators also decide how different user plane and control plane traffic types are prioritised. Should voice, messaging, signalling or operation and maintenance packets be dropped first? Different policies apply. Some guidance on the treatment of the different types of traffic in an IP backbone is given in Section 6.3.

6.1.5 QoS mechanisms in the circuit-switched core

In addition to the 3GPP QoS concept, the service quality experienced by end-users is affected by the way in which circuit-switched core call control works. When a call is set up the network reserves both radio and core network resources for the call. In a traditional time division multiplexing (TDM) based core this includes the circuits in the actual switching equipment and the time slots on the transmission links. In an MSC server system the reserved resources are ports in the media gateways.

The difference between the traditional TDM environment and the MSC server system is that in the TDM environment resource reservation is done for a circuit between the point where the call enters the network and the point where it leaves, whereas the MSC server system only reserves media gateway resources at ingress and egress. The availability of transport capacity between these points has to be ensured by some other means.

Reading the above one may think that the TDM solution is more complete and thus better than the MSC server solution. Unfortunately, the TDM network has to be constructed on a 2-Mb/s or 64-kb/s basis. Capacity allocations are rigid and to avoid the overwhelming complexity of a full mesh even smaller networks have to be built in a hierarchical architecture. So, network operators prefer MSC server-based solutions because of the simplified network architecture, lower capital expenditure and smaller operational cost.

The major QoS question in MSC server networks is how to make sure that the transport capacity between the media gateways is available. A second important question is to guarantee the availability of signalling links.

At this stage some may despair and decide that a signalled ATM network with a rigid capacity allocation is needed between mobile core network elements. This is supported by 3GPP specifications. When correctly operated it most likely solves the problems at hand. Unfortunately, it would in many cases mean the construction of a new dedicated network parallel to the IP network anyway, needed for both consumer and corporate data services. The need for major investments and the task of building lots of competences in technology that does not seem to have a future might be reason enough to take a second look at the QoS technologies available for IP-based networks.

First, it may be a good idea to think about the extent of the issues. Remember that QoS mechanisms do not create additional bandwidth. They just help in selecting which traffic to drop or to delay in an overload situation. Lost traffic equals lost revenue. A radio access network and mobile core are much more expensive to build and to operate than a standard IP or ATM backbone between major sites. Also, the network structure typically limits the amount of the backbone traffic as a core network site typically serves a distinct geographical region. Mobile subscribers tend to call people close to them, the kids at home or the boss in the office. This traffic normally does not need backbone transport.

Moderate overprovisioning of the backbone is recommended as it also makes it easier to adapt to changing traffic patterns. There is no point in building a network in which precious mobile network elements cannot be utilised to the full extent because of insufficient backbone transport capacity.

The QoS concerns that remain in a conservatively designed packet-based circuit-switched core are related to exceptional situations. What happens if the call distribution suddenly changes a lot or if some of the network resources become unavailable? In these cases a traditional TDM-based circuit-switched core will reject the calls of most subscribers and only accept calls from prioritised subscribers. This is done using the allocation retention priority (ARP), an attribute defined in the home location register (HLR) subscription. These mechanisms also work in a packet-based circuit-switched core, with the only difference that call control has no direct means to determine to what extent transport resources are available. So, instead of a busy tone the subscriber may get a call with unacceptable voice quality or no voice at all.

QoS mechanisms in the backbone help in avoiding situations where the packet loss rate for circuit-switched connections or related signalling becomes unacceptable. Conversational services have the highest priority in the differentiated services (DiffServ) QoS scheme used. Operation of the DiffServ scheme is described in more detail in Section 6.3.

6.2 Packet-switched core QoS

This section introduces a QoS mechanism for the PS core domain. More precisely, it describes PDP context QoS parameter control and provides examples of the traffic management features in the GGSN and SGSN.

6.2.1 Session management

The session management functionality allows operators to control flexibly how sessions are mapped onto different QoS profiles. The main elements involved in that process are the SGSN, HLR and GGSN (also sometimes called the 'intelligent edge').

The PDP context activation procedure was described in detail in Chapter 3.

6.2.1.1 Transmission-mode selection in 2G-SGSN

The 2G-SGSN selects the transmission mode for the PDP context as part of session management. In 2G-SGSN the values of the serving data unit (SDU) error ratio and residual bit error ratio affect the transmission mode used in different layers [2]. Table 6.1 shows the transmission mode specified in 3GPP R99 and later releases with a different SDU error ratio and residual bit error ratio combination.

If the MS is a 3GPP R97/98 one, the transmission mode is selected directly according to the reliability class value.

6.2.1.2 Mapping of 3GPP R97/98 QoS attributes onto 3GPP R99 attributes

Because network elements and mobiles may support various standard releases, 3GPP has specified how QoS profiles from different releases should map each other (see Chapter 3 for further details). In 3GPP R97/98 real time applications were not considered and therefore some of the current QoS parameters are not supported.

Attribute mapping from R97/98 to R99 is needed in the following cases [2]:

- Handover of PDP context from R97/98 SGSN to R99 SGSN.
- When the GGSN is R97/98 and SGSN is R99. In such a case, the activation PDP

Table 6.1 Selection of transmission mode according to R99 QoS attributes.

SDU error ratio	Residual bit error ratio	Resulting 2G-SGSN behaviour
$\leq 10^{-5}$	N/A	GTP buffer used, acknowledged LLC mode, LLC data-protected
$10^{-5} < x \leq 5 * 10^{-4}$ protocol	N/A	GTP buffer not used, unacknowledged LLC mode, LLC data-protected
$> 5 * 10^{-4}$ protected	$\leq 2 * 10^{-4}$	GTP buffer not used, unacknowledged LLC mode, LLC data-protected
$> 5 * 10^{-4}$ unprotected	$> 2 * 10^{-4}$	GTP buffer not used, unacknowledged LLC mode, LLC data-unprotected

context response from the GGSN accompanies the R97/98 attributes and the SGSN maps them onto R99.

• When the MS is an R99 one, but requests R97/98 attributes.

Attribute mapping from R99 to R97/98 is needed in the following cases:

• PDP context is handed over from GPRS/UMTS R99 to GPRS R97/98.
• When the GGSN is R97/98 and SGSN is R99. In such a case, the activation PDP context request from the SGSN shall be responsible for mapping the R99 attributes onto R97/98.
• A R99 HLR may need to map the stored subscriber QoS onto the R97/98 QoS attributes that are going to be sent to the R97/98 and R99 SGSN.

If the MS requests R99 QoS attributes, even if some network element (other than the SGSN) replies with R97/98 QoS attributes, the response to the MS should include R99 QoS attributes. Likewise, if the MS requests R97/98 QoS attributes, the response should include R97/98 QoS attributes.

6.2.1.3 Real time PDP context based admission control

Each network element may support a configurable, maximum, real time bandwidth dedicated for all real time PDP contexts. This bandwidth may be shared by two TCs, having different priorities. It may also be useful to have as a configurable parameter the maximum bandwidth per TC. In certain cases, the maximum overall bandwidth and the TC maximum bandwidth could be equal – for example, Max. overall bandwidth = 1 Gb/s and Max. TC streaming bandwidth = 1 Gb/s. Whenever a new real time PDP context request arrives at the network element, the element will check whether there is any bandwidth available from the combination of the remaining overall bandwidth and TC bandwidth. If not, a downgrade PDP context QoS profile procedure may take place, and it is up to the MS to accept or reject this new QoS profile. If there is enough bandwidth, the PDP context request will be accepted and the given GBR is taken away from the remaining overall and TC bandwidth.

In addition to using the above mechanism, it is recommended to perform admission control based again on network element utilisation. For instance, the central processing unit (CPU) load percentage, TC and ARP can be used as input for the admission control decision; as an example, a network element could be configured so that if the CPU load is above 60%, no new streaming PDP contexts with ARP3 are accepted. If the CPU load is above 80%, no new streaming PDP context with ARP2 or below and no new conversational PDP context with ARP3 are accepted. If the CPU load is above 90%, no new streaming PDP context and no new conversational PDP context with ARP2 or below are accepted. Finally, conversational PDP contexts with ARP1 may always be accepted. In this way the highest priority users and applications are served in highly loaded situations as well.

6.2.2 Intelligent edge concept (change for QoS control in packet core)

The intelligent edge concept was introduced to further improve QoS and charging control based on the actual services being used. This concept is described in the present section.

6.2.2.1 Service-based QoS differentiation

Broadband Internet access and mobile Internet access exhibit the following main differences:

- The network infrastructure is cheaper in broadband networks than in mobile networks. Base stations (BTS), BSC and RNC are expensive network elements that the operator will typically try to utilise optimally.
- In mobile networks, many users share the most commonly congested link (air interface), whereas in broadband access the most congested link is usually the private link.

These reasons among others are the drivers for advanced mobile network resource optimisation. As a most congested link is shared and mobile subscribers use applications with various QoS requirements, traffic differentiation is applied.

From the QoS viewpoint, the current 3GPP systems and/or specifications have the following limitations or constraints:

- Some MSs support only one PDP context at a time.
- Some MSs support only a limited number of APNs.
- Most currently available MSs do not request any QoS.
- Potential misuse of RT PDP contexts because of the open source APIs in the MS.
- Common subscriber QoS profile per APN for all access types.

In addition to these, some network element vendors may not support all parameters or TC combinations specified in 3GPP, which may cause extra signalling on the network.

Table 6.2   Relevant QoS parameters for traffic differentiation.

Traffic class	ARP	THP	MBR	GBR
Conversational	Yes (1, 2, 3)	No	Yes	Yes
Streaming	Yes (1, 2, 3)	No	Yes	Yes
Interactive	Yes (1, 2, 3)	Yes (1, 2, 3)	Yes	No
Background	Yes (1, 2, 3)	No	Yes	No

Finally, another issue is how to guarantee QoS when users connected to a GPRS access network move to UMTS coverage (and *vice versa*).

The most relevant QoS parameters used by network elements for prioritisation, scheduling and queuing are shown in Table 6.2.

There are three main approaches to traffic differentiation: subscriber-based differentiation, service-based differentiation and a combination of these two.

Subscriber-based differentiation may be done using the ARP that is stored in the HLR for different subscribers. For instance, for the same APN a different ARP is given for three types of subscribers depending on their importance or charging type: VIP, gold and low-priority users. This approach also has drawbacks since radio network resources may not be used optimally. Also, as the number of concurrent demanding services increases, gold and low-priority users will experience worse service quality.

Service differentiation as opposed to subscriber differentiation may require more intelligence in packet core elements, especially at the edge of the network. As explained above, 3GPP standards do not take into account potential terminal or equipment limitations and, therefore, alternative (i.e., non-standard) solutions are needed to deal with them. Also, because the majority of currently available MSs do not support simultaneous PDP contexts and support only a small number of APNs, an HLR-based solution in which a unique APN is assigned for each service type is not suitable.

In UMTS networks, QoS differentiation in the core is based on the PDP context QoS parameters that are mapped onto the transport QoS (see Section 6.2.4 for more details). In this respect, the GGSN – being the entry point of cellular networks – plays an essential role. The proposed idea for service differentiation is that the GGSN will identify which of the subscriber's services is in use and will select the adequate QoS accordingly. One possible way is by looking inside the IP flow using a Layer 4/7 lookup mechanism. After the IP flow is matched to the right service, the corresponding PDP context QoS profile is compared with the maximum QoS profile for that service. If the PDP context QoS profile is higher than what the service requires, a downgrade PDP context procedure is initiated. If the service requires a higher QoS profile than the PDP context currently has, an upgrade PDP context procedure is initiated (the upgraded QoS profile cannot exceed the negotiated QoS that results from the combination of the MS-requested QoS and the subscriber QoS in the HLR). If there are several active IP flows associated with one PDP context, the QoS profile suitable for the most demanding flow should be selected.

As MS vendors are introducing an open API for application developers, the requested QoS profile from an MS cannot be fully trusted. For example, if the end-user uses a peer-to-peer (P2P) application for file download, there is no guarantee that this application will request an NRT QoS profile, as it probably should. To prevent the misuse of QoS profiles, Layer 4/7 lookup is a very useful asset.

(E)GPRS and WCDMA have very different characteristics in terms of capacity and maximum throughput per user. Thus, knowing the access type when allocating resources in the core network may also be useful. For this reason, 3GPP R6 [4] has introduced the radio access technology (RAT) field in the PDP context activation and PDP context update messages between the SGSN and GGSN. Furthermore, a cell ID was also added to these messages. These parameters allow the edge of the network to modify the PDP context depending on RAT and access network capabilities.

6.2.2.2   Roaming and QoS-related issues

GPRS roaming standards define two main alternatives of connecting to a GGSN when the MS is connected to a roaming SGSN:

- The MS may be connected to a visitor network using the home GGSN.
- The MS may be connected to a visitor network using the visitor GGSN.

In practice, most operator networks are configured so that roaming users connect to their home network using their own home GGSN. The reasons for that are, among others, charging issues, APN configuration, languages used by different countries, etc.

As radio resources are expensive and limited, the operators might want to give the best resources to home subscribers instead of giving them to roaming users. Again, one important issue is how to avoid possible misuse of the real time PDP context (see Section 6.2.2.1) for roaming users, when the HLR, GGSN and everything behind the GGSN is in the home network. The most suitable place for solving this is the SGSN. The SGSN gets the IMSI with the public land mobile network (PLMN) Id during attach and routing area update (RAU) from the mobile station. The PLMN Id consists of the combination of the mobile national code (MNC) and the mobile country code (MCC). These two parameters (MNC and MCC) are unique per operator. Thus, the SGSN could maintain a table of PLMN Ids that the operator wants to restrict and at the same time configure the maximum QoS profile for these PLMN Id users. In this way the operator takes control over their network resources. That is, based on the PLMN Id, the SGSN can control the maximum QoS profile for roaming users.

6.2.3   Packet core and high-speed downlink packet access (HSDPA)

High-speed downlink packet access (HSDPA) provides a performance boost for WCDMA that is comparable with what EDGE does for GSM. It brings a two-fold air interface capacity increase and a five-fold downlink data speed increase. HSDPA also shortens the round trip time and reduces downlink transmission delay variance.

For packet core elements, HSDPA implies an extension of the QoS profile up to 16 Mb/s maximum bit rate (MBR) and guaranteed bit rate (GBR) for a real time PDP Context.

3G SGSN and GGSN will add three new octets in the GPRS tunnelling protocol (GTP)-C negotiated QoS profile (known as 'QoS2-negotiated') according to [5] change request (CR) 492 and [6] CR 796. Three new octets also need to be added to GTP for charging information according to [7] CR 031. Other 3G SGSN-related changes are: three new octets added to mobile application part (MAP) for communication between the 3G SGSN and HLR according to [8] CR 688 and to camel application part (CAP) for communication with charging when customised applications for mobile network enhanced logic (CAMEL) are used according to [9] CR 374. Finally, another change required for the GGSN is in the radius interface where the QoS profile is sent/received from/to a radius server.

As a summary, it can be said that the changes required from packet core network elements to support HSDPA are pretty minor compared with those needed in radio network elements.

6.2.4 Traffic management

Packet core traffic management functions include packet classification and marking, queuing, scheduling and congestion avoidance mechanisms.

There are three different QoS levels in packet core network elements, as shown in Figure 6.2:

• The first is UMTS QoS that is related to PDP context specific QoS management.
• The second is Gn and Iu/Gb transport QoS that includes, for instance, DiffServ edge functionality towards the radio and mobile packet core backbone.
• The third is the user layer QoS that is a DiffServ edge functionality towards external IP networks.

The SGSN and GGSN mark the Diffserv code point (DSCP) field of the transport IP header according to the PDP context type. Table 6.3 provides an example of the mapping between the PDP context type and DSCP field in the transport IP header.

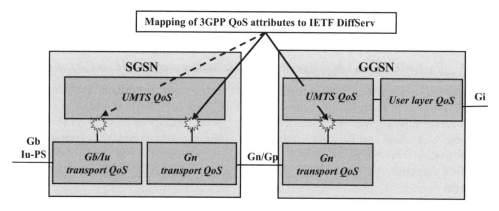

Figure 6.2 QoS function blocks in packet core elements.

Table 6.3   3GPP to DiffServ QoS mapping example.

	Classifier			Action	
Traffic class	THP	ARP	PHB	DSCP	
Conversational	–	ARP1	EF	101110	
Conversational	–	ARP2	EF	101110	
Conversational	–	ARP3	EF	101110	
Streaming	–	ARP1	AF41	100010	
Streaming	–	ARP2	AF42	100100	
Streaming	–	ARP3	AF43	100110	
Interactive	THP1	ARP1	AF31	011010	
Interactive	THP1	ARP2	AF32	011100	
Interactive	THP1	ARP3	AF33	011110	
Interactive	THP2	ARP1	AF21	010010	
Interactive	THP2	ARP2	AF22	010100	
Interactive	THP2	ARP3	AF23	010110	
Interactive	THP3	ARP1	AF11	001010	
Interactive	THP3	ARP2	AF12	001100	
Interactive	THP3	ARP3	AF13	001110	
Background	–	ARP1	BE	000000	
Background	–	ARP2	BE	000000	
Background	–	ARP3	BE	000000	

6.2.4.1   Traffic management in 3G SGSN

In case network element load exceeds the service rate, a single queue at each internal congestion point is no longer sufficient. Instead, a different queue for each type of service (PDP context type) is needed to which independent latency, jitter and packet loss characteristics apply. Figure 6.3 shows the QoS traffic management functions in the 3G-SGSN.

When IP packets arrive at the 3G SGSN from the GGSN, IP input scheduling is performed. IP input scheduling prioritises and schedules packets from ingress packet queues based on a DSCP. In Figure 6.4, an example of IP scheduling is given.

In this example six queues are shown: one expedited forwarding (EF), four assured forwarding (AF) and one best-effort (BE). For more details on these IETF schemes see Section 6.3.

When the packet arrives at the hardware driver, the access control list classifies the packet based on the DSCP field in the IP header. Then, the packet is sent to the proper queue. Each queue has a queue management function which is responsible for establishing and maintaining queue behaviour within the 3G SGSN and involves four basic actions:

- Add a packet to the queue.
- Drop the packet if the queue is full.
- Remove the packet when requested by the scheduler.
- Monitor queue occupancy.

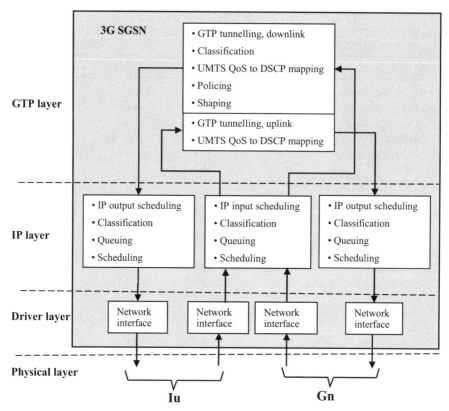

Figure 6.3   QoS traffic management functions in 3G SGSN.

Depending on the queue the packets belong to, queue management uses different congestion avoidance mechanisms. Figure 6.4 shows two of the most popular active queue management schemes: random early detection (RED) and weighted random early detection (WRED).

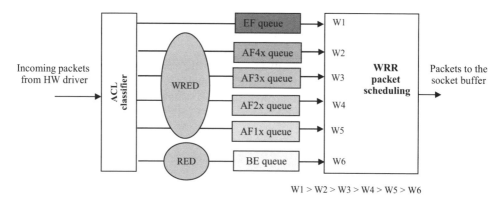

Figure 6.4   IP scheduling example.

RED uses the average queue occupancy as an input to decide whether congestion avoidance mechanisms ought to be triggered (the common action being packet drop). As the average queue occupancy increases, the probability of dropping a packet also increases.

- For occupancy up to a lower threshold min_{th} all incoming packets are accepted (drop probability is 0).
- Above min_{th} the probability of packet drop rises linearly towards a probability of max_p reached for a max_{th} occupancy.
- At max_{th} all incoming packets are dropped.

Average occupancy is calculated every time a new packet is received; it is based on a low-pass filter, or the exponential weighted moving average (EWMA), of instantaneous queue occupancy. The formula is:

$$Q_{avg} = (1 - W_q) \times Q_{avg} + Q_{inst} \times W_q \qquad (6.1)$$

where Q_{avg} is average occupancy, Q_{inst} is instantaneous occupancy and W_q is the weight of the moving average function. These values are typically set so that RED ignores short-term transients without inducing packet loss, but reacts before overall latency is affected or multiflow synchronisation of TCP congestion avoidance is experienced.

Dropping incoming packets randomly and at an early stage increases the likelihood of smoothing out transient congestion before queue occupancy gets too high. Randomising drop distribution at early stages also reduces the chances of simultaneously subjecting multiple flows to packet drops.

Queue managers are not limited to providing a single type of behaviour on any given queue. Additional information from the packet context may be used to select one of multiple packet discard functions. A precedence field can be used for the multiple packet discard function as is the case in WRED. The idea is to give different min_{th}, max_{th} and max_p parameters for each RED instance.

There are other congestion control mechanisms such as RED with in/out, adaptive RED (ARED) and Flow RED (FRED), described in [10].

The next step after prioritisation is scheduling, which dictates the temporal character-istics of packet departures from each queue. Since the type of service determines which queue the packet is placed in, the scheduler is then the main enforcement point of relative priority, latency bounds or bandwidth allocation between different traffic types. Scheduling mechanisms may be classified in two groups: simple scheduling and adaptive scheduling.

The simple scheduling group includes strict priority and round robin (RR) scheduling. A strict priority scheduler orders queues by descending priority and serves a queue of a given priority level only if all higher priority queues are empty. RR scheduling, on the other hand, avoids local queue starvation by cycling through the queues one after the other, transmitting one packet before moving on to the next queue.

From a service provisioning perspective, being able to maintain pre-defined bandwidth allocations to various traffic types that share a common link (CPU link or outbound link) is very often needed. Neither strict priority nor RR schedulers take into account the number of bits transmitted each time a queue is served. A number of scheduling

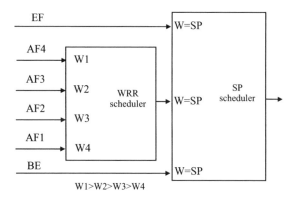

Figure 6.5 Cascade scheduler.

algorithms have been developed to meet this need – such as deficit round robin (DRR), weighted round robin (WRR), fair queuing (FQ) and weighted fair queuing (WFQ).

In general, these two types of scheduling alone are not enough for the 3G SGSN, as some of the flows may have very tight QoS requirements that can only be met with SP scheduling. On the other hand, AF queues typically do not have such tight QoS requirements and therefore WRR is a more suitable scheduler for them. Figure 6.5 shows an example of cascade scheduling. In this case a combination of WRR and SP scheduling is used to accommodate EF and AF classes.

The GTP process classifies the packet according to which PDP context it belongs to. For all traffic the DSCP field in the IP header may be changed according to PDP context attributes and the router configuration at the Gn or Iu interfaces. Also, for roaming subscriber PDP contexts, metering and policing is done. Metering and policing functionalities are described in Section 6.2.4.2.

After the packet has been processed by the GTP layer, it is sent to the IP stack that forwards it to the right interface using a similar type of scheduling to that presented earlier. Note that outbound scheduling can be done per physical or logical interface. If the outbound interface is an ATM, then different QoS mechanisms can be used as explained in Section 6.3.4.

6.2.4.2 Traffic management in GGSN

The scheduling, queuing and prioritisation of IP traffic in the GGSN are typically done in a similar way to that done in the 3G SGSN (see Section 6.2.4.1). Since the GGSN is the edge element for GPRS and UMTS PS services, metering and policing for downlink traffic are key functionalities.

The GTP level classification identifies the PDP context the packet belongs to. PDP context specific QoS attributes are then used for QoS-related traffic management functions on the GTP layer. The metering function ensures that downlink traffic conforms to the negotiated bit rate at the PDP context level. The traffic conditioner function, which is part of the metering function, is the actual component providing the conformance of downlink user data traffic.

An algorithm for bit rate conformance definition was presented in [10]. The algorithm is known as a 'token bucket'. In this context 'token' represents the allowed data volume (e.g., a byte). 'Tokens' given at a constant 'token rate' by a traffic contract are stored temporarily in a 'token bucket' and are consumed by accepting the packet. This algorithm uses the following parameters:

- Token rate r (as a *maximum bit rate/guaranteed bit rate*).
- Bucket size b (combination of the *maximum bit rate/guaranteed bit rate* and the *maximum SDU size*).
- *Token bucket counter* (TBC): the number of given/remaining tokens at any time.

The TBC is usually increased by r in each time unit. However, the TBC has an upper bound b (token bucket size) and the value of the TBC shall never exceed b. When packet p_1 with length l_1 arrives, the receiver checks the current TBC. If the TBC value is equal to or larger than l_1, packet arrival is judged compliant – that is, the traffic is conformant. At this moment tokens corresponding to the packet length are consumed and the TBC value decreases by l_1. The same happens to packet p_2 with length l_2 in our example. However, for packet p_3, the TBC is below l_3 and packet arrival is considered non-compliant (the traffic is not conformant). In this case, the value of TBC is not updated and the p_3 is either dropped or forwarded to the shaper.

If the packet is not compliant and belongs to a NRT PDP context, it is buffered until it becomes compliant or until the time to live (TTL) has expired. Non-compliant packets belonging to an RT PDP context are dropped.

The last function of the GTP layer is marking the IP header DSCP field according to the PDP context QoS profile. Also, for uplink traffic the DSCP field can be marked in order to enable consistent traffic differentiation behind the Gi interface.

6.2.4.3 Traffic management in 2G-SGSN

This section describes traffic management in the 2G-SGSN and, more specifically, active queue management techniques that may be implemented in that element (see [11] for more details).

In addition to playing a central role in, say, session, mobility and charging management procedures, the 2G-SGSN acts as a buffer for the radio access network. That is, the 2G-SGSN shall temporarily hold downlink packets (instead of forwarding them immediately) if the BSC is not able to receive them due to, say, lack of own-buffer space. The main benefit of this approach is to avoid placing too high memory requirements at the BSC. This flow control procedure between the 2G-SGSN and the BSC is specified in 3GPP [3] and [12] (see Figure 6.6).

There are three different flow control levels. The first is the BSSGP virtual connection (BVC) flow control, which refers to the cell level. In case the available buffer space in the BSC reserved for a particular BVC drops below a certain threshold, the BSC will signal the 2G-SGSN to reduce its sending rate for the traffic accessing that BVC. The second level is MS-specific flow control. Again, if the available memory in the BSC reserved for a particular MS gets too low, the 2G-SGSN will reduce the sending rate for that particular

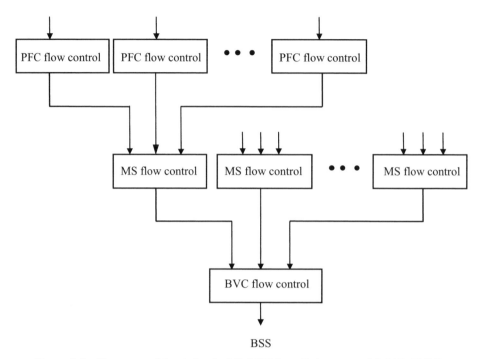

BSS

Figure 6.6 Flow control levels in the 2G-SGSN applied to every LLC-PDU [12].

MS. The last (optional) level is the PFC that handles flows within a certain MS that have specific QoS requirements.

As specified in [12], the 2G-SGSN will apply these flow control tests to every logical link control–packet data unit (LLC-PDU): flow control is performed on each LLC-PDU first by the PFC flow control mechanism (if applicable and negotiated), then by the MS flow control mechanism and last by the BVC flow control mechanism.

This flow control approach has the following benefits:

- It prevents downlink traffic overflow (i.e., packet drops) in the BSC.
- It ensures that a certain congested cell, MS or PFC, will not create unnecessary downlink buffer delay in the 2G-SGSN for other flows accessing non-congested cells, MSs or PFCs.

So, in high-load conditions, the 2G-SGSN will be the main element in charge of handling potential excess downlink traffic in the BSS. Or, in other words, the 2G-SGSN downlink buffer is a potential traffic bottleneck since overload may not only be caused by 2G-SGSN or Gb capacity limitations but also by cell or even MS congestion which are indeed more common scenarios. This is illustrated in Figure 6.7.

One way to deal with potential buffer delay is to prioritise the traffic based on how delay-sensitive it is, as explained in detail for the 3G-SGSN in Section 6.2.4.1. Likewise,

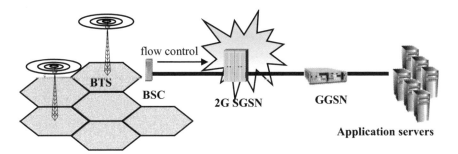

Figure 6.7 2G-SGSN is a potential traffic bottleneck in loaded conditions.

in the 2G-SGSN, the traffic (LLC packets) from different TCs may be handled in separate buffers (as shown in Figure 6.8). A weighted fair queuing scheduler may then allocate a certain share of the output capacity to each buffer.

Although QoS-based queuing and scheduling may lower or even eliminate buffer delays for the highest priority classes (i.e., real time traffic), the lowest priority classes (i.e., non-real time traffic) are then even more likely to experience long delays (depending on the traffic mix). Measurements performed in live (E)GPRS networks typically confirm that end-to-end latency grows with network load. Thus, some other mechanisms to control or reduce buffer delays for non-real time TC are needed in 2G-SGSN to optimise both end-user experience and spectral efficiency. It should be noted that, although the same applies to any other core network element (e.g., GGSN, 3G-SGSN or backbone

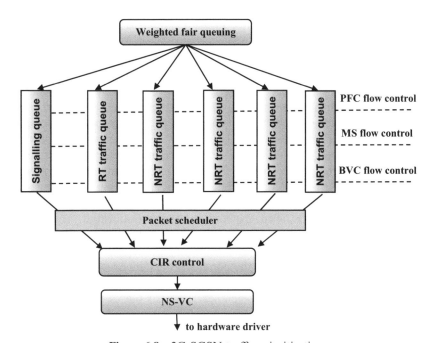

Figure 6.8 2G-SGSN traffic prioritisation.

routers), the buffer delay issue is typically most acute and also in a way specific to the 2G-SGSN because of the standard flow control between the radio or core network domains. Therefore, specific non-classical approaches to solve this buffer delay problem are also worth investigating. The first way to control buffer delays in the 2G-SGSN is to introduce a pre-defined lifetime for LLC frames. The idea is very simple: after having spent a certain pre-defined time in the 2G-SGSN and/or BSC buffers, the LLC frame will be discarded. Such a mechanism is available by default in most router-like network elements in order to ensure that packets that are too old are removed. In the 2G-SGSN, this scheme can also help to guarantee a certain maximum buffer delay depending on the TC considered. The objective is to find the right trade-off between high resource utilisation and optimised end-user throughput. For instance, network utilisation may be affected if the packet lifetime is set too low. On the other hand, a too large lifetime may degrade end-user throughput because of high latency. In order to avoid unnecessary packet drops at the BSC, LLC frames successfully sent from the 2G-SGSN to the BSC should be given at least a pre-defined minimum lifetime – that is, the total LLC frame lifetime should be split between the 2G-SGSN and the BSC. Moreover, since flow control cannot provide any delay bounds (and there is no active queue management at the BSC), we should also introduce a pre-defined maximum lifetime for LLC frames at the BSC.

Another way to limit buffer delays is simply to limit the buffer size. It very much resembles the previous approach although in this case the output interface speed shall be known in order to predict the maximum buffer delay. What complicates things in this respect in the 2G-SGSN is the multilayer flow control presented earlier. For instance, although the output link speed of the 2G-SGSN would allow immediate forwarding of received packets, some packets may have to be buffered because the BSC is unable to accept them. Thus, extracting a maximum buffer delay out of the 2G-SGSN buffer size is not easy.

A third, more advanced approach is to randomly drop packets before the buffer gets full or before the packet lifetime expires. The RED algorithm [16] (already described in detail in Section 6.2.4.1), which does exactly this, is probably the most popular active queue management scheme used nowadays. As mentioned, the RED algorithm drops arriving packets probabilistically. The probability of packet drop increases as the estimated average queue size grows. RED responds to a time-averaged queue length, not an instantaneous one. Thus, if the queue has been mostly empty in the 'recent past', RED is not likely to drop packets (unless the queue overflows). On the other hand, if the queue has recently been relatively full, indicating persistent congestion, newly arriving packets are more likely to be dropped.

An improvement of RED called 'explicit congestion notification' (ECN) was later introduced. As stated in [13], explicit congestion notification allows a TCP receiver to inform the sender of congestion in the network by setting the ECN-Echo flag upon receiving an IP packet marked with the congestion experienced (CE) bit(s). The TCP sender will then reduce its congestion window. Thus, the use of ECN is believed to provide performance benefits [14]. Reference [15] also places requirements on intermediate routers – for example, active queue management and setting of the CE bit(s) in the IP header to indicate congestion. Therefore, the potential improvement in performance can only be achieved when ECN-capable routers are deployed along the path.

We also note that numerous variants of RED and ECN have been proposed [16].

RED and ECN implementation could, in principle, take into account the 2G-SGSN multilayer flow control mechanism. That is, one instance of RED or ECN could be applied to each independent flow control entity. As an illustration, if the RED threshold in a 2G-SGSN buffer is exceeded mostly due to a few congested cells (BVC), it does not necessarily mean that packets accessing other non-congested cells – buffered in the 2G-SGSN for such other reasons as Gb capacity limitation – should be randomly dropped by the same rules. However, a multilayer RED (or ECN) approach in the 2G-SGSN would add significant complexity and require extra CPU and memory, while practical performance gains are not so clear. This is because, although packets may not be buffered in the 2G-SGSN for the same reasons (e.g., MS vs. Gb vs. BVC congestion), they all indicate some sort of congestion as well as a potential significant buffer delay (e.g., BSC buffers are full and cannot accept any more data). The state of the art [16] recommends buffer delay to be only a fraction of the round trip time, and thus it is probably a good idea to allow only a relatively small buffer delay in the 2G-SGSN. As a conclusion, it seems the potential performance gains of applying one separate instance of RED (or ECN) to each independent flow control entity do not justify the required extra complexity.

Another alternative to the 2G-SGSN is to follow a TTL-based RED approach since, as explained above, it is not straightforward to relate 2G-SGSN buffer occupancy and buffer delay. There are two possible implementations for a TTL-based RED approach:

- In the first the packet is checked periodically and if its age (current time less timestamp) exceeds a threshold, the packet is randomly dropped (or marked if ECN is used and the flow supports ECN).
- The second implementation is a bit simpler. Here the packets are given a random lifetime (in addition to the deterministic lifetime that is used in both the 2G-SGSN and BSC) when they enter the 2G-SGSN. As in the first implementation, the packet is periodically checked to see whether it should be discarded – that is, if the age of the packet (current time less timestamp) exceeds its lifetime. If that is the case, the packet is dropped (or marked if ECN is used and the flow supports ECN).

The last congestion control alternative that we consider here is called 'window pacing'. With this scheme, the router can decrease the TCP-advertised window value in uplink TCP acknowledgements if the defined buffer filling level threshold for a specific TC is reached. A decreased, advertised window value forces the sending TCP to slow down the transmission speed, since the sending window is the minimum of congestion window and advertised window.

Simulations were performed in order to evaluate the efficiency of some the aforementioned schemes in decreasing buffer delay and improving global throughput. The simulator is described in Figure 6.9.

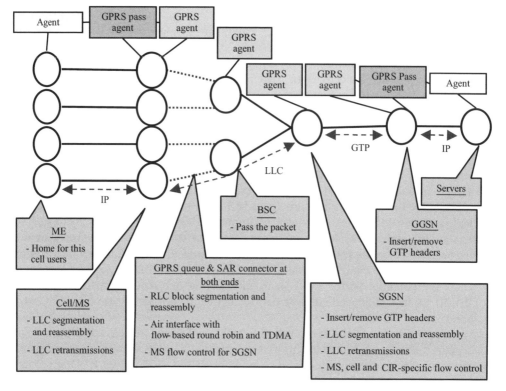

Figure 6.9   The end-to-end (E)GPRS simulator [11].
Reproduced by permission of © IEEE 2006.

The GPRS agent has four different instances: GGSN, SGSN, BSC and cell. Moreover, a 'pass agent' is needed for each TCP/UDP/sink agent. The main features of the GPRS model are:

- MS/cell/committed information rate (CIR) flow control in 2G-SGSN downlink queuing.
- Dynamic BSSGP flow control (BSC → 2G-SGSN) for MS and BVC flows.
- LLC retransmissions (if needed).

The simulator makes use of publicly available ns-2 [17] modules such as TCP (NewReno) and traffic sources.

Figures 6.10 and 6.11 illustrate the end-to-end delay and TCP throughput (i.e., goodput) experienced by end-users with various congestion control schemes.

In this scenario, we have:

- Five active streaming (mean bit rate of 40 kb/s, maximum bit rate of 80 kb/s, UDP used as transport protocol) users using the streaming TC.

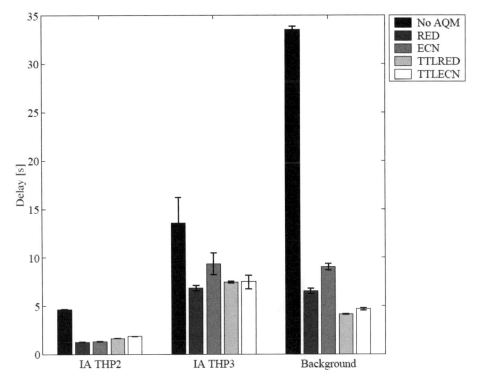

Figure 6.10   End-to-end delay (95th percentile) [11].
Reproduced by permission of © IEEE 2006.

- Fifteen active push to talk over cellular (PoC, constant bit rate of 8 kb/s, UDP used as transport protocol) users using the interactive THP1 TC.
- Thirty active web browsing (main page and the 30 inline items per page all have a size of 4.91 kB, which results in a total page size of 152 kB; page reading time is 1.0 s; four persistent TCP connections are utilised) users using the interactive THP2 TC.
- Thirty active web browsing (same as above) users using the interactive THP3 TC.
- Thirty active file downloading (four files, each having a size of 875 kB, which results in a total content size of 3.5 MB; time between two downloads is 1.0 s; four persistent TCP connections are utilised) users using the background TC.
- A total of 48 EGPRS time slots available for packet-switched traffic (equivalent to, e.g., five sites having three sectors each, where each sector allocates three (E)GPRS timeslots for packet-switched traffic).
- The users are distributed evenly in two cells – that is, the cells have the same traffic mix. The cells are under the same Gb 'pipe'.
- Gb capacity of 2048 kb/s.
- The RED/ECN parameters are the following: $min_{th} = 15\,\text{kB}$, $max_{th} = 45\,\text{kB}$, $max_{DP} = 0.2$.
- The TTLRED/TTLECN (the second, simpler implementation is used) parameters are the following: $min_{th} = 0.75\,\text{s}$, $max_{th} = 2.25\,\text{s}$ and $max_{DP} = 0.2$.

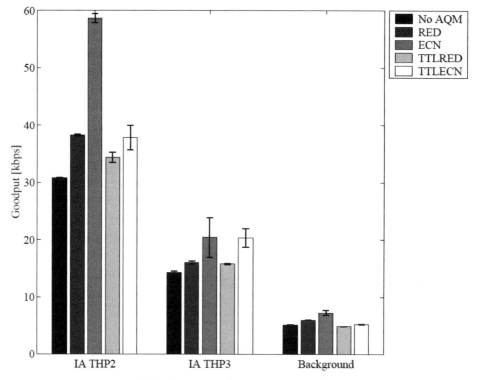

Figure 6.11 Average end-user TCP throughput [11].

Reproduced by permission of © IEEE 2006.

From the simulation results we can observe that:

- First of all, end-to-end delay (95th percentile) can be fairly high without any congestion control scheme: up to 5 s for the interactive THP2 TC, 14 s for the interactive THP3 TC and 34 s for the background TC. These are definitely too high for interactive-type applications such as web browsing.
- In most cases, RED and ECN (including their TTL-based variants) seem to be able to reduce average buffer delays dramatically (e.g., down to 1 s from 5 s).
- RED seems to be able to increase average throughput by 10–20% depending on the TC while ECN can do even better (40–90%).

It should be noted that RED/ECN (as well as TTLRED/TTLECN) implementation in the simulator followed the basic approach recommended in the literature. In other words, there was only a single instance of RED/ECN per TC-specific BSSGP buffer. Some obvious conclusions are that:

- Appropriate congestion control mechanisms are very useful in the 2G-SGSN for handling non-real time traffic in loaded scenarios.

- RED and ECN seem to be suitable active queue management schemes for the 2G-SGSN.

6.3   Backbone QoS

This section introduces QoS mechanisms in the backbone for both IP and ATM.

6.3.1   QoS is an end-to-end issue

In the mobile environment quality of service is an end-to-end issue. This principle has to be respected in the packet backbone as well. Figure 6.12 shows the full extent of the issue. 'End to end' means in the worst case *across many different administrative domains* combined with *using many different technologies.*

In Figure 6.12 there are three operators involved in the connection. In the example a global carrier connects two UMTS operators to each other. Each of the operators may have networks using a different backbone technology and a technology-dependent QoS scheme. UMTS operators themselves have two or more QoS schemes in use, as 3GPP-specified mechanisms only apply from the terminal to the GGSN. So inter-working between different backbone technologies is needed. Note that the end-user service experiences the sum of QoS budgets. In order to keep delay, delay variation and loss at an acceptable level, typically service level agreements (SLAs) are used for generating commercial incentives for service quality.

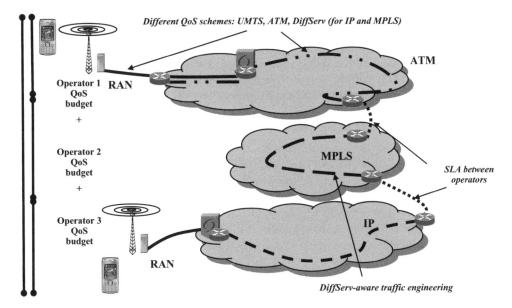

Figure 6.12   QoS in a multioperator environment.

6.3.2 Choice of backbone technology

3GPP does not specify the backbone network service. So, implementation of the backbone network and provision of sufficient QoS mechanisms is left to network operators. This means that the network operator may choose freely what technology to use when implementing the backbone for the UMTS network. In practice, the operator has to decide whether the packet backbone is built using ATM or IP or IP/MPLS or traditional TDM. The TDM option is not discussed here as QoS is not an issue in TDM networks.

ATM is a good choice for operators already running a large ATM network. Also, in environments where transmission links with speeds above 2 Mb/s are not available using ATM, inverse multiplexing for ATM (IMA) may be the only viable option. In all other cases the most likely technology choice is an IP as it is today the *lingua franca* of networking. The comparably low cost of equipment, availability of skilled personnel and commercial services make IP a safe and future-proof choice.

Small networks that do not grow beyond two to four sites can be built with routed IP backbones. In larger networks router-based MPLS has become the backbone of choice. MPLS is typically implemented to allow traffic engineering or VPN services. Recently, improved possibilities to carry non-IP traffic have also contributed to the success of MPLS.

6.3.3 QoS in IP networks

For IP, IETF has specified differentiated service (DiffServ) and integrated service (IntServ) QoS models. The IntServ model is based on signalled resource reservations and provides guaranteed bandwidth. In native IP backbone environments the IntServ model faces severe scalability problems and is not in use.

The DiffServ model is based on marking codepoints noting the per-hop behaviour of individual packets at the network edge and treating the packets accordingly in the network. It is much more scalable than IntServ. The downside is that DiffServ only provides soft QoS. It defines how a packet is to be treated in comparison with other packets. Excessive high-priority traffic in a DiffServ network leads to an increase in delay, jitter and packet loss for all.

In IETF the following DiffServ per-hop behaviours (PHBs) have reached RFC status:

- Expedited forwarding (EF) PHB Group [RFC 3246] [18], for providing a low-loss, low-latency, low-jitter, assured bandwidth, end-to-end service (virtual leased line). Codepoint 101110 is recommended for EF PHB.
- Assured forwarding (AF) PHB Group [RFC 2597] [19] defines four AF TCs each containing three drop precedence levels. AF codepoints are shown in Table 6.4.

The most obvious way of QoS inter-working between the mobility layer and an IP-based backbone is using DiffServ. The UMTS TC of a connection is mapped onto a DiffServ codepoint (e.g., EF for conversational class, selected AF codepoints for interactive/streaming and best effort for background). This is the default inter-working scheme.

Table 6.4 Assured forwarding (AF) PHB Group [RFC 2597].

	Low drop precedence	Medium drop precedence	High drop precedence
Class 1	001010	001100	001110
Class 2	010010	010100	010110
Class 3	011010	011100	011110
Class 4	100010	100100	100110

6.3.4 QoS in ATM networks

ATM is based on fixed-length packets and is a connection-oriented technology. End-to-end connections across the ATM network consist of virtual circuits. The connections are provisioned or set up using signalling. For different types of services, different adaptation layers (AALs) are available.

QoS in ATM networks is based on ATM service classes. The following service classes have been standardised:

- Available bit rate (ABR):
  - provides rate-based flow control;
  - provides minimum guaranteed cell rate.
- Unspecified bit rate (UBR):
  - best effort.
- Constant bit rate (CBR):
  - emulates circuit switching;
  - for jitter-sensitive applications.
- Variable bit rate–non-real time (VBR-NRT):
  - allows sending of traffic at variable rate;
  - statistical multiplexing in the network.
- Variable bit rate–real time (VBR-RT):
  - for jitter-sensitive applications;
  - like VBR-NRT.

In ATM networks the QoS is guaranteed with strict admission control and policing.

6.3.5 QoS in MPLS networks

Multi-protocol label switching (MPLS) is a technology that allows multiservice networking in an all-IP environment. It was originally designed to combine the benefits of IP and ATM and to act as glue between these two network technologies. Today, MPLS is primarily used for providing traffic engineering and virtual private network capabilities to IP backbones. Increasingly, MPLS is also used for the transport of non-IP traffic such as ATM cells, frame relay frames or TDM circuits across a router-based backbone network.

MPLS provides a variety of traffic engineering tools. In MPLS, QoS information can be carried in the experimental (EXP) bits of the MPLS shim header or implicitly mapped onto the label. The latter method is used in cell-based interfaces, where the EXP field is not present.

MPLS provides both soft DiffServ-type QoS and hard QoS services that use resource reservation. The three bits of the EXP field in the MPLS shim header allow a simple extension of DiffServ to the MPLS domain. For most applications and networks this is the preferred way to implement QoS in the MPLS backbone. As an alternative, resource reservations are possible using traffic engineering. The IntServ QoS scheme can also be extended to the MPLS backbone.

6.3.6   Deriving backbone QoS needs

A basic idea of the 3GPP QoS concept is that the UMTS TC for the user application is either requested by the terminal or derived from user data. For transport over IP networks the UMTS TC is mapped onto DiffServ codepoints in the SGSN and GGSN as explained in Section 6.2.4. Mappings in a 3G network are also outlined in Figure 6.13. Note that circuit-switched user traffic is conversational.

The selection of QoS scheme and applicable class of service (CoS) in the backbone is up to the network operator. In practice, IP, MPLS and ATM backbones are used.

As shown in Figure 6.13, mapping the UMTS TC onto the backbone CoS can be conveniently done via DiffServ codepoints (although direct mappings are also possible). In Table 6.5, example DSCP mappings for circuit-switched traffic are given.

Using the aforementioned mappings it is possible to construct a working QoS scheme for a UMTS core network that includes an IP backbone.

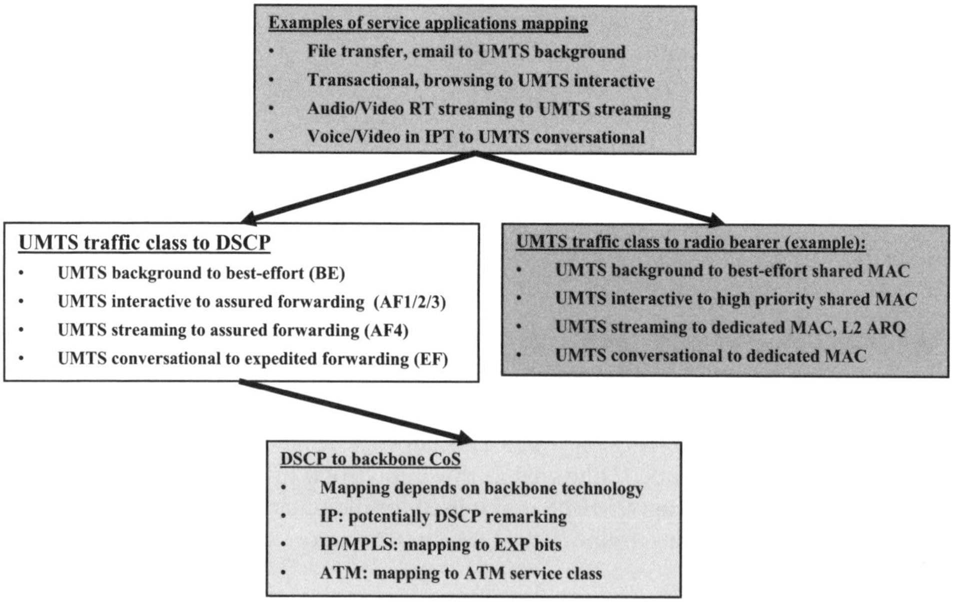

Examples of service applications mapping
- File transfer, email to UMTS background
- Transactional, browsing to UMTS interactive
- Audio/Video RT streaming to UMTS streaming
- Voice/Video in IPT to UMTS conversational

UMTS traffic class to DSCP
- UMTS background to best-effort (BE)
- UMTS interactive to assured forwarding (AF1/2/3)
- UMTS streaming to assured forwarding (AF4)
- UMTS conversational to expedited forwarding (EF)

UMTS traffic class to radio bearer (example):
- UMTS background to best-effort shared MAC
- UMTS interactive to high priority shared MAC
- UMTS streaming to dedicated MAC, L2 ARQ
- UMTS conversational to dedicated MAC

DSCP to backbone CoS
- Mapping depends on backbone technology
- IP: potentially DSCP remarking
- IP/MPLS: mapping to EXP bits
- ATM: mapping to ATM service class

Figure 6.13   QoS mappings.

Table 6.5 DiffServ mapping for circuit-switched core traffic types.

Traffic type	Per-hop behaviour (PHB)
Voice (conversational user plane)	Expedited forwarding
3GPP control protocols	Assured forwarding Class 4, low drop precedence
Urgent operation and maintenance traffic	Assured forwarding Class 4, low drop precedence
Non-urgent O&M traffic	Assured forwarding Class 1, low drop precedence

Note that the relative priorities of the selected AF PHB are commonly used by ISPs and other network operators in order to ensure interoperability with other DiffServ-aware applications running in the same network, Note also that low drop precedence is indicated for all the AF PHBs. In case of congestion, drop precedence determines the importance of the packet within the AF class. A congested node first drops packets with high drop precedence. If this is not enough to resolve the congestion, then packets with medium drop precedence and, finally, also those with low drop precedence value will be dropped.

6.3.7 Need for QoS in IP backbones

For efficient network utilisation the mobile operator should apply QoS differentiation in the IP network. Voice packets should not experience queuing delay, and delayed voice packets should be dropped as they are of no use for the application. On the other hand, data packets can be delayed but should not be dropped, as in most cases the application will eventually need to ask for a retransmission of the lost data. Note that the additional delay tolerated by a data application depends on many factors such as the connection speed, round trip delay and configuration of connection endpoints. If a packet is queued too long the application will consider it as lost.

Meeting the requirements of all services with a uniform best-effort service and heavy overprovisioning is feasible as long as all networking equipment is concentrated on one site or a small number of sites that can be cost-efficiently connected to high-speed links. In a large network with numerous sites and expensive wide-area links, plain over-provisioning becomes too expensive. Another issue is the need to cope with rapidly changing traffic patterns and exceptional situations such as loss of transmission or routing capacity because of a failure. Under unexpected conditions even a heavily overprovisioned best-effort network may fail as mission-critical applications face extensive loss and delay. In QoS-enabled networks, prioritised packets survive while less important traffic is dropped. QoS helps in coping with unexpected situations.

6.3.8 Queuing and scheduling

A QoS scheme only affects the traffic flow when the traffic offered to a specific link exceeds the transport capacity of the channel. In packet-based systems this means that packets are queued for further transport.

The way a QoS scheme works in practice depends on implementation of the queuing and scheduling in network elements. These mechanisms sort out the order in which packets are sent to the link. They also determine which packets are to be dropped in case of congestion.

State-of-the-art routers have several output queues per link, at least one for real time traffic and one for non-real time traffic. In current systems the number of queues can be very high. In practice, however, it may be difficult to find meaningful differentiation for more than four queues.

The scheduler picks the packets from the different queues in the output line using a pre-defined algorithm. Often, a strict priority scheduler is used for allowing real time traffic to pass any other packets. Alternatively, a modified round robin scheme can be used. The benefit of using round robin based schemes is that it is possible to limit the amount of the highest priority traffic to, say, 75% of link capacity. In exceptional situations where the offered real time traffic exceeds link capacity, this will prevent starvation of all the other flows because many TCP-based non-real time applications can adapt to low-speed transport. Especially in voice networks it is important to protect other traffic types. Voice connections will not survive long if signalling links are lost.

The ability of TCP-based applications to slow down is widely utilised in queue management. The idea is to react to queues becoming full by dropping random packets in the hope that end-user applications will reduce the offered traffic. This feature is called 'random early detection'.

Note that from the end-user perspective only short queues are useful for real time traffic. Voice packets that face an extra end-to-end delay of 20–30 ms will be discarded by the end-user application. For non-real time traffic, especially packet core user traffic, a queuing delay of tens or even some hundreds of milliseconds is much less critical than packet loss. When TCP is used, even one lost packet will cause TCP not just to ask for a retransmission but to slow down and so reduce data throughput for a while.

6.3.9   Implementing QoS interworking

A typical core network site hosts a large number of network elements. It serves a clearly defined geographical area. Typically, most of the traffic handled on such a core site is routed back to the radio network or to a local interconnect or to a local server. Mobile subscribers mostly communicate with people nearby and use local services. So, wide area transport is often not needed. In such a networking environment it is useful to feed all the traffic into a cost-efficient LAN network and to provide the more expensive wide area interfaces centrally. Readers familiar with LAN technology may start wondering about inter-working of UMTS QoS and IEEE's 802.1p frame prioritisation. In practice, this LAN QoS scheme is not needed as carrier grade LAN switches can also base packet priorisation on DiffServ codepoints.

In addition to the large core network sites, UMTS core network equipment can be placed as stand-alone items. In such configurations it may be meaningful to implement wide-area interfaces and WAN QoS interworking in mobile network elements.

The different backbone technologies set different requirements for QoS inter-working. For IP backbones, no special inter-working is needed if the DiffServ codepoints used in the UMTS domain can also be used in the backbone. Otherwise, the codepoints will have to be re-marked.

QoS interworking with the mobility layer and an IP/MPLS backbone is implemented using IP QoS mechanisms. DiffServ can be used for priority-based forwarding and MPLS for traffic engineering. This is not a major issue as codepoint re-marking is supported on carrier grade routers without significant performance impacts.

MPLS traffic engineering allows constraint-based routing of IP traffic. In practice, this means that packets can be routed through the backbone based on the available bandwidth and other resource requirements in addition to the simple hop count. DiffServ-aware traffic engineering extends MPLS traffic engineering to enable guaranteed bandwidth services.

ATM is quite widely deployed in operator backbones. Although most of the new services are based on IP technology, there are still many benefits in using such existing cell-based networks as 3G backbones. The key issues are use of ATM in 3GPP R99 specifications, availability of a reliable hard QoS scheme and traffic management capabilities that are superior to native IP.

In some cell-based networks, MPLS is used as a control plane instead of or in addition to the ATM control plane. QoS is provided according to the following principles:

- ATM CoSs are available for ATM traffic;
- specific ATM MPLS CoSs are available for IP traffic.

For IP over ATM the DSCP can be used for selecting a PVC with the desired CoS when more than one PVC is configured (e.g., available bit rate for non-real time traffic and constant bit rate for real time traffic).

References

[1] 3GPP, R5, TR 26.975, Performance Characterization of the AMR, v. 5.0.0.

[2] 3GPP, R6, TS 23.107, Quality of Service (QoS) Concept and Architecture, v. 6.1.0.

[3] 3GPP, R6, TS 23.060, General Packet Radio Service (GPRS); Service Description, v. 6.10.0.

[4] 3GPP TS, R6, 29.060, GPRS Tunnelling Protocol (GTP) across the Gn and Gp Interface, v. 6.10.0.

[5] 3GPP, R6, TS 24.008, Mobile Radio Interface Layer 3 Specification; Core Network Protocols, v. 6.10.0.

[6] 3GPP, R5, TS 32.215, Telecommunication Management; Charging Management; Charging Data Description for the Packet Switch (PS) Domain, v. 5.9.0.

[7] 3GPP, R6, TS 29.002, Mobile Application Part (MAP) Specification, v. 6.11.0.

[8] 3GPP, R6, TS 29.078, Customised Applications for Mobile Network Enhanced Logic (CAMEL) Phase 4; CAMEL Application Part (CAP) Specification, v. 6.4.0.

[9] 3GPP, R6, TS 29.061, Interworking between the Public Land Mobile Network (PLMN) Supporting Packet Based Services and Packet Data Networks (PDN), v. 6.6.0.

[10] G. Armitage, *Quality of Service in IP Networks: Foundations for a Multi-Service Internet*, 1st Edition, April 2000, Macmillan Technical Publishing.

[11] R. Cuny and J. Lakkakorpi, Active queue management in EGPRS, submitted to *IEEE 63rd Vehicular Technology Conference, 7–10 May 2006.*

[12] 3GPP, R6, TS 48.018, General Packet Radio Service (GPRS); Base Station System (BSS)-Serving GPRS Support Node (SGSN); BSS GPRS Protocol (BSSGP), v. 6.11.0.

[13] IETF, RFC 3481, TCP over Second (2.5G) and Third (3G) Generation Wireless Networks, 2003.

[14] IETF, RFC 2884, Performance Evaluation of Explicit Congestion Notification (ECN) in IP Networks, 2000.

[15] IETF, RFC 3168, The Addition of Explicit Congestion Notification (ECN) to IP, 2001.

[16] Sally Floyd's RED page, available at *http://www.icir.org/floyd/red.html*

[17] UCB/LBNL/VINT, Network Simulator – ns (version 2), September 2004, available at *http://www.isi.edu/nsnam/ns/index.html*

[18] IETF, RFC 3246, An Expedited Forwarding PHB (Per Hop Behavior), 2002.

[19] IETF, RFC 2597, Assured Forwarding PHB Group, 1999.

7

Service and QoS Aspects in Radio Network Dimensioning and Planning

David Soldani, Carolina Rodriguez and Paolo Zanier

The cellular network planning process consists of network dimensioning, detailed network planning, network operation and optimisation.

Lots of material covering the basics of radio, transmission and core network planning is available in the literature. For example, a comprehensive description of processes and tools for WCDMA radio network planning and optimisation can be found in [1] and [2]. Several methods for dimensioning, and performance improvement of GSM, GPRS and EDGE networks were presented in [3]. Planning and optimisation aspects for 2G, 2.5G, 3G, and evolution to 4G radio transmission and core are well covered in [4].

However, the published analytical methods and tools typically do not show enough flexibility to provide answers to what kind of service the operator is planning to provide, how these services will be implemented and how much money is needed for the total roll-out. The theoretical approaches, based on the circuit-switched type of communications, are very limited in their scope of application, and radio network planning tools do not take into consideration the possibility of handling radio resources according to offered traffic mix characteristics, such as priorities and quality of experience (QoE). On the other hand, dynamic simulators typically run with far too high time resolution, and thus need lengthy simulation times to design access networks and/or to analyse thoroughly the deployment of application services.

The ability to model voice and data services, as well as the deployment of innovative applications integrating voice, data and multimedia, such as presence, multimedia chat, PoC and conferencing, is one of the key required ingredients for service planning. The traffic mix and volume need to be modelled differently depending on carried application service characteristics. The radio resource management (RRM) functions deployed in the real network, such as power control (PC), admission control (AC), load control (LC) and handover control (HC) with QoS differentiation, are essential features for planning tools.

QoS and QoE Management in UMTS Cellular Systems
Edited by David Soldani, Man Li and Renaud Cuny © 2006 John Wiley & Sons, Ltd

Also, for each of the analysed services, customised user satisfaction criteria need to be defined and the performance of each service separately monitored.

This chapter addresses some of the radio dimensioning and detailed planning issues that may arise when deploying multimedia services in WCDMA and (E)GPRS cellular systems – note that the service optimisation aspects thereof are presented in Chapter 10. This includes the possibility of taking into account the offered service characteristics, discriminated treatment across the radio access network and service-related performance monitoring.

The proposed methods and tools allow network planners to assess the provision of the QoS negotiated for each of the deployed services and maintenance of the data transfer characteristics, as well as to find an optimum trade-off between the quality, capacity and coverage requirements for any of the services in a network operator's service portfolio.

7.1 WCDMA radio dimensioning and planning

This section introduces some possible methods and tools for WCDMA radio dimensioning and planning.

7.1.1 Radio dimensioning aspects of UTRAN FDD

This section addresses the radio dimensioning issue that arises from the introduction of novel services in WCDMA networks [5]. To investigate the impact and entailed limitations of the deployment of new services on the performance of offered ones, we describe two useful methods and a simple simulator for radio interface dimensioning. The tool (based on throughput estimates and snapshots of the system status) supports essential radio resource management functions, such as admission control and packet scheduler, with QoS differentiation. Using this tool, it is possible to define customised user satisfaction criteria for each of the implemented services, and performance results, at a given QoS, can be presented as the maximum number of subscribers a WCDMA cell can satisfactorily accommodate. Such a simulator, in its simplicity, provides an appropriate platform for studying quickly the adverse effects on already deployed services in WCDMA networks due to the provisioning of a novel application. In the following, the proposed solution is used for analysing the deployment of push to talk over cellular (PoC) by means of four case studies.

7.1.1.1 Simulator description

In this section, we present a modified version of the virtual time simulator described in [6], where computational time is significantly reduced. The tool supports simplified models for packet-switched services and works on snapshots of the system status, upon which performance statistics are derived. Load estimates are based on throughput. In this chapter, the simulator is used to evaluate rapidly the influence of the deployment of PoC on the maximum number of subscribers a WCDMA cell can accommodate within the constraints defined by a minimum percentage of satisfied users. Since the system in high-traffic situations is downlink capacity limited (see, e.g., [1] and/or [2]), the simulator supports only this direction.

7.1.1.2 Simulation structure

The simulation flow chart is illustrated in Figure 7.1. As can be noticed in the figure, the simulator consists of four modules and two loops, denoted by 'inner' and 'outer' loop in the following. The first module is the call generator, which is described in Section 7.1.1.3, followed by an admission control (AC) and packet scheduler (PS) function, which are presented in Section 7.1.1.4. The last block has been designed to collect performance statistics, as explained in Section 7.1.1.5. The inner loop constitutes a computational procedure in which a cycle of operations (new configuration of active users) is repeated to approximate the desired result more closely. In other words, the greater the number of iterations allowed, the more accurate the performance results are. In fact, the tool, being event-based (from the active connection viewpoint), during each iteration analyses a snapshot of the cell status, where more users are simultaneously active, depending on the related probability to generate a call (use a certain service). The outer loop may be enabled to increase the offered traffic (load) till at least one of the thresholds defined by the percentage of dissatisfied users of that particular service is exceeded. When such a

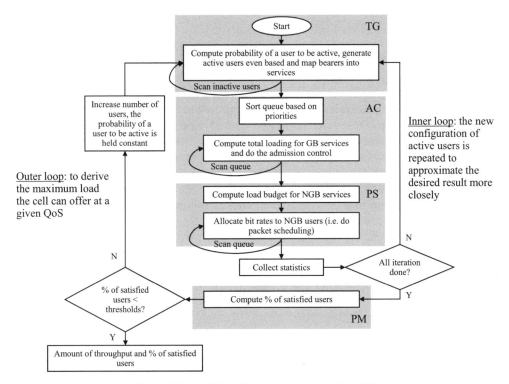

Figure 7.1 Cell-based simulation flow chart [5].

condition is satisfied, the simulation ends and the offered and served traffic, together with average cell throughput, are collected as described in the following sections.

7.1.1.3 Call generator

The simulator supports several application services, namely: circuit-switched (CS) speech and video calls, video sharing (VS), push to talk over cellular (PoC), streaming, multimedia messaging (MMS), WAP browsing and Dialup connections (see Chapter 2 for more details). In the simulator, for each of the above services, it is possible to define a different *QoS profile* – traffic class (TC), traffic handling priority (THP), bit rates and allocation retention priority (ARP), see Chapter 3 and [7] – *mean service time, mean arrival rate* and *share of subscription*. On this basis, a user is supposed to make a call every mean arrival rate seconds, and the corresponding connection may last on average mean service time seconds. The ratio between the two corresponding parameters yields the Erlangs per user for that particular service type, whereas the share of subscriptions defines the percentage of users that have contracted to receive and pay such a service. Hence, during a snapshot of the cell status, the probability P_i of a user making use of a certain service type i, when the subscribers are scanned for the first time, is given by:

$$P_i = \frac{T_i}{A_i} \frac{S_i}{100} \qquad (7.1)$$

where T_i is the mean service time, A_i the mean arrival rate (reciprocal of the mean arrival intensity, denoted by λ in the literature) and S_i the share of subscription to that particular service type. The possibility of a PoC user communicating simultaneously in the downlink to more peers is taken into account by introducing a group factor in (7.1). In this work, we assume the group factor following a geometric distribution with a cut-off: the average group size is supposed to be composed of four people, and the minimum and maximum group size is assumed to be 1 and 25 users, respectively. Further, during inner iterations, the probability of a user making a new call is conditioned by the fact that such a subscriber may have already established one or more connections. Hence, (7.1) is accordingly corrected. Ultimately, when the simulation is over – for example, the condition to exit the outer loop of Figure 7.1 is satisfied – for each of the services deployed in the cell, the maximum offered load is estimated as follows:

$$N_i = U_i \cdot \frac{A_i}{T_i} \qquad (7.2)$$

where U_i is the average number of active bearers carrying service type i; A_i and T_i are as defined in (7.1); and N_i denotes the offered traffic in number of subscribers per cell.

7.1.1.4 Radio resource management functions

The simulator supports only an admission control (AC) and packet scheduler (PS) function, as illustrated in Figure 7.1. Connection requests are arranged in a queue and served following the strict priority principle. The lower the associated priority value, the higher the priority of the service in question to be processed by the AC and PS functions. Guaranteed bit rate (GB) services are not admitted if either the total load in the cell L_{Total} exceeds the *overload threshold*, $L_{Target} + Offset$ – that is:

$$L_{Total} = L_{NGB} + L_{GB} > L_{Target} + Offset \qquad (7.3)$$

or the load increase ΔL_{GB} due to the bearer is question (if admitted), plus the actual non-controllable (GB) load L_{GB} in the cell, goes beyond the target (planned) load L_{Target} – that is:

$$L_{GB} + \Delta L_{GB} > L_{Target} \qquad (7.4)$$

non-guaranteed bit rate (NGB) traffic is admitted if (7.3) is not satisfied, and bit rates are allocated using the fair throughput (FT) scheduling algorithm presented in [6]. The load budget LB_{NGB}, which is shared fairly in terms of allocated bit rates between peers, is given by:

$$LB_{NGB} = L_{Target} - (L_{NGB} + L_{GB}) \qquad (7.5)$$

The load increase and decrease computations are based on the downlink fractional load equation presented in [1], which for one radio bearer service (k) reduces to:

$$\eta_{DL}^k = \frac{1 + SHO}{W} \rho_k R_k \nu_k ((1 - \alpha_k) + i_{k,DL}) \qquad (7.6)$$

where SHO is the soft-handover overhead, W is the chip rate, ρ_k is the required E_b/N_0 of the service in question, R_k the transport channel user bit rate, ν_k the service activity factor (AF), $i_{k,DL}$ the other-to-own-cell-interference ratio and α_k is the orthogonality factor, which ranges from 0 to 1, depending on multipath conditions ($\alpha = 1$ means perfect orthogonality). The downlink load factor (L_{DL}) is given by [1]:

$$L_{DL} = \sum_k \eta_{DL}^k = P_{TxTotal}/P_{TxMax} \qquad (7.7)$$

where $P_{TxTotal}$ and P_{TxMax} denote the actual and maximum transmission power of the cell.

7.1.1.5 Performance monitoring

The performance indicators collected by the simulator are call block ratio and throughput, based on which the following user satisfaction criteria are defined. CS speech and video calls are satisfied if they do not get blocked. Their guaranteed bit rates (GBs) are 12.2 and 64 kb/s, respectively. The same criterion is applied to video sharing, whose target bit rate (GB) is 64 kb/s. For NGB services, in addition to the above condition, user satisfaction criteria take into account average allocated bit rates. In particular, PoC, streaming, MMS, Dialup and WAP users are satisfied if the average bit rate during iterations (inner loop) is higher or equal to 8 (or 16), 64, 32, 64 and 32 kb/s, respectively.

If the outer loop is enabled, the simulation is over when at least one of the following conditions turns out to be true (see Figure 7.1):

- Less than 90% of speech users are satisfied.
- Less than 90% of video users are satisfied.
- Less than 90% of streaming users are satisfied.

- Less than 70% of MMS users are satisfied.
- Less than 90% of PoC users are satisfied.
- Less than 90% of VS users are satisfied.
- Less than 50% of Dialup users are satisfied.
- Less than 70% of WAP users are satisfied.

(*Note*: The above values are used in the case study presented in Section 7.1.1.6 and, in general, are parameters for the radio network planner to set.)

This means that the maximum load that can be served in the cell, at a given QoS and share of subscriptions, is achieved and the offered traffic from (7.2) can be accordingly derived. In this case, the main output of the simulator is thus the maximum number of subscribers a WCDMA cell can accommodate satisfactorily.

Moreover, the simulator may be used for studying the impact of a new service on existing subscriber satisfaction. This can be done by gradually increasing the subscription level for the new service while keeping the input load of the other services constant.

7.1.1.6 Simulation assumptions, results and discussion

In this section, four case studies concerning the impact of PoC deployment on the quality of offered services in a WCDMA cell are presented. The services are mapped onto QoS profiles with different priority values – that is, speech calls have top priority, followed by CS video and VS applications. Within NGB traffic, the bit rate to PoC users is allocated (scheduled) first, followed by streaming, WAP and MMS, and Dialup connections that are served last. The most important parameters are listed in Table 7.1 and the traffic models and mix are reported in Table 7.2.

Table 7.1 Most important cell-based parameters.

Parameter	Value			
Number of iterations (inner loop)	1000			
Downlink load target	70%			
Overload offset	10%			
Orthogonality (α)	0.5 (ITU Vehicular A)			
Soft handover overhead (*SHO*)	20%			
Other-to-own-cell-interface ratio (i)	0.55			
Chip rate (W)	3.84 Mchip/s			
Offered services/Traffic class	Downlink DCH bit rate (kb/s)	Priority	E_b/N_0 (dB)	Activity factor (v)
Speech/CS conversational (GB)	12.2	1	7	0.67
Video/CS conversational (GB)	64	2	6	1
SWIS/Streaming (GB)	64	3	6	1
PoC/Interactive THP1 (NGB)	0, 8, 16	4	6	1
Streaming/Interactive THP2 (NGB)	0, 64	5	7	0.6
WAP-MMS/Interactive THP3 (NGB)	0, 64, 128, 144, 256, 384	6	5/5.5	1/0.6
Dialup/Background (NGB)	0, 64, 128, 144, 256, 384	7	5.5	0.8

Table 7.2   Adopted traffic models and mix.

Offered service	Share of subscriptions (%)	Mean service time (s)	Mean arrival intensity (Hz)
Speech (CS)	100	90	1/4800
Video (CS)	3	120	1/24000
Streaming	10	600	$1/(5*3600)$
MMS	10	10	$1/(2*3600)$
SWIS (RTVS)	3	180	$1/(2*3600)$
Dialup	1	1200	$1/(2*3600)$
WAP browsing	20	600	$1/(4*3600)$
PoC	Varies*	60	$1/(2*3600)$

*The volume is increased from 0 to 100%, whereas the average PoC group size is held constant: one user in Cases 1 and 2, and four users in Cases 3 and 4.

For Cases 1–3, the served PoC traffic is monitored in terms of absolute and relative cell throughput, whereas the effects of PoC deployment on the other existing services are characterised by the maximum number of satisfactory non-PoC subscriptions in the cell and the variation thereof in percentage. In Case 4, we display the course of average cell throughput and the percentage of satisfied users as a function of the number of PoC subscribers.

7.1.1.7   Case 1

This case focuses on the deployment of PoC carried by a bearer with a maximum bit rate of 8 kb/s. The average PoC group size in the downlink is one, which means that one user is allowed to call only one person at a time. Different priorities are only allocated to distinct services. Simulation results are illustrated in Figures 7.2 and 7.3. The former

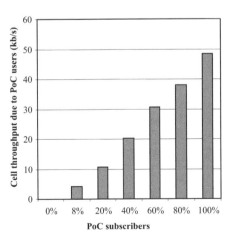 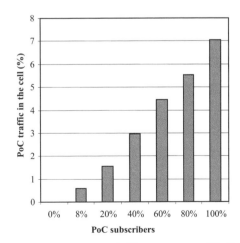

Figure 7.2   Case 1 – (a) Average cell throughput due to PoC users, and (b) percentage of PoC traffic in the cell as a function of PoC subscribers [5].

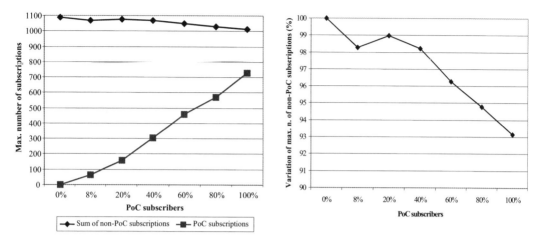

Figure 7.3 Case 1 – (a) Maximum number of satisfactory subscriptions, and (b) variation in percentage thereof as a function of the PoC subscribers in the cell [5].
Reproduced by permission of © IEEE 2004.

depicts the served PoC traffic in the cell, whereas the latter shows the impact of PoC on other services. From Figure 7.2(a) and (b), we conclude that if all subscribers subscribe to the PoC service, less than 50 kb/s would be the average cell throughput due to PoC traffic (7% of the total traffic in the cell). Hence, as depicted in Figure 7.3, the impact of PoC usage on other services is really small: performance deterioration in the cell would be noticed only when more than 40% of end-users subscribed to PoC, and only 7% of non-PoC subscriptions would be unsatisfactory if all users made use of PoC.

7.1.1.8 Case 2

This case uses the same parameter settings as Case 1, but PoC is offered at 16 kb/s. Simulation results are illustrated in Figures 7.4 and 7.5. The former depicts the served PoC traffic in the cell, whereas the latter shows the impact of PoC on other services. Figure 7.5 reveals that even in this case PoC slightly deteriorates the performance of other services. If all users subscribed to PoC the corresponding generated traffic would require on average only ∼13% of cell capacity; and the maximum number of non-PoC subscribers satisfactorily handled in the cell would decrease by ∼13%, which is, as expected, almost twice as much as the corresponding traffic served in Case 1.

7.1.1.9 Case 3

Parameters are set as in Case 2, but now the PoC group size is on average composed of four users. It is further assumed that a given PoC user will press the PoC talk key as often and speak as long as in Case 2. As a result, a PoC user will generate on average four times more traffic in the cell than in Case 2, and eight times more than in Case 1. Case 3

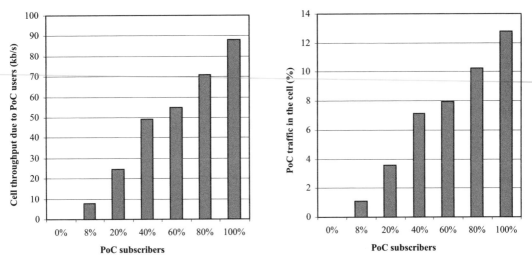

Figure 7.4 Case 2 – (a) Average cell throughput due to PoC users, and (b) percentage of PoC traffic in the cell as a function of PoC subscribers [5].

Reproduced by permission of © IEEE 2004.

investigates the 'worst scenario', where the average PoC group size is fairly high (four) and the group members are all located in the same cell. Figure 7.6 depicts the served PoC traffic in the cell. Figures 7.7 and 7.8 display the influence of PoC on other services. Here, PoC has a significant impact on the maximum number of non-PoC subscribers the cell can accommodate. If less than 20% of the subscribers subscribed to PoC, the corresponding generated traffic would require ∼10% of cell capacity, and the maximum

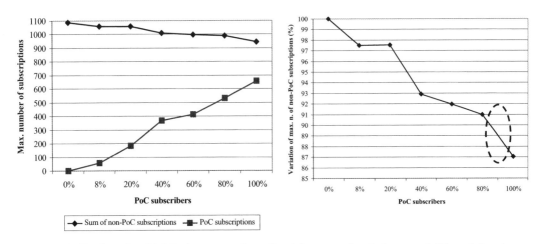

Figure 7.5 Case 2 – (a) Maximum number of satisfactory subscriptions, and (b) variation in percentage thereof as a function of the PoC subscribers in the cell [5].

Reproduced by permission of © IEEE 2004.

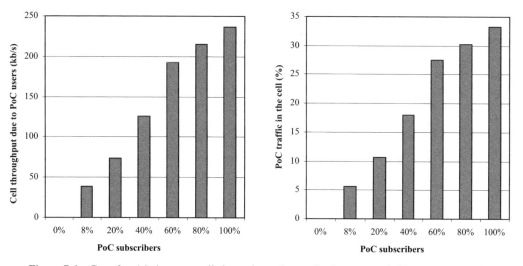

Figure 7.6 Case 3 – (a) Average cell throughput due to PoC users, and (b) percentage of PoC traffic in the cell as a function of PoC subscribers [5].
Reproduced by permission of © IEEE 2004.

number of non-PoC subscribers satisfactorily handled in the cell would decrease by only 10%. If all users subscribed to PoC, on average one-third of cell capacity would be exploited by PoC traffic. In Figure 7.8, the course of the curves displayed separately for each of the deployed services is rather similar to the slope of the graph in Figure 7.7. This is due to the fact that the simulation ends when any of the dissatisfaction criteria are verified.

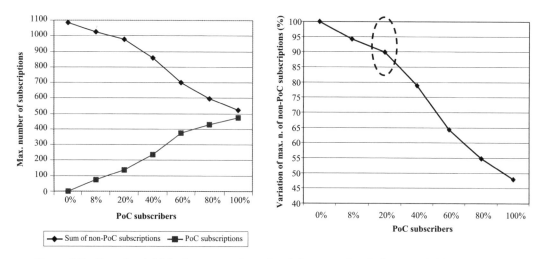

Figure 7.7 Case 3 – (a) Maximum number of satisfactory subscriptions, and (b) variation in percentage thereof as a function of the PoC subscribers in the cell [5].
Reproduced by permission of © IEEE 2004.

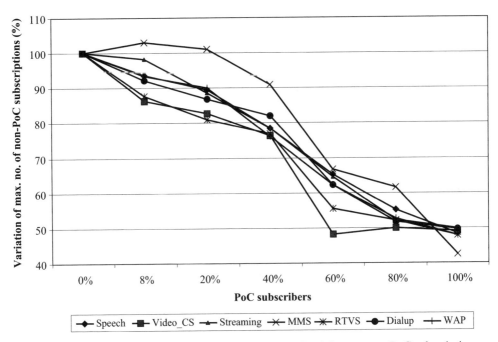

Figure 7.8 Case 3 – Variation in the maximum number of satisfactory non-PoC subscriptions as a function of the percentage of the PoC subscribers in the cell [5].

Reproduced by permission of © IEEE 2004.

7.1.1.10   Case 4

Parameters are set as in Case 3, but the input load for PoC is gradually increased, whereas for all other services the offered traffic volume in the cell is fixed at 500 users. This case illustrates the services that are likely to be affected by PoC penetration. As already pointed out, while in previous cases the goal was to compute the maximum number of subscribers a cell can properly accommodate with or without PoC presence, the main objective of this case is to quantify the impact of PoC penetration when the number of subscribers of other services is kept constant. In general, the traffic volume in the cell without PoC is for the operator to determine, yet, in this study, it was chosen such that all percentages of satisfied users of the offered services were above the quality thresholds defined in Section 7.1.1.5. Figure 7.9 depicts the average cell throughput of GB, NGB and PoC traffic as a function of PoC subscribers (in percentage). As expected, when PoC traffic increases, NGB load decreases, whereas the load due to GB services remains constant. This is due to the fact that PoC has the highest priority among NGB services (see Table 7.1) and that from (7.4) it has no means to affect the admission of GB services. However, from Figure 7.9 we cannot draw any conclusion on the influence of PoC on the performance of other services, as the quality experienced by the users of NGB applications is not shown. These effects can be seen in Figure 7.10, where the percentage of satisfied users is depicted separately for each of the offered services in the cell. When the percentage of PoC subscribers goes beyond 50%, the bit rate allocated to WAP

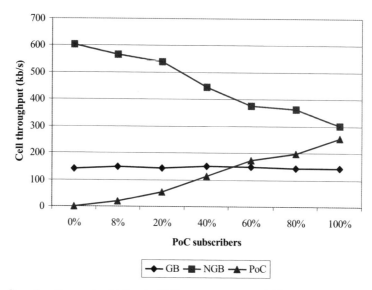

Figure 7.9 Case 4 – Guaranteed bit rate (GB), non-guaranteed bit rate (NGB) (without PoC) and PoC average cell throughput (kb/s) as a function of the percentage of PoC subscribers in the cell [5].

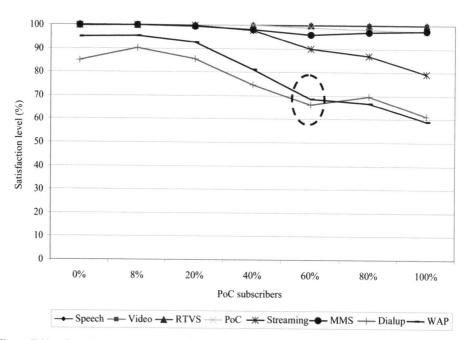

Figure 7.10 Case 4 – Satisfaction levels per service as a function of the percentage of the PoC subscribers in the cell [5].

connections yields too many dissatisfied users. This would tell the operator to add extra radio capacity in the cell – for example, introduce one more carrier.

Simulation results show that the maximum number of subscribers a WCDMA cell can accommodate, at a given QoS, depends heavily on the assumed subscription percentage for each of the offered services. Nevertheless, the radio interface does not appear to be a bottleneck when the PoC service is launched. In the worst case – that is, the PoC group size is fairly high and group members are all located in the same cell – additional radio resources may be required when the number of PoC subscriptions exceeds 20% of existing customers. Using the proposed tool, the influence of the penetration of any service can be rapidly evaluated by increasing the traffic volume of the application in question while the subscription levels of all other services are held constant.

7.1.2   A virtual time simulator for UTRAN FDD

This section describes a virtual time simulator for studying the provisioning of QoS before the deployment of new services or RRM functions throughout UTRAN. This tool is a good trade-off between the complexity of advanced *dynamic* simulators and the straightforwardness of *quasi-static* tools [1]–[4]. The proposed solution supports essential modules for service-driven radio network planning, such as multicell propagation scenarios and realistic RRM functions [8]. Also, for more accurate analyses, the tool includes user-definable traffic models for packet-switched (PS) and circuit-switched (CS) connections, and the possibility of monitoring system and service performances as recommended in [9].

As shown in the following sections, this solution has the potential of investigating any QoS management algorithm and multimedia service provisioning in UTRAN before its deployment throughout a real WCDMA network [10].

7.1.2.1   Simulator structure

The simulator consists of a modular structure with clear interfaces. Each module is implemented independently so that each entity may be straightforwardly replaced by an alternative solution. The tool includes the following functions: traffic and path loss generators, admission control (AC), load control (LC), packet scheduler (PS), power control (PC), process calls (PrCs) and performance monitoring (PM). AC, LC and PS are cell-based algorithms, whereas PC, PrC and PM are system-based functions. The statistically large number of items of user equipment (UEs) in the system does not make it necessary for them to be on the move: mobility effects may be taken into account by, say, speed-dependent E_b/N_0 requirements. Soft handover (SHO) affects mainly AC and the PS. In the former, diversity (DHO) branches are processed first, followed by the main branches. In the latter, the bit rate assigned to the radio link set (UE) is the minimum of the bit rates allocated separately (for each cell) to all radio links of the active set. SHO gains may be taken into account in the E_b/N_0 requirements based on SHO condition. Since the system in heavy-traffic situations is downlink capacity limited [1]–[4], the presented simulator only supports this direction. The structure to simulate the uplink could be easily implemented using exactly the same concept, except that the transmission power levels at the base station (BS) would need to be replaced by received power levels.

The maximum resolution of the tool is one radio resource indication (RRI) period – that is, the time needed to receive the power levels from the base stations. A simulation flow chart is illustrated in Figure 7.11.

7.1.2.2 Traffic generator, models and mix

Call and session arrivals are generated following a Poisson process [9], and mapped onto the appropriate QoS profiles [7], depending on the type of traffic carried. Circuit-switched (CS) speech and video calls are held for an exponentially distributed service time, and their inter-arrival periods follow exactly the same type of distribution [9]. Packet-switched services are implemented as an ON/OFF process with truncated distributions [11]. The duration of the ON period depends mainly on the allocated bit rate and object size, which is modelled differently depending on the application carried [12]. Different distributions are also used to model the related OFF time behaviors and session lengths. The utilised traffic models and the adopted traffic mix (in *share of calls*) in the case study presented in Section 7.1.2.8 are listed in Table 7.3.

All calls/sessions (generated at the beginning of each simulation) are subsequently processed (played back) taking into account the corresponding arrival times, service activities and priorities, hence the name *virtual time* simulator.

7.1.2.3 Path loss generator

Across the simulation area, each call gets assigned a random position, but also other mobile distributions are possible. For each mobile location, the power levels received from all cells are calculated first and then the cells satisfying SHO conditions are assigned as active. Each cell can be configured separately. For path loss calculations, the Okumura–Hata model described in [13] and the models defined in [14] are supported. By implementing an appropriate interface it is also possible to import the propagation calculated by another radio network planning tool. Correlated slow fading can be overlaid as described in [9]. Figure 7.12(a) and (b) show the path losses and the cell dominance areas for the simulation scenario (downtown Helsinki) adopted in the case study discussed in Section 7.1.2.8.

7.1.2.4 QoS management functions in UTRAN

The cell-based radio resource management functions supported – namely, admission control, packet (bit rate) scheduler, and load control with QoS differentiation (priorities and differentiated parameters for different QoS profiles, or radio bearer attributes) – were described in Section 5.2.

7.1.2.5 Process call function

All active calls in the system are processed once each radio resource indication period, as illustrated in Figure 7.11. If the ongoing connection is CS, the simulator collects its throughput, increases the active connections counter and releases the call in case it lasted longer then the corresponding call duration period (see Table 7.3). For each packet-

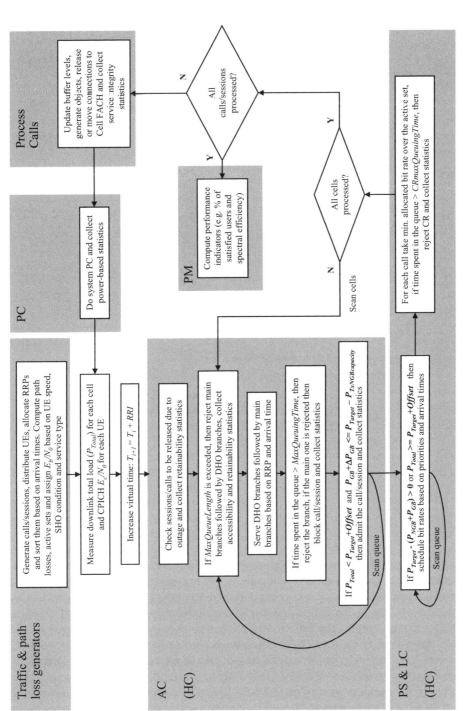

Figure 7.11 Simulation flow chart [10].

Reproduced with kind permission of Springer Science and Business Media.

Table 7.3 Adopted traffic models and mix [10].

Service	Data rate (kb/s)	Buffer size (s)	Object size (kB)	Off time (s)	Session length (objects)	Mix (%)
PoC	8	1	Exponential 6 mean, 0.5 min, 40 max	Exponential 60 mean, 1 min, 1200 max	Geometric 8 mean, 1 min, 30 max	18
Streaming	64	8	Uniform 160 min, 3200 max	—	1	12
MMS	Best effort	—	Exponential 20 mean, 3 min, 200 max	—	1	5
Dialup	Best effort	—	Log-normal ($\mu = 5$, $\sigma = 1.8$) 0.1 min, 20 000 max	Pareto ($k = 2$, $\alpha = 1$) 1 min, 3600 max	Inv. Gaussian ($\mu = 3.8$, $\lambda = 6$) 1 min, 50 max	15
SWIS	64	1	Exponential 80 mean, 32 min 2400 max	—	1	10
WAP	Best effort	—	Log-normal ($\mu = 2$, $\sigma = 1$) 0.1 min, 50 max	Exponential 20 mean, 1 min, 600 max	Geometric 3 mean, 1 min, 50 max	13
Speech	12.3	—	—	—	Exponential 90 s	20
Video	64	—	—	—	Exponential 120 s	7

switched connection, we check first whether either the RNC buffer or the source buffer is not empty. Then, if there are data to transmit, active session throughput is collected and the active connection counter accordingly increased. Besides this, the status of the corresponding buffer in the UE is monitored and updated. If during the ON period the buffer becomes empty, the rebuffering procedure is activated and the user of the service is considered unsatisfied. When the user is reading (or the connection is in idle mode in the case of PoC), the transfer delay of the delivered object is calculated, the connection is marked as inactive and the inactivity period monitored. If the dwelling time of the DCH in question lasted longer than the corresponding inactivity timer, the terminal is moved to Cell_FACH state, and the corresponding allocated resources are released. When the reading time is over, either a new object for download and the corresponding reading time are regenerated, or ongoing packet communication is released, depending on the corresponding session length (see Table 7.3). In the former

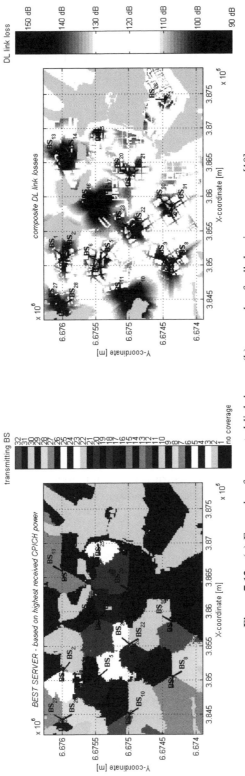

Figure 7.12 (a) Example of supported link losses, (b) example of cell dominance areas [10].

Reproduced with kind permission of Springer Science and Business Media.

case, if the time needed to fill up the buffer in the terminal is more than the corresponding *buffering delay*, the user of the service in question is recorded as unsatisfied.

7.1.2.6 Power control function

The simulator supports an ideal power control function that includes the effects of a large-scale propagation channel (see Section 7.1.2.3), but not fast fading. Multipath fading and SHO effects are taken into account in the service E_b/N_0 requirement. Interference is realistically modelled: at any simulation time step, the received power from all cells except that from the best server is counted as interference, and hence the corresponding coupling effect is fully taken into account. (*Note*: From the best server only that fraction of power determined by non-orthogonality is considered as interference.)

The power control system of equations can be written as:

$$\frac{Wp_{i_m}/L_{m,i_m}}{R_{im}P_m/L_{m,i_m}(1-\alpha_{im}) + \sum_{n,n\neq m} P_n/L_{n,i_m} + N_{i_m}} = \rho_{i_m}, \quad i_m \in I(m), m = 1, \ldots, M \quad (7.8)$$

where the symbols in (7.8) are explained in Table 7.4.

Equation (7.8) simply equates the received E_b/N_0 with given transmission powers to the required E_b/N_0 for sufficient quality of the connection. Taking into account that:

$$P_m = \sum_{i_m \in I(m)} p_{i_m} + p_{c,m} \quad (7.9)$$

where $p_{c,m}$ is the sum of common channel powers from BS_m, (7.8) can be rewritten in compact form as an $M \times M$ linear system of equations of type $Ax = b$, where the unknowns are total BS powers. During the simulations, we first resolve this linear system and then from (7.8) we derive individual radio link powers. The solutions are

Table 7.4 Symbols in the PC system of equations [10].

Symbol	Explanation
i_m	Index of a UE served by BS m
m, n	Indices of BSs
$I(m)$	Set of UE indices served by BS m
M	Number of cells
p_{i_m}	BS transmitted power for UE i_m
P_m, P_n	Total transmit power of BS m and BS n
L_{m,i_m}	Path loss from BS m to UE i_m served by BS m
L_{n,i_m}	Path loss from BS n to UE i_m served by BS m
R_{i_m}	Bit rate used by UE i_m
α_{i_m}	Orthogonality factor for UE i_m
N_{i_m}	Noise power (thermal plus equipment) of UE i_m
ρ_{i_m}	Required E_b/N_0 for UE i_m

then used to estimate the transmission powers of GB and NGB services. AC and PC ensure that BS total transmission power is kept below its maximum and WCDMA pole capacity is not exceeded. The existence of solutions to this type (7.8) of equations was studied in [15], among others. The power of common channels is a cell-based management parameter (see Section 7.1.2.8).

7.1.2.7 Performance monitoring function

Several performance indicators can be collected during the *measurement period* – for example, *call block ratio* (CBR) caused by queuing and/or buffer overflow; *call drop ratio* (CDR) due to power outage; *active session throughput* (AST); *capacity request rejection ratio* (CRRR) for NGB traffic; *object transfer delay* for browsing; MMS and Dialup connections; and *rebuffering* for streaming, PoC and VS applications. *Link-* and *cell-based powers*, as well as common pilot channel (CPICH) E_c/N_0 values are also computed during the simulated time. From these measurements we can derive the *geometry factor* (*G*), defined as the ratio between power received from the serving cell and power received from the surrounding cells plus noise [1].

System and service performance can be assessed as recommended in [9], and for this purpose tailored *user satisfaction* criteria can be input to the simulator. In this work, a speech or video user is satisfied if the call neither gets blocked nor dropped. In addition to this criterion, for PoC, VS and streaming users no rebuffering is allowed during the communication, and the time to fill up the related buffer needs to be reasonably short; for Dialup (HTTP, email, FTP, etc.), WAP browsing and MMS, AST has to be higher than 64 kb/s, 32 kb/s and 8 kb/s, respectively. Furthermore, none of the capacity requests of NGB services must be rejected. *Spectral efficiency* is computed as the system load (cell throughput normalised with respect to the chip rate, 3.84 Mchip/s) at which a certain percentage of users of the worst performing service are satisfied. Different thresholds can be set for distinct bearers, though 90% is the default value for all applications.

7.1.2.8 Case study

In this case study, we investigate the benefits to the network administrator of offering speech, video and VS with a guaranteed bit rate (GB), and all other services (including streaming and PoC which are idle for most of the session duration) on a best-effort (NGB). Besides this, we analyse service performance deterioration at heavy-traffic volume when the bearers start to compete for radio resources.

Speech and video calls are served as CS conversational, whereas VS is carried on the PS streaming class. PoC, streaming and WAP/MMS are mapped onto the PS interactive. Dialup connections – which comprise, for example, FTP, email and HTTP traffic – are carried on the PS background class. Radio resource priority (RRP) values (see Section 5.2) are set such that speech calls have top priority, followed by video and VS services. Within interactive class, using different traffic handling priorities (THPs), PoC is handled first, followed by streaming and WAP/MMS. Dialup is served in the end.

Differentiated parameter values, which further improve PoC and streaming performance at the expense of lower priority services, and the mapping of services onto distinct QoS profiles are reported in Table 7.5.

Table 7.5 Mapping of services onto QoS classes and parameter values [10].

QoS profile	Service	Bit rate (kb/s)	RRP	Min. all. bit rate (kb/s)	AC max. queuing time (s)	Granted min. DCH Alloc. time (s)	Granted min. DCH Alloc. time in overload (s)	Buffering delay (s)	Inactivity timer (s)	CR max. queuing time (s)
CS-conv.	Speech	12.2	1	GB	5	—	—	—	—	—
	Video	64	2	GB	10	—	—	—	—	—
PS-stream.	SWIS	64	3	GB	10	—	—	5	—	—
PS-int. THP1	PoC	0, 8	4	8	15	15	10	4	60	4
THP2	Streaming	0, 64	5	64	15	10	5	16	5	10
THP3		0, 16, 32, 64, 128, 144, 256, 384	6	32	15	5	0.2	—	10	10
PS-backg.	Dialup	0, 16, 32, 64, 128, 144, 256, 384	7	16	15	1	0.2	—	5	5

Reproduced with kind permission of Springer Science and Business Media.

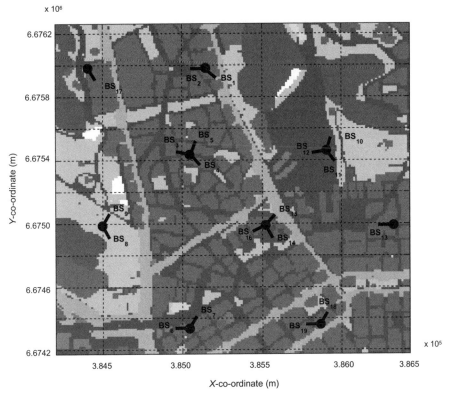

Figure 7.13 Simulation scenario used in the case study discussed in Sections 5.2.6 and 7.1.2.8 [10].

Reproduced with kind permission of Springer Science and Business Media.

Performance results were analysed using a macrocellular network located in downtown Helsinki (see Figure 7.13), where terminals were uniformly randomly distributed on the land area. The simulation was performed over a period of 2 h using a time step of 200 ms (RRI period). Traffic mix and traffic intensity were held constant – that is, two call/session attempts per second. The corresponding offered traffic was about 750 users per cell over the simulated time. Table 7.6 reports the most important network parameters.

The simulation results are shown in Figures 7.14–7.19 and in Table 7.7. The offered traffic mix over the entire simulated period is illustrated for each of the simulated cells in Figure 7.14(a). Figure 7.14(b) shows the average cell throughput as a function of deployed cells and services.

A snapshot (1000 s) of the load status in Cell 11 (see Figure 7.13), where most of the users turned out to be unsatisfied, is depicted in Figure 7.15. The values plotted against the simulation time are transmission powers, the power budget (PB) and scheduled capacity (SC). From this figure, the dynamics of supported QoS management functions can be observed. At a given NGB traffic volume and available bit rate, the curve of the bit rates (SC) allocated by the PS closely follows that of the power budget (PB). Besides this,

Table 7.6 Most important system-based parameters [10].

Parameter	Value
Call/session mean arrival rate	0.5 s
Radio resource indication period (RRI)	0.2 s
Simulation time (s)	7200 s
Power target for downlink AC	3 dB below BTS total power
Overload offset for downlink AC	1 dB above power target
Orthogonality (α)	0.5
Period for load control actions	0.2 s (1 RRI)
Period for packet scheduling	0.2 s (1 RRI)
E_b/N_0 requirements	
Speech	7 dB
SWIS	6 dB
Streaming	6 dB
PoC	7 dB
MMS/WAP	5/5.5 dB
Dialup	5.5 dB
Maximum BTS Tx power	43 dBm
P-CPICH Tx power	33 dBm
Sum of all other CCH Tx powers	30 dBm
Length of AC queue	10 Radio bearers
Dedicated NGB capacity	0 dB, i.e. not used
Power weight for inactive NGB traffic (k)	0.5

when the total load $P_{TxTotal}$ in the cell (i.e., $P_{GB} + P_{NGB}$) reaches the overload threshold, denoted by $P_{TxTotal} + Offset$, NGB-allocated bit rates are accordingly reduced by LC and in turn immediately resumed when the P_{GB} decreases and the system returns to its normal state of operation. Ultimately, the measured GB load (P_{GB}) hardly ever exceeds the target threshold ($P_{TxTarget}$) and, following the input traffic mix $P_{TxTotal}$, thanks to the AC and PC functions always preserves its point of equilibrium ($P_{TxTarget}$).

Let us now examine the performance experienced by users of the deployed services. As shown in Table 7.7, in terms of call block ratio (CBR), none of the calls/sessions were rejected due to buffer overflow and almost none of the speech, video and VS calls were blocked due to the time spent in the AC queue. This is due to the fact that the offered traffic hardly ever exploited available cell capacity. As a result, almost all users of GB services were satisfied. The favourable effects of prioritisation between GB services can be seen in Figure 7.16(a), where the percentage of satisfied users for each of the deployed services is illustrated as a function of simulated cells. In Cells 10 and 11, the quality experienced by speech users, in terms of CBR, is better than the accessibility offered to VS and video users. In the figure, more relevant is the differentiated treatment of NGB traffic, which reflects the provisioned discrimination between the real time (RT) and non-real time (NRT) services of interactive and background classes. In fact, in each of the simulated cells, the percentage of satisfied users of Dialup is the lowest, followed by WAP and MMS; and for RT services, PoC performance is always better than streaming

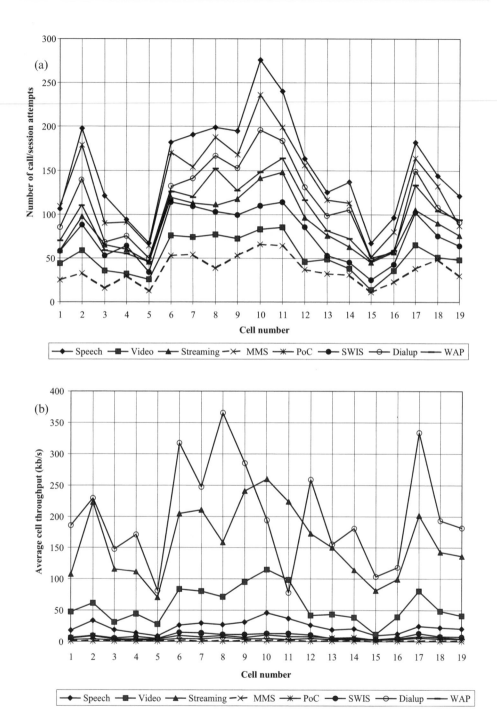

Figure 7.14 Traffic distribution over the 19 cells: (a) offered load in call arrivals; (b) average cell throughputs (service-based values in kb/s for each of the simulated cells) [10].

Table 7.7   System-based measurement results [10].

Service type	CBR (%)	CDR (%)	CRRR (%)	RBR (%)	DBR (%)	Median AST (kb/s)	Median object size (kB)	Calculated object delay (s)	SU (%)
Speech	0.05	0.00	—	—	—	12.2	—	—	99.95
Video	0.16	0.00	—	—	—	64.0	—	—	99.84
Streaming	0.00	0.00	1.93	6.35	0.05	63.4	1682	212.2	91.67
MMS	0.00	0.00	2.29	—	—	70.5	15	1.7	97.55
PoC	0.00	0.03	1.08	2.49	0.19	8.0	4	4.0	96.31
SWIS	0.32	0.00	—	0.00	0.00	64.0	89	11.1	99.68
Dialup	0.00	0.00	8.44	—	—	51.4	120	18.7	59.94
WAP	0.00	0.05	2.57	—	—	66.0	48	5.8	94.24

Note: RBR = Re-Buffering Ratio, DBR = Delay Buffering Ratio; SU = Satisfied Users.
Reproduced with kind permission of Springer Science and Business Media.

performance. Analogous results for NGB services are depicted in Figure 7.17, where the position of dissatisfied users is marked by differently tinted asterisks depending on the service used. This performance overview enables the operator to capture the benefit provided by service differentiation and its limitations in high-traffic scenarios. In

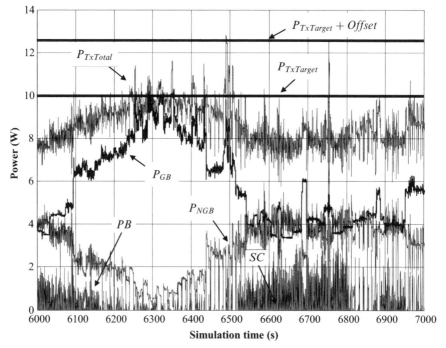

Figure 7.15   Cell 11: Snapshot of the simulation period [10].
Reproduced with kind permission of Springer Science and Business Media.

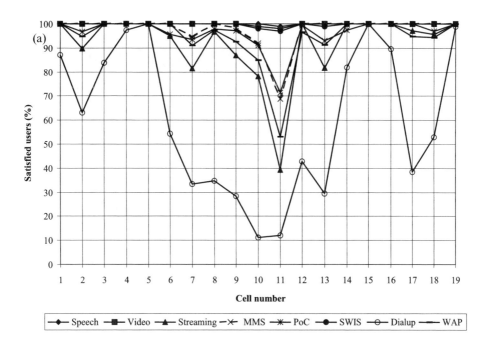

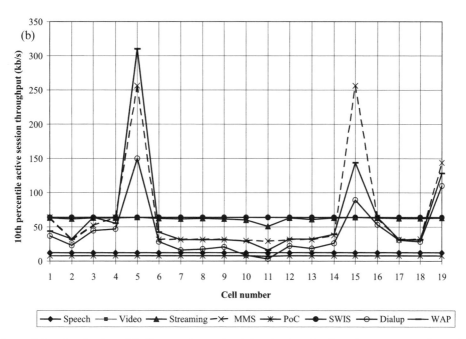

Figure 7.16 Service-based indicators for each of the simulated cells: (a) percentage of satisfied users; (b) 10th percentile of average active session throughput during the simulated time [10].

Reproduced with kind permission of Springer Science and Business Media.

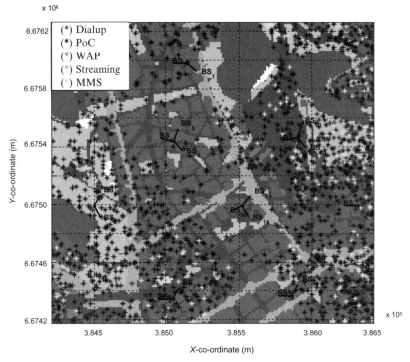

Figure 7.17 Dissatisfied users of Dialup, PoC, WAP, streaming and MMS services [10].
Reproduced with kind permission of Springer Science and Business Media.

addition, the most critical areas in the network where more capacity is needed may be identified. In this case, NGB QoE is definitely unacceptable in Cell 11; can be tolerable in Cells 7, 10 and 13; and ends up more or less satisfactory in all other cells. A more detailed analysis of the reason users of NGB services were not satisfied is possible based on the raw performance indicators illustrated in Figures 7.16(b), 7.18 and 7.19(a). In these figures, the intended differentiation between services in terms of the metrics characterising the QoE of each of the deployed services is also visible. In particular, Figures 7.16(b) and 7.18(a) show, respectively, the 10th percentile of the average active session throughput (AST) and the capacity request rejection ratio (CRRR) collected for each of the above services during the measurement period, as a function of deployed cells. The throughput experienced by Dialup users is lower than the corresponding one offered to WAP/MMS users who underwent the same treatment. Conversely, the accessibility offered to PoC and streaming services, while requesting capacity from the PS, is better than the corresponding blocking experienced by WAP/MMS and Dialup users. Figure 7.18(b) reveals how throughput deterioration adversely affects PoC and streaming performance in terms of rebuffering in the UE. As expected, the rebuffering ratio is higher for streaming, whereas dissatisfaction with the excessive time needed to refill the buffer, depicted in Figure 7.19(a), is higher for PoC. This is due to the fact that the tolerance for streaming users (up to 16 s) was higher than that for PoC users, who were not supposed to wait for more than 4 s (see Table 7.5). The differentiation between WAP/

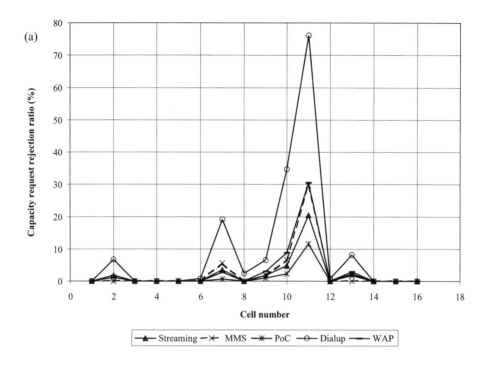

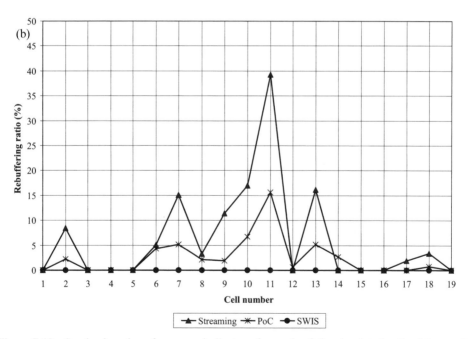

Figure 7.18 Service-based performance indicators for each of the simulated cells: (a) capacity request rejection ratio; (b) rebuffering ratio [10].

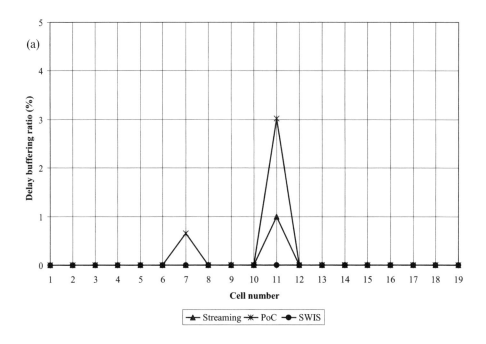

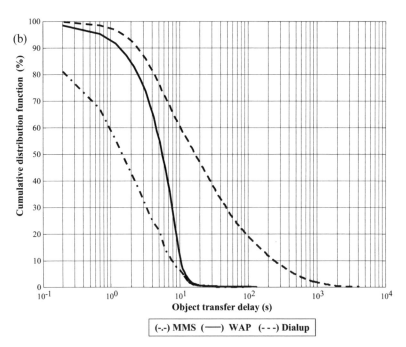

Figure 7.19 (a) Excessive time needed for the rebuffering ratio (service-based performance indicators for each of the simulated cells); (b) MMS, WAP and Dialup object transfer delays (system-based statistics over all simulated time) [10].

Reproduced with kind permission of Springer Science and Business Media.

MMS and Dialup connections and the benefit thereof is also shown in Figure 7.19(b) and Table 7.7, where the transfer delays (system-based statistics upon all simulated time) of the WAP, MMS and Dialup objects during the simulation period are presented.

7.2 High-speed downlink packet access (HSDPA) dimensioning

The basic functionality and QoS differentiation capability of HSDPA were described in Chapters 4 and 5. HSDPA is a 3GPP R5 concept introduced to support new multimedia services requiring improved QoS and high peak data rates in the downlink. These services will account for the biggest source of traffic load in 3G networks in the years to come, resulting in new revenue opportunities for the operator. But nothing comes for free. HSDPA has to be deployed and, thus, dimensioning methods for this new radio technology are of great importance.

In the following sections, we present the downlink radio dimensioning issue that arises from the introduction of HSDPA in an existing operating WCDMA network, which reflects the most probable deployment scenario [18]. This analytical approach allows the network administrator to derive maximum HSDPA throughput, or HSDPA power if throughput is provided as an input, as a function of actual cell load (due to dedicated and common channels), admission control and congestion control thresholds. The required HSDPA throughput may be the average value over the cell area or the minimum value associated with a certain coverage probability. The results of this process can be used to find out whether resources already allocated for WCDMA are sufficient for a satisfactory HSDPA service, or whether it is necessary to add new carriers or sites. The effects on coverage may be evaluated using the proposed changes to the radio link power budget calculation.

7.2.1 Relevant radio resource management

In this section, we present the radio resource management (RRM) aspects relevant to the HSDPA dimensioning process. It is important to remember that in the following sections we assume there to be no QoS supported in HSDPA (e.g., HSDPA power allocation is static and streaming class is not supported).

7.2.1.1 Admission and load control

When HSDPA is not active in the cell, admission and load control actions are based on two parameters for the operator to set, namely:

- The power target, denoted by P_{Target}.
- The overload threshold, denoted as $P_{Target} + Offset$.

Guaranteed bit rate (GB) bearer services are not admitted if either the total transmission power in the cell exceeds the overload threshold, or the power increase due to the bearer service in question (if admitted), plus the actual non-controllable power (NC, common channel and GB traffic associated power) in the cell, goes beyond the target threshold. Non-guaranteed bit rate (NGB) traffic is not admitted only in the case of overload.

Two more thresholds are introduced for HSDPA traffic control: $P_{TargetHSDPA}$ and $P_{TargetHSDPA} + Offset_{HSDPA}$. The first HSDPA user is admitted only if the total transmission power in the cell, denoted as P_{Total}, is below the target threshold; that is:

$$P_{Total} \leq P_{TargetHSDPA} \tag{7.10}$$

As a consequence of HSDPA activation there is an increase in downlink interference, which in turn causes a power rise in the ongoing dedicated connections (DCHs). Hence, the total power at the transmitting end is given by:

$$P_{Total} = P_{CCH} + P_{DCH} + \Delta P_{DCH} + P_{HSDPA} \tag{7.11}$$

where P_{CCH} is the common channel (CCH) power and P_{DCH} is the transmission power needed for DCH traffic. HSDPA power (P_{HSDPA}), which is a parameter for the operator to set (i.e., it has a fixed power level), is used for high-speed physical downlink shared channel (HS-PDSCH) and shared control channel (HS-SCCH) transmission:

$$P_{HSDPA} = P_{HS\text{-}PDSCH} + P_{HS\text{-}SCCH} \tag{7.12}$$

The DCH power rise is estimated as follows:

$$\Delta P_{DCH} = P_{HSDPA} \cdot \frac{\eta_{DCH}}{1 - \eta_{DCH}} \tag{7.13}$$

The actual loading due to DCH traffic η_{DCH} is defined as [1]:

$$\eta_{DCH} = \frac{P_{DCH}}{P_{MaxCell}} \tag{7.14}$$

where $P_{MaxCell}$ is the maximum cell transmission power.

When there is at least one HSDPA user in the cell, overload control actions are started if the following condition is true:

$$P_{NonHSDPA} \geq P_{TargetHSDPA} + Offset_{HSDPA} \tag{7.15}$$

where $P_{NonHSDPA}$ is the transmitted carrier power of all codes not used for HS-PDSCH or HS-SCCH transmission. Overload control actions are primarily targeted at NGB traffic carried on dedicated channels (DCHs). If the target power level, defined by $P_{TargetHSDPA}$, cannot be reached, overload control actions are then targetted at HSDPA traffic.

7.2.1.2 Definition of SINR

For dedicated channels (DCHs), one of the important link budget parameters is the target E_b/N_0 (specified for the transport channel) that corresponds to a certain BLER for a given data rate (R, also specified for the transport channel). Now, the basic WCDMA E_b/N_0 equation for one user looks like [1]:

$$\frac{W}{R} \cdot \frac{C}{I} \geq \left(\frac{E_b}{N_0} \right)_{target} \tag{7.16}$$

where W is the WCDMA chip rate, C and I are, respectively, received carrier and interference power.

However, since HSDPA relies on adaptive modulation and coding every TTI (physical layer), we are forced to move to Layer 1 and use slightly different definitions for the HS-DSCH [2]:

- HS-DSCH SINR is the narrowband signal-to-interference-plus-noise ratio after despreading the HS-PDSCH. Hence, the SINR includes the HS-PDSCH processing gain (spreading factor) for the HS-PDSCH and the effect of using orthogonal codes.
- 'Instantaneous required HS-DSCH SINR' is the per-TTI required SINR on the HS-DSCH to obtain a certain BLER target for a given number of HS-PDSCH codes and a modulation and coding scheme.
- 'Average HS-DSCH SINR' is the HS-DSCH SINR experienced by a user averaged over fast fading.

Now, in the case of HSDPA, (7.16) will be written as [2]:

$$SF_{HS\text{-}PDSCH} \cdot \frac{C}{I} \geq SINR_{target} \qquad (7.17)$$

where $SINR_{target}$ and SF refer to the physical layer.

We can also derive the relationship between per-TTI SINR and per-TTI E_s/N_0 as:

$$\frac{E_s}{N_0} = \frac{SINR}{M} \qquad (7.18)$$

where M is the number of HS-PDSCH codes used during the TTI.

7.2.1.3 Power control headroom and HSDPA power

The power control headroom is defined for dedicated physical channels as the margin required in the downlink power budget for inner-loop power control to compensate for fast fading [2]. Note that in the following we assume that HSDPA transmission power is a configuration parameter.

Figure 7.20 shows how the maximum total transmission power in the cell can be divided between the different transport channels without and with HSDPA, respectively. In the former case, the power available for DCH transmission (denoted as 'variable power' in Figure 7.20) may reach $P_{Target} + Offset$, whereas in the latter case, with HSDPA active, it cannot exceed $P_{TargetHSDPA} + Offset_{HSDPA}$. Hence, when HSDPA is used, the power control headroom for dedicated traffic is reduced.

When there is just DCH traffic, the maximum downlink loading factor is defined as follows:

$$\eta_{DCH_{max}} = \frac{P_{Target} + Offset}{P_{MaxCell}} \qquad (7.19)$$

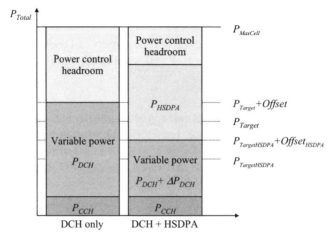

Figure 7.20 Power control headroom definition [18].

Reproduced by permission of © IEEE 2005.

If we want to have a consistent power control headroom setting, without and with HSDPA, we can set:

$$\frac{PCheadroom_{DCH}}{VariablePower_{DCH}} = \frac{PCheadroom_{HSDPA}}{VariablePower_{HSDPA}} \tag{7.20}$$

Using (7.20) and the power budget of Figure 7.20, it is possible to derive the maximum downlink loading factor when HSDPA is active:

$$\eta_{HSDPA_{max}} = \frac{P_{TargetHSDPA} + Offset_{HSDPA} + P_{HSDPA}}{P_{MaxCell}}$$

$$= \eta_{DCH_{max}} + \frac{P_{HSDPA}}{P_{MaxCell} - P_{CCH}}(1 - \eta_{DCH_{max}}) \tag{7.21}$$

Equation (7.21) can be rearranged to calculate the HSDPA power corresponding to the maximum utilisation of the available resources:

$$P_{HSDPA} = \frac{\eta_{DCH_{max}} - \dfrac{P_{DCH} + P_{CCH}}{P_{MaxCell}}}{\dfrac{1}{P_{MaxCell} - P_{DCH}} - \dfrac{1 - \eta_{DCH_{max}}}{P_{MaxCell} - P_{CCH}}} \tag{7.22}$$

Now that we are able to derive HSDPA power, starting with the existing data on DCH traffic we want to know what the corresponding average HSDPA cell throughput will be.

7.2.2 HSDPA power vs. throughput

The relationship between HSDPA power and average cell throughput is disclosed in three steps. First, we define how actual throughput relates to the SINR at the UE (defined in Section 7.2.1.1). Then, we describe how the SINR can be expressed as a function of HSDPA transmission power and user location. Finally, we derive the cumulative dis-

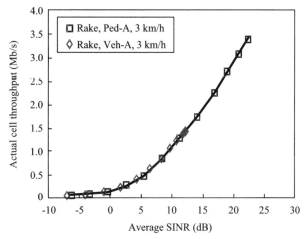

Figure 7.21   Actual cell throughput vs. average SINR [18].
Reproduced by permission of © IEEE 2005.

tribution function (CDF) of actual HSDPA throughput over the cell area and the corresponding mean value (i.e., average cell throughput).

7.2.2.1   Actual cell throughput vs. SINR

Actual cell throughput can be computed as a function of the average SINR, given the maximum number of HS-PDSCHs (see Section 4.3). Figure 7.21 shows the performance results attained for five HS-PDSCH codes using a link level simulator, where the HS-DSCH was 100% utilised. When more HSDPA users are active in the cell, throughput per connection depends on such factors as packet data transfer activity and the scheduling algorithm. In the case of round robin scheduling, the average cell throughput of Figure 7.21 is equally divided among active users.

7.2.2.2   Average SINR vs. HSDPA transmission power

Looking back to at the definition of SINR given by (7.17), it is easy to see that the SINR with respect to HSDPA transmission power can be expressed as follows:

$$SINR = SF_{HS\text{-}PDSCH} \cdot \frac{C}{I}$$

$$= SF_{HS\text{-}PDSCH} \cdot \frac{P_{HSDPA} - P_{HS\text{-}SCCH}}{P_{Total} \cdot \left(1 - \alpha + \frac{1}{G}\right)} \tag{7.23}$$

where $SF_{HS\text{-}PDSCH}$ is the spreading factor of the HS-PDSCH (equal to 16); C and I are, respectively, the received carrier power and interference level at the UE; P_{Total} is total averaged cell transmission power when the HSDPA is active, calculated using (7.11); α is the orthogonality factor; and G the geometry factor, defined as the ratio between the

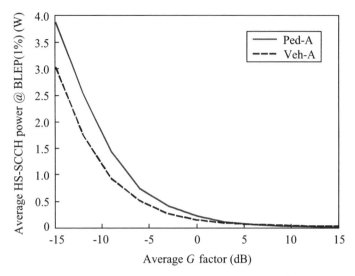

Figure 7.22 Average HS-SCCH power vs. average G factor [18].
Reproduced by permission of © IEEE 2005.

power received from the serving cell and the power received from the surrounding cells plus noise.

The power used to transmit the HS-SCCH $P_{HS\text{-}SCCH}$ depends on the G factor (distance from the base station). Average HS-SCCH power as a function of the geometry factor G, for ITU Pedestrian A and Vehicular A multipath channel profiles, with a block error probability (BLEP) of 1%, is depicted in Figure 7.22. From the figure we see that the average HS-SCCH power variation for a Vehicular A profile is less than 0.5 W within the typical range for G, which is from -5 dB to 20 dB, where -5 dB is at the cell edge.

7.2.2.3 G cumulative distribution function

User location (geometry factor G) is one of the key inputs to calculate HSDPA transmission power (P_{HSDPA}) using (7.23). The G distribution may be derived from real network measurements or simply from simulation results. The CDF adopted to derive the numerical results of Section 7.2.4, valid for Vehicular A and Pedestrian A multipath fading, is illustrated in Figure 7.23.

7.2.2.4 Actual throughput cumulative distribution function

The actual HSDPA throughput distribution (CDF) over the cell area for a certain HSDPA transmission power is computed as follows. For each value of G in Figure 7.23, using (7.23), we calculate the corresponding SINR. Then, the related values of actual cell throughput are derived from the SINR using a look-up table resulting from interpolation of the data shown in Figure 7.21. For instance, the CDF in Figure 7.24 was obtained assuming HSDPA transmission power (P_{HSDPA}) equal to 4 W, a Vehicular A

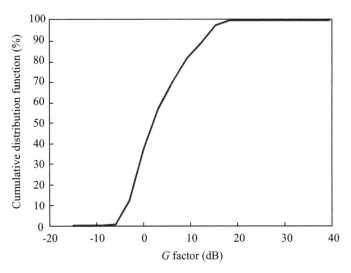

Figure 7.23   *G* factor CDF [18].

Reproduced by permission of © IEEE 2005.

profile, round robin scheduling and input power for dedicated and common channels ($P_{DCH} + P_{CCH}$) equal to 7.5 W. Changing any of the above settings leads to a different curve.

Average cell throughput (*AvTput*) is computed from the discrete distribution of actual cell throughput (*ActualTput*); that is:

$$AvTput = E[ActualTput] = \sum_{n=-\infty}^{+\infty} ActualTput(n) \cdot pdf(n) \tag{7.24}$$

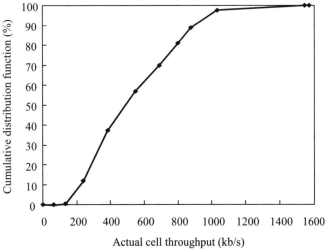

Figure 7.24   Actual cell throughput CDF [18].

Reproduced by permission of © IEEE 2005.

Using this approach, the probability that actual HSDPA throughput is higher than a pre-defined threshold equates to the probability that geometry factor G is higher than a certain value. Thus, given a required minimum HSDPA throughput and the probability of such throughput being higher than this value, we get the SINR value from the throughput, G from the probability and $P_{HS-SCCH}$ from G (Figure 7.22). The corresponding HSDPA power can be derived from (7.23) as:

$$P_{HSDPA} = \frac{\dfrac{SINR}{SF_{HS-PDSCH}} \cdot \left(1 - \alpha + \dfrac{1}{G}\right) \cdot (P_{DCH} + P_{CCH})}{1 - \dfrac{SINR}{SF_{HS-PDSCH}} \cdot \left(1 - \alpha + \dfrac{1}{G}\right) \cdot \dfrac{1}{1 - \eta_{DCH}}} \qquad (7.25)$$

7.2.3 Dimensioning assumptions, inputs and flows

The HSDPA dimensioning method described here relies on the following assumptions:

- HSDPA will be deployed in operating WCDMA networks, thus the goal of the dimensioning process is to find out whether there are enough resources to meet operator constraints or add more capacity.
- Only downlink radio interface dimensioning is of interest. In the uplink only the presence of HSDPA return channels is taken into account.
- Radio link budget: uplink-limited, this limiting factor is the highest bit rate of an HSDPA return channel due to the HS-DPCCH overhead (see Section 7.2.5).

The inputs are:

- Power needed to carry the DCH traffic (P_{DCH}).
- Common channel power (P_{CCH}).
- Target and overload thresholds (P_{Target}, Offset).
- G cumulative distribution function.
- Average HSDPA cell throughput or minimum HSDPA cell throughput and probability of being higher.

Dimensioning consists of the following steps:

- Based on DCH inputs and the HSDPA throughput requirement, we can calculate the HSDPA transmission power (P_{HSDPA}) and the actual loading (η_{HSDPA}), defined as the ratio between the total transmission power – given by (7.11) – and the maximum cell power ($P_{MaxCell}$).
- Using (7.21), we derive the new maximum downlink loading.
- If $\eta_{HSDPA} \leq \eta_{HSDPAmax}$, the process ends.
- If $\eta_{HSDPA} > \eta_{HSDPAmax}$, then the number of carrier frequencies must be increased.

There are two possible strategies to distribute traffic between two frequencies:

- Input traffic (DCH and HSDPA) is evenly distributed (users can be moved from one carrier to the other based on load).
- One frequency is dedicated mainly to HSDPA traffic, the other to DCH (users can be moved from one carrier to the other based on service).

If only one carrier is available, then the number of sites must be increased:

- Should the operator wish to limit the costs, we can calculate HSDPA power as power not used by the DCH and the corresponding average cell throughput.
- The impact on coverage can be evaluated using the changes in the link budget described in Section 7.2.5.

7.2.4   Numerical results

In this section, we present some use cases and related numerical results attained following the dimensioning process introduced in Section 7.2.3.

7.2.4.1   Available average HSDPA throughput

Non-HSDPA power ($P_{DCH} + P_{CCH}$) is provided as an input. The maximum downlink DCH loading factor is assumed to be 41 dBm. Using (7.22), we derive the maximum HSDPA power that can be used without additional resources (adding a frequency) and the corresponding average throughput. Results for non-HSDPA power in the range from 5 to 10 W (half the maximum transmission power in the cell) are depicted in Figure 7.25.

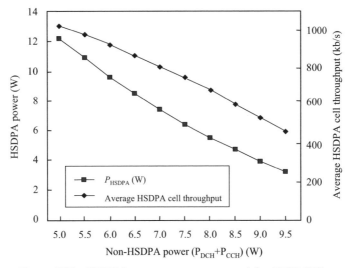

Figure 7.25   HSDPA power as power unused for DCH [18].

Reproduced by permission of © IEEE 2005.

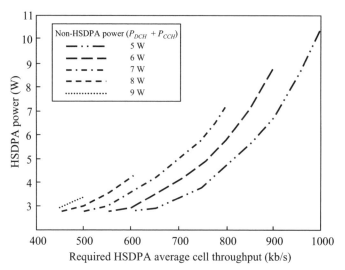

Figure 7.26   HSDPA power vs. required average cell throughput [18].
Reproduced by permission of © IEEE 2005.

The corresponding average HSDPA cell throughput varies from about 1 Mb/s to 400 kb/s. Under this assumption, in the worst traffic scenario, communication would still be satisfactory for more than ten users of the most common application services offered on best effort.

7.2.4.2   Required average HSDPA throughput

Figure 7.25 shows the upper limit in terms of HSDPA power and average throughput for different non-HSDPA transmission power values. Figure 7.26, instead, provides an estimate of average throughput in the cell if not all the available power is allocated to HSDPA – for example, if some margin on loading is taken.

7.2.4.3   Minimum throughput at cell edge

A performance requirement (constraint) for HSDPA may be minimum throughput at the cell edge. Figure 7.27 shows the HSDPA power needed to obtain certain minimum throughput values as a function of total non-HSDPA power ($P_{DCH} + P_{CCH}$). For a given value of non-HSDPA power and minimum throughput, with a 99.5% probability to be the cell edge condition, we get the corresponding required HSDPA transmission power. Then, using Figure 7.26, we can derive average cell throughput.

7.2.5   Impact on radio link budget

Very few things affect the radio link budget calculation presented in [2]. In the uplink, the overhead introduced by the HS-DPCCH, which carries feedback information for

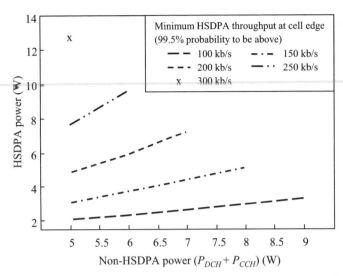

Figure 7.27 HSDPA power as a function of minimum HSDPA throughput at the cell edge and non-HSDPA power [18].

retransmissions and rate control, can be taken into account as an offset to be added to the target DCH E_b/N_0; that is:

$$\left(\frac{E_b}{N_0}\right)_{HSDPA} = \left(1 + \frac{\Delta_{ACK} \cdot r^2}{1 + r^2}\right) \cdot \left(\frac{E_b}{N_0}\right)_{DCH} \tag{7.26}$$

where r is the ratio between gain factor for the dedicated physical control channel (β_c) and gain factor for the dedicated physical data channel (β_d), and Δ_{ACK} denotes the power offset of the ACK/NACK part of the HS-DPCCH with respect to the associated DPCCH [2]. Δ_{ACK} depends on soft handover (SHO) conditions. Its recommended value is 6 dB for all bit rates of the uplink-associated DCH.

In the downlink, estimation of path loss is derived as in the case of DCH, taking into account that:

- The SHO gain is equal to 0.
- Planned E_b/N_0 is replaced by the SINR calculated at the cell border – that is, $G = -5$ dB.
- The processing gain is $10 * \log_{10}(SF_{HS\text{-}PDSCH}) = 12$ dB.

7.3 (E)GPRS dimensioning

(E)GPRS radio and transmission dimensioning needs a good understanding of service requirements in such terms as delay, bandwidth and network capability. This section introduces some important aspects that should be considered in that respect.

7.3.1   (E)GPRS dimensioning procedure for CS and PS traffic

This section proposes a practical procedure to calculate the radio capacity needed to support a determined amount of CS and PS traffic in (E)GPRS networks.

The network dimensioning procedure utilises several inputs:

- CS traffic to be supported during the busy hour (denoted as *ErlangsCS*).
- PS traffic to be supported during the busy hour (kb/s/cell).
- (E)GPRS layer and its characteristics: BCCH (frequency reuse), non-hopping (frequency reuse) or hopping layer (number of frequencies per BTS).
- Average MS time slot capability (maximum number of time slots an MS can support in the uplink and downlink).

Additionally, if a certain QoS needs to be guaranteed, the following performance indicators have to be considered as dimensioning targets:

- Average PS load supported (kb/s/cell).
- Maximum possible PS load supported (without CS load, kb/s/cell).
- Minimum guaranteed PS traffic (kb/s/cell).
- PS average throughput per MS (kb/s).
- Throughput (kb/s) for 90% of user connection time in poor radio link conditions (at the border of the cell).

The dimensioning procedure includes the steps described in the following sections.

7.3.1.1   Dimensioning for CS traffic

Step 1   CS traffic dimensioning
Assume that call arrivals follow a Poisson distribution and call durations are exponentially distributed. Hence, the Erlang B formula may be utilised for calculating load from the blocking probability (BP) and the number of channels (CS resources).

The number of time slots, denoted by N_{CS}, needed to support a CS load equal to *Erlangs CS* with defined blocking probability can be derived using the following formula:

$$BP_{Erlang\ B}(ErlangsCS, N_{CS}) = \frac{ErlangsCS^{N_{CS}}/N_{CS}!}{\sum\limits_{i=0}^{N_{CS}} ErlangsCS^{i}/i!} \qquad (7.27)$$

The BP for CS traffic is typically 2% (operator-selected value) for (E)GPRS radio interface dimensioning.

7.3.1.2   Dimensioning for PS traffic

Step 2   Calculate the Erlangs PS in first iteration
The PS load is the average data amount transmitted per cell per hour (kb/s/cell/hour). Erlangs due to PS traffic, denoted by *ErlangsPS*, is calculated as follows:

$$ErlangsPS = \frac{T_{PS}}{TSLC} \qquad (7.28)$$

Table 7.8 Typical values for time slot capacity.

Layer	GPRS (CS-1 to CS-2) (kb/s)	GPRS (CS-1 to CS-4) (kb/s)	EGPRS (MCS-1 to MCS-9) (kb/s)
BCCH	11	20	45
Non-hopping	11–10	20–14	40–20
Hopping	12–10	10–18	55–20

where T_{PS} is the average PS traffic (throughput in kb/s) to be supported during busy hours, and $TSLC$ is the average time slot capacity in kb/s, which depends on the EGPRS layer.

In the first iteration, $TSLC$ is set to a typical value (e.g., 11 kb/s for GPRS and 40 kb/s for EGPRS). Table 7.8 reports typical time slot capacity values for different configurations.

Step 3   Time slot capacity to support PS traffic

Since the $TSLC$ depends on the (E)GPRS layer where the PS services are deployed, a $TSLC$ should be calculated for each layer and a weighted value should be derived depending on the traffic share between layers. For instance, if there are six *ErlangsPS* in the BCCH layer (two time slots are dedicated for signalling) and four *ErlangsPS* in the hopping layer, the weighted $TSLC$ is given by:

$$\frac{60 \cdot TSLC_{BCCH} + 40 \cdot TSLC_{HOPPING}}{100} \tag{7.29}$$

In the TSLC estimate, the influence of CS traffic must then be taken into account. This is possible by defining an *Equivalent ErlangsPS* that includes this effect; that is:

$$Equivalent\ ErlangsPS = ErlangsPS_{PS} + ErlangsPS_{CS} \tag{7.30}$$

where $ErlangsPS_{PS}$ is calculated using (7.28) and corrected by existing PS features, as detailed below, and $ErlangsPS_{CS}$ are the additional erlangs due to the influence of CS traffic. In the following paragraphs, we explain how the latter quantity may be estimated for particular network configurations.

In the hopping layer, we know by experience (i.e., trial and simulation results) that 1 erlang of CS traffic (*ErlangsCS*) creates less interference than 1 erlang of PS traffic (*ErlangsPS*). Table 7.9 shows the CS load needed to generate the same interference as one *ErlangPS* for a drop call rate (DCR) of 2%, as a function of features activated. This translation is needed only when PS services are deployed in the hopping layer. If they are

Table 7.9 Equivalence between *ErlangsCS* and *ErlangsPS*.

Features activated for CS	PS load (*ErlangsPS*)	Translation factor (*ErlangsCS*)
—	1	1
DLPC	1	1.6
DLPC and DTX	1	2

deployed in the non-hopping layer, CS and PS loads are added to calculate an equivalent number of *ErlangsPS*.

The *Equivalent ErlangsPS* is the PS load that generates the same interference as the CS plus PS loads; that is:

$$Equivalent\ ErlangsPS = ErlangsPS + \frac{ErlangsCS}{TF_{CS}} \qquad (7.31)$$

where TF_{CS} is the translation factor due to the CS features (DTX and/or PC) reported in Table 7.9, *ErlangsPS* is calculated using (7.28) and *ErlangsCS* is derived from (7.27).

This value needs to be recalculated if downlink power control for (E)GPRS is applied and a PS is deployed in the hopping layer. In fact, if (E)GPRS is used with full power (i.e., no downlink PC for a PS is applied), then there is a back-off of 2 dB in power due to the non-linearity of the amplifiers. This is equivalent to having downlink power control for PS activated with a power offset from the maximum output level of 2 dB, assuming 8-PSK modulation coding schemes are used most of the time. When downlink PC is implemented, the PS load generates less interference with PS and CS traffic. This effect is modelled with a new *Equivalent Erlangs PS* value reduced by back-off gain. The value of back-off gain as a function of PS load share is shown in Figure 7.28.

If PS traffic is allocated in the hopping layer, independently of whether downlink PC is applied or not, in order to estimate the *TSLC* the effective frequency load (EFL) needs to be calculated first. The EFL is defined as 'PS load per frequency per TSL', and can be calculated as:

$$\frac{EFL = Equivalent\ ErlangsPS}{8 \cdot F} \qquad (7.32)$$

where F is the number of frequencies per cell (BTS) and 8 is the number of TSLs per TRX (frequency).

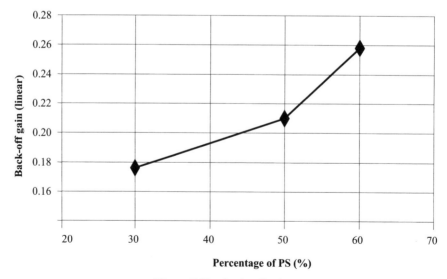

Figure 7.28   Back-off gain.

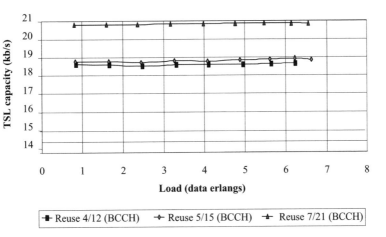

Figure 7.29   Time slot capacity for GPRS in the BCCH layer.

If PS traffic is handled in the non-hopping or BCCH layer, the *TSLC* is obtained using Figures 7.29–7.32 – that is, 'TSL capacity vs. data erlang' figures – which are valid for GPRS and EGPRS. In this case, the *Equivalent ErlangsPS* is used in the x-axis, and the *TSLC* can be read from the y-axis for a particular reuse factor.

If PS traffic is transmitted over a hopping layer, Figures 7.33 and 7.34 – that is, 'TSL capacity vs. EFL' figures – are used. These figures return the time slot capacity (in the y-axis) for a given EFL (in the x-axis).

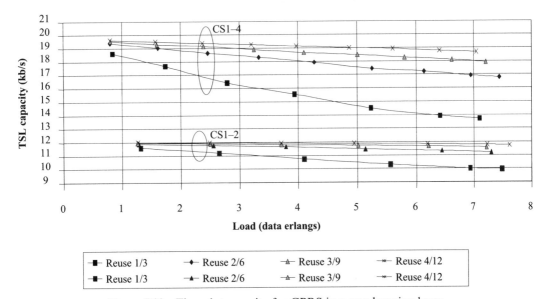

Figure 7.30   Time slot capacity for GPRS in a non-hopping layer.

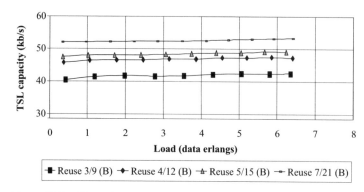

Figure 7.31 Time slot capacity for EGPRS in the BCCH (B) layer.

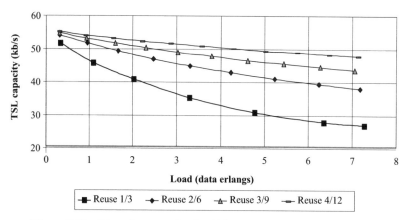

Figure 7.32 Time slot capacity for EGPRS in a non-hopping layer.

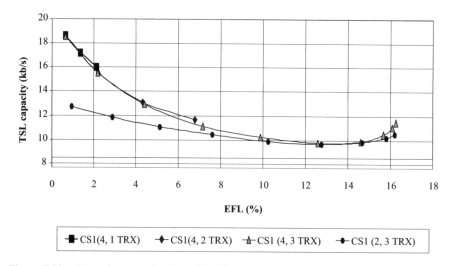

Figure 7.33 Time slot capacity for GPRS in a hopping layer (x frequencies, y TRXs).

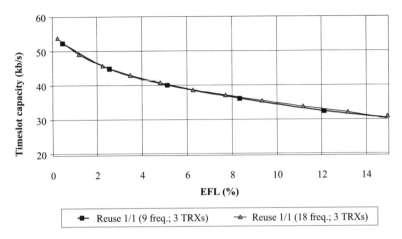

Figure 7.34 Time slot capacity for EGPRS in a hopping layer (x frequencies, y TRXs).

Step 4   Compute accurate values

Several iterations at Steps 2 and 3 are needed to obtain accurate values for *ErlangsPS* and *TSLC*. The procedure should be stopped when the differences between consecutive values are below a certain threshold (e.g., 5%). The typical number of iterations needed to obtain accurate results is 2 or 3. When deploying (E)GPRS in the BCCH TRX, *TSLC* does not depend on load, hence only the initial calculation is needed.

Step 5   Minimum number of TSLs needed for average PS load

The minimum number of time slots N_{PS} needed for PS services is given by:

$$N_{PS} = Roundup(ErlangsPS) \qquad (7.33)$$

where *ErlangsPS* is computed using (7.28) and the accuracy of *TSLC* is improved by the above-described procedure.

Step 6   Total number of TSLs needed for average CS and PS load

The total number of TSLs needed to support CS and PS load is given by:

$$TotalTSL = (N_{CS} + GuardTSL + N_{PS} - DedicatedTSL) + DedicatedTSL \qquad (7.34)$$

where N_{CS} is estimated using (7.27), N_{PS} is given by (7.33), *GuardTSL* is a parameter for the operator to set in order to maintain a safety guard between CS and PS territories that may depend on the number of TRXs (therefore, recalculation may be needed at the end of the dimensioning process) and *DedicatedTSL* is territory size exclusively for PS, which can be configured by the operator based on the minimum required average PS load or MS throughput, as explained in Step 9. The term ($N_{PS} - DedicatedTSL$) represents PS territory needs that are beyond dedicated resources.

The number of transceivers in the cell is thus the minimum to support *TotalTSL*.

Step 7   Calculate the reduction factor

The reduction factor (RF) is defined as the portion of the time slot used by a particular subscriber. The full *TSLC* may not be available for each subscriber due to the fact that PS resources are shared between users and CS traffic may pre-empt PS traffic. The RF ranges from 0 to 1, where '0' means that the user does not have any bit rate and '1' means that the user has the whole *TSLC* at his/her disposal. The RF is used to estimate user throughput in the next step. Several tables are available in the literature that report RF values as a function of used technology, PS usage and CS load in the cell.

Step 8   Evaluate the performance of the (E)GPRS network

This step calculates different quality indicators to estimate the performance that would be achieved in a network if it was dimensioned using the above procedures. These formulas will also be used in the following step to guarantee minimum system capacity and connection quality. In general, it is up to the operator to decide whether to guarantee these minimum values or to deploy PS services in two layers (BCCH and hopping/non hopping). In this case, a weighted *TSLC* should be used to derive the quality indicators, depending on PS load.

The maximum possible PS load (in kb/s), denoted as *MPSL*, is achieved on average when all the available time slots in the cell are effectively used by PS services – that is, there is no CS load and thus all TSLs are made available to PS traffic. This value can be calculated as:

$$MPSL = (N_{PS} + N_{CS}) \cdot TSLC \tag{7.35}$$

where N_{PS} is the number of TSLs available to PS traffic, calculated using (7.33), and N_{CS} is the number of TSLs calculated using (7.27) for an average CS load, and *TSLC* is the capacity of a TSL determined in Step 3. (*Note*: In (7.35), it is assumed that *GuardTSL* is kept free by the operator.)

The average supported PS load (in kb/s), denoted as *APSL*, is the capacity available to PS taffic when CS load is also the average load (initial requirement). In this case, the time slots available to PS traffic are given by:

$$APSL = N_{PS} \cdot TSLC \tag{7.36}$$

The minimum supported PS load (in kb/s), denoted as *mPSL*, is the load available when CS traffic utilises all the resources shared by CS and PS traffic. Therefore, only dedicated time slots for PS traffic are available in this situation:

$$mPSL = DedicatedTSL \cdot TSLC \tag{7.37}$$

Average user throughput (kb/s), denoted as *AUT*, is given by the following expression:

$$AUT = AverageMScapability \cdot RF \cdot TSLC \tag{7.38}$$

where *AverageMScapability* is the average number of time slots an MS can use to receive data, *RF* is the reduction factor calculated in Step 7 and *TSLC* is the time slot capacity obtained after Step 4.

Minimum user throughput (kb/s), denoted as *MUT*, is defined as throughput that is exceeded at least 90% of the time when the MS has poor radio link conditions (near the

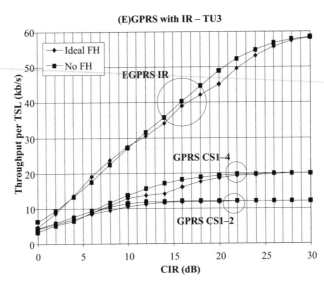

Figure 7.35 Throughput vs. CIR for EGPRS (all MCS) with incremental redundancy and with/ without frequency hopping (FH) and GPRS (CS1-2 and CS1-4).

border of a cell):

$$mUT = Average MScapability \cdot RF \cdot mTSLC \qquad (7.39)$$

Minimum TSL capacity, denoted as $mTSLC$, can be obtained using Figure 7.35 or Table 7.10 (or the $TSLC$ at the border of the cell if it is known by the operator). Figure 7.35 shows throughput per TSL for different CIR levels and for different technologies. Thus, once the CIR at a cell border is known, the $mTSLC$ can be read from the graphs. Table 7.10 summarises the typical values of throughput at the border of a cell, assuming that the radius of the cell is designed for a CIR equal to 9 dB.

7.3.2 (E)GPRS dimensioning with capacity and bit rate guarantees

Step 9   Capacity and bit rate guarantee
When a certain capacity and average throughput per user must be guaranteed, additional calculations are required:

1. *Guarantee average user throughput.* This is computed using (7.38). If the average user throughput to be guaranteed is higher than that obtained with this formula, it is

Table 7.10 Typical values for the minimum time slot capacity, $mTSLC$, at cell border with CIR = 9 dB.

Layer	GPRS (CS-1 to CS-2)	GPRS (CS-1 to CS-4)	EGPRS (MCS-1 to MCS-9)
Any	10 kb/s	12 kb/s	25 kb/s

necessary to increase the number of TRXs and perform a new iteration from Step 7 onwards until this requirement is fulfilled.
2. *Guarantee throughput for 90% of user connection time for a user in poor radio link conditions (at the border of a cell).* This throughput is calculated using (7.39). If the value obtained is lower than the target value, another TRX must be added and the procedure repeated from Step 7 onwards until the target value is reached.
3. *Guarantee minimum PS load (throughput).* This requirement can be fulfilled using a certain number of dedicated time slots for packet-switched services. *DedicatedTSL* can be derived using (7.28) or (7.33), depending on the level of accuracy, with the minimum desired load (throughput) and rounding up the attained *ErlangsPS* value to the minimum integer. If the number of dedicated time slots for PS services plus the number of time slots needed for CS services (with a blocking probability according to the dimensioning criterion) exceeds the number of time slots available for traffic with the current number of TRXs, the number of TRXs has to be accordingly increased.

Networks can be dimensioned to guarantee the support of minimum PS load and average MS throughput in a statistical manner. However, throughput and dedicated resources per user cannot be guaranteed without admission control and quality control functions.

7.3.3   (E)GPRS dimensioning with QoS guarantees

With the support of streaming class, the BSS must be capable of delivering a guaranteed bit rate to the application by ensuring that sufficient resources are allocated. Furthermore, CS traffic in the cell may no longer have priority over PS traffic. CS traffic pre-emption on PS territory may only be allowed if resource reallocation is possible by maintaining the QoS negotiated. If reallocation is not possible, downgrade may be denied and CS traffic may be blocked.

The dimensioning process requires initial understanding of the variety of services (web browsing, WAP, PoC, audio/video streaming, etc.) available to end-users and their requirements. Estimation of the actual service mix will determine the radio resources needed to support each service application. Another factor to consider is the average radio conditions present in the network, because that will determine radio bearer capability and performance.

7.3.3.1   Estimation of radio and dynamic Abis resources

AC estimates the necessary radio resources and dynamic Abis resources (i.e., transmission resources on Abis dedicated to support EGPRS bit rates) based on the guaranteed bit rate of the aggregate BSS QoS profile (ABQP), which can either be negotiated with the SGSN or given by BSC parameterisation (see Chapters 3 and 4, and [19]).

In order to estimate the radio resources needed to allocate an RT service with GBR or NRT service with a nominal bit rate (NBR), several things have to be considered. Resources will depend not only on the GBR or NBR of the service, but also on PDCH average throughput capacity (*TSLC*) – that is, CS/MCS usage – which further depends on radio conditions.

In order to cope with varying radio conditions, a certain margin, denoted the *ACmargin*, is included in calculation of the necessary number of radio TSLs; that is:

$$RadioCapacity = \left(1 + \frac{ACmargin}{100}\right) \cdot \frac{\sum BRforService}{TSLC} \qquad (7.40)$$

where *BRforService* is the required GBR for RT services or the NBR for NRT services (in kb/s), *TSLC* is average PDCH throughput for the average CIR in the cell or cluster, as introduced before – average CIR determines the CS/MCS employed on average and, therefore, expected average PDCH throughput (kb/s) – and *ACmargin* (%) is the margin to assure that any variation in radio conditions will not affect service performance.

In dynamic Abis, the needed resources for a bearer service (bit rate) are given by:

$$AbisCapacity = \frac{BRforService}{TSLC} \cdot NslaveTS \qquad (7.41)$$

where *NslaveTS* is the number of slave channels required when sending an RLC/MAC block with certain CS/MCS (slave channels/radio block). This can also be determined from the average CIR in a cell or cluster.

Dimensioning procedures should reserve enough PS capacity to guarantee a bit rate for RT services mapped onto streaming class, since such services have higher constraints in terms of required QoS. If any of the NRT services mapped onto interactive class require a target NBR, this needs to be derived using (7.40) and (7.41). Otherwise, these services will be assigned the remaining available capacity.

7.3.3.2 Review of the main factors for throughput guarantees

When dimensioning PS resources for RT and NRT services with certain guarantees, the following aspects need to be considered:

- *Guarantee bit rate/Nominal bit rate*. These are service-dependent. Real time services will require a guaranteed bit rate, while non-real time services will set a targetted nominal bit rate. Fulfilment of these requests will depend on network capacity.
- *Interference level* in the area (e.g., average CIR in the cell). The actual CIR sets the CS/MCS to use and therefore the throughput capability of the PDCH time slot. CIR estimation must exceed, say, 95% of the cell coverage area in order to support RT services over the cell area.
- *ACmargin* is a safety margin to guarantee the service even with varying radio conditions.
- Average *MS multislot capability* will determine the allocation needed for radio resources in order to guarantee the service requirements for certain types of MSs.
- *EGPRS/GPRS ratio* in the network also has an impact on dimensioning – that is, the percentage of EGPRS-capable MSs over GPRS-capable MSs – since their capabilities in the PDCH are different.
- *EGPRS/GPRS multiplexing* influences the PS capacity dimensioning process, because when GPRS and EGPRS territories are combined in the same PS resources, the rate reduction impacting EGPRS users must be considered in PS services.

7.3.3.3 Steps for (E)GPRS dimensioning with QoS

The formal steps for dimensioning with QoS may be as follows:

1. Estimation of service types and traffic mix.
2. Service requirements: throughput, delay, maximum BLER and the QoS attributes generally associated with each service.
3. Available radio technology (e.g., GPRS and EGPRS) and multiplexing effects. If GPRS and EGPRS PS territories are combined, performance restrictions should be taken into account.
4. Average MS multislot capability and type of services available and their penetration in the network.
5. Radio and dynamic Abis capacity estimation to support each service using formulas (7.40) and (7.41).
6. Amount of default or dedicated PS territory to enable the service.
7. Required CS capacity (note that blocking may increase due to PS services).

7.3.4 (E)GPRS dimensioning example

Assume we have the traffic mix of Table 7.11 during the busy hour (BH). In order to minimise the number of resources needed, services such as web browsing, WAP and MMS are treated as best-effort traffic. Capacity necessary for these services is calculated as the last step.

Assumptions on the user profile:

- Number of users in the cluster (denoted by U_{Total}): 1000.
- Percentage of active streaming users during the BH (denoted by $S_{ActiveBH}$): 10%.
- Percentage of active PoC users during the BH (denoted by $S_{ActiveBH}$): 20%.

Assumption on MS capability:

- Average EGPRS MS capability: $2 + 1$.
- Percentage of EGPRS-capable MSs: 60%.
- Average GPRS MS capability: $3 + 1$.
- Percentage of GPRS-capable MSs: 40%.

Table 7.11 Service mix defined and bit rate mapping.

Service type	RAB attributes	GBR/NBR
Video streaming	Streaming/ARP $= 2$	GBR $= 32$ kb/s
PoC	Interactive/THP $=$ ARP $= 1$	NBR $= 8$ kb/s
Web browsing	Interactive/THP $=$ ARP $= 3$	NBR $= 0$ kb/s
WAP	Background/ARP $= 3$	NBR $= 0$ kb/s
MMS	Background/ARP $= 3$	NBR $= 0$ kb/s
Others	Background/ARP $= 3$	NBR $= 0$ kb/s

Radio environment assumptions:

- The average CIR in the cluster provides an average throughput of 25 kb/s.
- *ACmargin* is 10%.

Let us now calculate the radio capacity needed per service during the BH.

Video streaming

The GBR is assumed to be 32 kb/s. Therefore, the number of radio TSLs per MS to guarantee the service are:

$$RadioCapacity_{EGPRS} = (1 + 0.1) \cdot \frac{32}{25(MCS\text{-}6)} \approx 1.4 \, TSL/MS \qquad (7.42)$$

if the service is requested by an EDGE MS; and:

$$RadioCapacity_{GPRS} = (1 + 0.1) \cdot \frac{32}{12(CS\text{-}2)} \approx 2.7 \, TSL/MS \qquad (7.43)$$

if the service is requested by a GPRS MS.

If we assume 60% of the terminals are EDGE and 40% are GPRS, the number of TSLs to support this traffic in the cluster, denoted as $TSL_{Service}$, can be calculated using the following formula:

$$TSL_{Service} = U_{Total} \cdot S_{ActiveBH} \cdot \sum_{T=EGPRS, GPRS} (RadioCapacity_T \cdot ShareMS_T) \qquad (7.44)$$

where U_{Total} is the total number of users, $S_{ActiveBH}$ is the number of active services during the busy hour and $ShareMS$ is the share of mobile terminals using that particular technology. Equation (7.44) in this case yields:

$$1000 \cdot 0.1 \cdot \left(\frac{1.4 \, TSL}{MS \cdot MS_{EDGEcapable}} + \frac{2.7 \, TSL}{MS \cdot MS_{GPRScapable}} \right) = 410 \, TSL \text{ per cluster} \quad (7.45)$$

When EGPRS and GPRS traffic is multiplexed on the same TSL, the rate reduction factor that affects each user due to PDCH sharing and modulation restrictions must be considered in (7.40) – for example, how multiplexing affects the TSLC must be modelled.

Push to talk over cellular (PoC)

The nominal bit rate requirement is assumed to be 8 kb/s. Therefore, the required number of radio TSLs per MS to guarantee the service during the BH are:

$$RadioCapacity_{EGPRS} = (1 + 0.1) \cdot \frac{8}{25(MCS\text{-}6)} \approx 0.4 \, TSL/MS \qquad (7.46)$$

if the service is supported by an EDGE MS; and:

$$RadioCapacity_{GPRS} = (1 + 0.1) * \frac{8}{12(CS\text{-}2)} \approx 0.7 \, TSL/MS \qquad (7.47)$$

if the service is supported by a GPRS MS.

If we assume 60% of the terminals are EDGE and 40% are GPRS, the number of TSLs to support this traffic in the cluster would be:

$$1000 \cdot 0.2 \cdot \left(\frac{0.4\,TSL}{MS \cdot MS_{EDGEcapable}} + \frac{0.7\,TSL}{MS \cdot MS_{GPRScapable}} \right) = 104\,TSL \text{ per cluster} \quad (7.48)$$

If we assume that users are evenly distributed in the cluster, we can dimension the PS territory homogeneously among the cells. Note that RT services may be very demanding resource-wise and a successful deployment strategy may require dedicating more PS resources to every cell. The number of resources will be defined by the dimensioning process as well as network configuration details.

Traffic with no bit rate guarantees
Service such as email, WAP, web browsing and MMS share the available resources with no guarantees. Network statistics on number of events per user per month are used for dimensioning purposes. Users will share the residual capacity if they are mapped onto interactive/background class with no nominal bit rate target.

Dimensioning for services with no guarantees is based on the following approach. Let *UserBHdata* be the amount of data delivered to a particular user during the BH. The *UserBHdata* is given by:

$$UserBHdata = \frac{MonthlyUserData}{30} \cdot BHusage \quad (7.49)$$

where *MonthlyUserData* is the kB – or number of events, if the average size of each event is provided, see (7.50) – that a user generates during a month on average, *BHusage* is the percentage of total daily traffic that is generated during the BH (in %) and 30 the average number of days in a month. As an example, if we consider that 80% of daily traffic occurs 20% of the time, as a BH definition (each operator can make its own assumptions), then *BHusage* is approximately 17%.

Bit rate requirements during the BH can be calculated based on the average size of the WAP page, webpage download, MMS size, etc. An example of total PS load during the BH, as well as the minimum throughput to support such a load, due to the assumed traffic mix is reported in Table 7.12.

Required total throughput per user during the BH is 0.78 kb/s. Assuming that subscribers are evenly distributed in the cluster it is possible to estimate the needed capacity to support the generated load for all users in the cell at a total rate of 349 kB during the BH.

For all users in the cluster, the *TotalDataVolume* of traffic generated during the BH is:

$$TotalDataVolume = U_{Total} \cdot \sum_{S=Services} (ServicePenetration_s * UserBHdata_s * Size_s) \quad (7.50)$$

where *Services* includes all available services (e.g., WAP, web browsing, etc.), *ServicePenetration* is the percentage of active users for service *S*, *UserBHData* is the calculated value using (7.49) per service *S*, *Size* is the size (in kB) of the events of each service *S* and U_{Total} is the total number of users in the area.

As an example, Table 7.13 summarises *UserBHData* and *TotalDataVolume* during the BH for several services.

Table 7.12   User bit rate and total data generated during the BH.

Services	Monthly user data (number of events per user per month)	Busy hour (BH) usage (%)	User BH data (events per user in BH)	User BH data rate (kb/s)	Total data per user per BH (kB)	Size (kB)	Units
MMS	50	17	0.28	0.04	17	60	MMS
Email	100	17	0.57	0.03	11.33	20	Email
Interactive WWW	62.5	17	0.35	0.09	42.5	120	Webpage download
Media services download	50	17	0.28	0.06	28.33	100	Media file download
WAP	750	20	5	0.56	250	50	WAP pages
Total				0.78	349.17		

Table 7.13   Total data bit rate and volume for all users.

Services	Monthly user data (number of events per user per month)	Service penetration (%)	User BH data (events per user in BH)	Total data volume in BH of all users (MB)	Bit rate of all users BH (kb/s)	Size (kB)	Units
MMS	50	80	0.28	13.6	30.22	60	MMS
Email	100	50	0.57	5.67	12.59	20	Email
Interactive WWW	62.5	50	0.35	21.25	47.22	120	Webpage download
Media services download	50	50	0.28	14.17	31.48	100	Media file download
WAP	750	70	5	175	388.89	50	WAP pages
Total				229.68	510.41		

According to these calculations, the network must support a load of 229.7 MB and a throughput of 510 kb/s.

Support for these services can be satisfied with no dedicated PS resources in the cell provided CS load allows usage of shared time slots. In other words, this is when CS traffic has enough capacity in the cell and does not interfere with PS services. This may be ensured through the dimensioning procedure presented in Section 7.3.1.2.

References

[1] J. Laiho, A. Wacker and T. Novosad (eds), *Radio Network Planning and Optimisation for UMTS*, 2nd Edition, John Wiley & Sons, 2006, 630 pp.

[2] H. Holma and A. Toskala (EDS), *WCDMA for UMTS*, 3rd edition, John Wiley & Sons, 2004, 450 pp.

[3] T. Halonen, J. Romero and J. Melero (eds), *GSM, GPRS and EDGE Performance*, 2nd Edition, John Wiley & Sons, 2003, 615 pp.

[4] A. R. Mishra, *Fundamentals of Cellular Network Planning and Optimisation: 2G/2.5G/3G ... Evolution to 4G*, John Wiley & Sons, 2004, 286 pp.

[5] D. Soldani and R. Cuny, On the deployment of multimedia services in wireless networks: Radio dimensioning aspects for UTRAN FDD, *IEEE, 1st International Conference on Quality of Service in Heterogeneous Wired/Wireless Networks, QShine 2004, Dallas, TX*, pp. 189–196.

[6] D. Soldani and J. Laiho, A virtual time simulator for studying QoS management functions in UTRAN, *Proceedings of Vehicular Technology Conference, Fall, IEEE, 2003*, Vol. 5, pp. 3453–3457.

[7] 3GPP, TS 23.107, QoS Concept and Architecture.

[8] D. Soldani, A. Wacker and K. Sipilä, An enhanced virtual time simulator for studying QoS provisioning of multimedia services in UTRAN, *Management of Multimedia Networks and Services* (Lecture Notes in Computer Science 3271), Springer-Verlag, pp. 241–254.

[9] ETSI, TR 101 112 (UMTS 30.03), Selection Procedures for the Choice of Radio Transmission Technologies of the UMTS, v. 3.2.0.

[10] D. Soldani, QoS management in UMTS terrestrial radio access FDD networks, dissertation for the degree of Doctor of Philosophy, Helsinki University of Technology, October 2005, 235 pp. See *http://lib.tkk.fi/Diss/2005/isbn9512278340/*

[11] N. Shankaranarayanan, Z. Jiang and P. Mishra, User-perceived performance of web-browsing and interactive data in HFC cable access networks, *Proceedings of ICC, IEEE, 2001*, Vol. 4, pp. 1264–1268.

[12] C. L. Klemm and M. Lohmann, Traffic Models for Characterization of UMTS Networks, *Proceedings of GLOBECOM, IEEE, 2001*, Vol. 3, pp. 1741–1746.

[13] COST 231, TD(91)73, Urban Transmission Loss Models for Mobile Radio in the 900 and 1800 MHz Bands.

[14] ITU-R, M. 1225, Guidelines for Evaluation of Radio Transmission Technologies for IMT-2000, 1997.

[15] S. V. Hanly, Information capacity of radio networks, Ph.D. Dissertation, Kings College, University of Cambridge, 1993, 225 pp.

[16] 3GPP, TS 25.308, UTRA High Speed Downlink Packet Access (HSDPA); Overall Description; Stage 2, Release 5.

[17] T. Kolding, K. Pedersen, J. Wigard, F. Frederiksen and P. Mogensen, High speed downlink packet access: WCDMA evolution, *IEEE Vehicular Technology Society (VTS) News*, No. 1, 2003, pp. 4–10, Vol. 50.

[18] P. Zanier and D. Soldani, *A Simple Approach to HSDPA Dimensioning*, IEEE, Personal Indoor Mobile and Mobile Radio Communications, Berlin, 2005.

[19] 3GPP, R4, TS 23.060, General Packet Radio Service (GPRS); Service Description; Stage 2, v. 4.7.0.

8

QoS Provisioning

David Soldani, Man Li and Jaana Laiho

QoS provisioning is a process that deploys QoS in networks and mobile terminals. The process translates planning results into mechanisms and parameters understandable by network elements and mobile terminals and further configures them on the equipment or devices. It is of course not possible to explain all parameters affecting QoS, which need to be provisioned in all types of network elements and mobile terminals. Some were presented in Chapters 4–6. Instead, in this chapter we discuss the basic QoS provisioning concepts and then provide examples to further illustrate the basic ideas.

QoS provisioning can be classified in three categories:

- radio, core and transport (IP and data link layer, DLR) QoS provisioning, which configures QoS mechanisms inside the network elements
- service QoS provisioning that maps services onto bearer QoS attributes; and
- terminal QoS provisioning of application-specific QoS information to mobile terminals.

The following sections discuss each category in detail. Furthermore, from the network and service management viewpoint, service and QoS provisioning involve the network element layer, network management layer and service management layer, as introduced by the TeleManagement Forum (TMF) [1]. Business management layer decisions reflect the actual content for provisioning lower layers.

8.1 Hierarchy in QoS management

The telecommunications management network (TMN) model provides an insight into how the business of a service provider or an operator is managed. As already introduced, the TMN model consists of four layers – namely, business management at the top, then service management and network management, and finally element management at the

QoS and QoE Management in UMTS Cellular Systems
Edited by David Soldani, Man Li and Renaud Cuny © 2006 John Wiley & Sons, Ltd

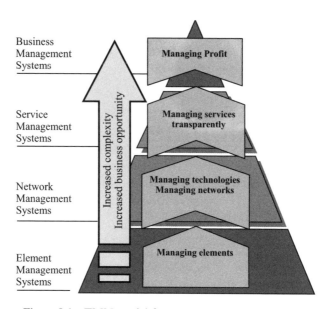

Figure 8.1 TMN model for management hierarchy [1].

bottom. Management decisions at each layer are different though related to each other. Working from the top down, each layer imposes requirements on the layer below. The TMN of the TMF sets the guidelines for the optimisation of functionalities and processes. In 3GPP, the same model was adopted (see [1], [2]). The scope of TMF is to:

- find a standardised way to define service quality;
- set requirements for networks in terms of QoS measurements; and
- make it possible to have QoS reports between providers and systems that implement the service.

The TMN model is depicted in Figure 8.1. Information flow from the business management layer all the way down to the service management and network management layers is essential since business aspects have to be considered carefully in the optimisation and network development process. The efficiency of the business plan can be measured in terms of capital and operational expenditure (CAPEX, OPEX) and revenue. The desired business scenario is then translated into offered services, service priorities and service QoS requirements. On the lowest (network element) level of the TMN model, business-related issues are converted into configuration parameter settings.

The following functions are supported by business management systems (BMSs):

- Creation of an investment plan.
- Definition of the main QoS criteria for the proposed network and its services.
- Creation of a technical development path (expansion plan) to ensure that the anticipated growth in subscriber numbers is provided for.

The functions supported by service management systems (SMSs) include:

- Management of subscriber data.
- Provisioning of services and subscribers.
- Accounting and billing operations for services offered.
- Creation, promotion and monitoring of services.

The following functions are supported by network management systems (NMSs):

- Planning the network.
- Collecting information from underlying networks.
- Pre-/post-processing of raw data.
- Analysis and distribution of information.
- Optimisation of network capacity and quality.

Element management systems (EMSs) can be considered as part of network element functionality that is responsible for:

- Monitoring the functioning of equipment.
- Collecting raw data (performance indicators).
- Providing a local GUI for site engineers.
- Mediating with the NMS.

How these management layers are in practice mapped onto overall, end-to-end QoS functions is depicted in Figure 8.2. Different layers and details on the above functions are discussed in Section 8.2.

The task of managing QoS and services is complex and thus proper mechanisms to ensure consistency and operational efficiency are needed. The high-level requirements for telecommunication management are introduced in [2]. Management system interactions are shown in Figure 8.3. In particular:

- Enterprise systems (ESs) are those information systems (ISs) that are used in tele-communication organisation but are not directly or essentially related to telecommunications aspects (call centres, fraud detection and prevention systems, invoicing, etc.). Enterprise systems are not standardised, since they involve many operator choices (organisational, etc.) and even regulatory. Also, enterprise systems are often viewed as a competitive tool. However, it is essential that the requirements of such systems are taken into account and interfaces to operationssystems (OSs) are defined, to allow for easy interconnection and functional support.
- A network management system (NMS) is a service that employs a variety of tools, applications and devices to assist human network managers in monitoring and maintaining operational networks. It can also be defined as the execution of the set of functions required for controlling, planning, allocating, deploying, co-ordinating and monitoring the resources of a telecommunications network. The processed defined in the enhanced telecom operations map (eTOM) is supported by network management systems [1], [2] (eTOM is an initiative by the TMF to deliver a business process model

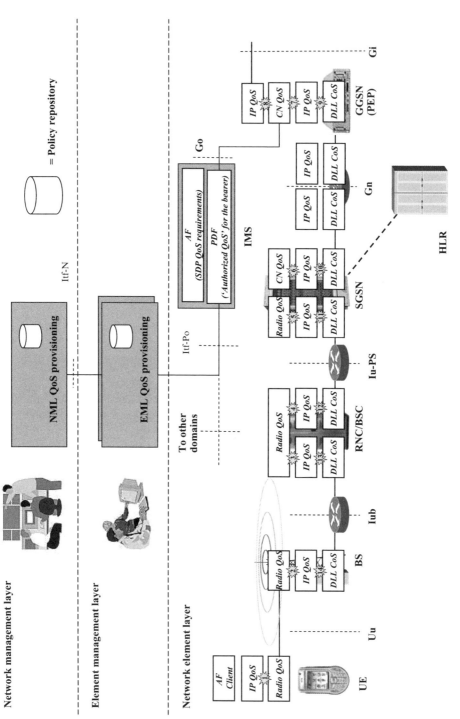

Figure 8.2 QoS mechanisms and management in different layers and network domains.

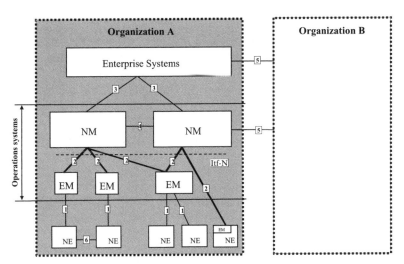

Figure 8.3 Management system interactions according to [2]. In this context the TMF management hierarchy is in the NM layer [1].

Possible interfaces: (1) between NEs and the element manager (EM) of a single PLMN organisation; (2) between the EM and NM of a single PLMN organisation; (3) between the NM and the ES of a single PLMN organisation; (4) between the NMSs of a single PLMN organisation; (5) between the ES and NM of different PLMN organisations; (6) between NEs.

or framework for use by service providers and others within the telecommunications industry; it describes all the enterprise processes required by a service provider and analyses them at different levels of detail according to their significance and priority for the business). The NMS contains the configuration and performance data repositories, and allows mass modifications in network element configurations. In the case of policy-based management, the NMS provides storage of all network policies for all domains, global policy distribution and conflict detection. Further, NMS is capable of accessing alarm information, contains the analysis and optimisation logic to operate the network, and provides support to provision and monitor mobile services. The logic to derive key performance indicators (KPIs) as described in Chapter 9 is located in the NMS.

- The element management (EM) system is a prerequisite for NMSs. EM is perceived as lower level management, targetted at specific network elements. It allows local and manual configuration changes. In the case of policy-based management, it provides storage of specific policies that apply to that domain, storage of general policies that apply across domains and local policy conflict detection. Also, EM is responsible for collecting and transferring performance measurements and generated alarms/events to NMS systems. More about QoS monitoring is in Chapter 9.

8.2   Radio, core and transport QoS provisioning

This section describes the network element functionalities that need to be configured and maintained with parameter settings and/or QoS policies.

The first objective of radio, core and transport (IP layer and DLL) QoS provisioning is to ensure that QoS differentiation mechanisms are provisioned at all network layers (shown in Figure 8.3). In 3GPP R5 and later release-compliant systems, the parameters and/or QoS policies of the following functions need to be configured and maintained. See Chapters 5 and 6 for a detailed description of the functions and parameters thereof.

Radio QoS functions
- RAB management (located in the UE, UTRAN and SGSN) – e.g., packet scheduler.
- Admission/Capability control (located in the UTRAN and SGSN).
- Classification, policing and shaping (located in the UE, UTRAN and SGSN).
- Translation (in the UE).
- Mapping (located in the UTRAN and SGSN).

CN QoS functions
- CN BS management (located in the SGSN and GGSN).
- Admission/Capability control (in the SGSN and GGSN).
- Classification (located in the SGSN and GGSN).
- Policing and shaping (in the SGSN and GGSN).
- Translation (in the GGSN).
- Mapping (in the SGSN and GGSN).

PDF function
- The most important policies to be configured in the IMS are:
  ○ allowed media types;
  ○ allowed codecs; and
  ○ maximum bandwidth on a per-TC basis.

IP QoS functions
- Classification.
- Queuing.
- Scheduling.

Data link layer (DLL) class of service (CoS)
- ATM QoS functions.
- MPLS QoS functions, etc.

Radio, core and transport QoS may have its own way of differentiating traffic using a different queuing structure, scheduling, access control, etc. For example, transport provisioning configures all the necessary parameters and mechanisms to make QoS differentiation function correctly at lower layers. In this section, we use core network QoS bearer provisioning as an example to illustrate the basic concepts.

The second objective of radio, core and transport QoS provisioning is to deploy QoS mappings between layers. One can view QoS provisioning as configuring 'bit pipes' that will treat traffic differently. It is hence not enough to only configure pipes at different layers. A proper mapping between the bit pipes at different layers must also be provisioned. In Section 8.2.2, we also provide examples of mapping of bit pipes between the different network layers (shown in Figure 8.3).

8.2.1 Core network bearer QoS provisioning

We focus on core networks that are based on IP technology. 3GPP 23.107 dictates that for an IP-based backbone, differentiated services as defined by IETF shall be used [3]. IP-based backbone QoS provisioning involves configuring DiffServ QoS policy on all network elements that support DiffServ QoS. DiffServ policy includes traffic filtering, metering, packet drop algorithms, scheduling and bandwidth for different QoS classes.

As there may be a large number of network elements that support DiffServ QoS, configuring QoS individually will be very time-consuming. Taking into account that QoS rules may be the same or very similar for groups of network elements, IETF has developed a framework for policy-based management [4].

The following subsections look at the different DiffServ policy components in detail.

Classifier
Typically, the first element of a DiffServ policy is a classifier that filters traffic into different streams. Each stream is then offered different treatment. Classifiers can be simple or complex. Combinations of the following parameters can be configured to set up classifiers:

- IP source address or prefix.
- IP destination address or prefix.
- IPv6 flow ID.
- Source protocol and port range.
- Destination protocol and port range.
- Differentiated service code point.

Any of the above parameters can also be a wild card meaning 'match all'.

Meter
A meter measures the rate at which the packets comprising a stream of traffic pass it and compares this rate with a set of thresholds. A given packet is said to 'conform' to the meter if, at the time the packet is being looked at, the stream appears to be within the meter's profile. There can be different types of meters that use different algorithms for measuring traffic. Token bucket is a typical way of metering. The parameters to be configured for a token bucket include average token rate and burst size.

Action
Actions are the things a differentiated services interface may do to a packet in transit. At minimum, such a policy might calculate statistics on traffic in various configured classes, mark it with a specific DSCP, drop it or queue it before passing it on for further processing. Actions may be cascaded to allow multiple actions to be applied to a single traffic stream.

Queuing

Queuing includes algorithmic droppers, queues and a scheduler that are inter-related in their use of queuing techniques. When a packet is destined to a specific queue, the queuing policy needs to determine whether the packet can be queued, depending on certain conditions (e.g., the average length of the queue). For example, if the RED queuing mechanism is as described in Section 6.2.4.1, min_{th}, max_{th}, max_p, and W_p must be provisioned.

Scheduling

Since there can be multiple queues in a single interface, the scheduler defines the algorithm to service these queues. The scheduler allows flexibility in constructing both simple and somewhat more complex queuing hierarchies from those queues. For a strict priority scheduler, the priority of each queue should be configured. For a weighted scheduling method – for example, weighted fair queuing or weighted round robin – the weight of each queue should be provisioned.

For all the DiffServ policy elements described above, each element also has a pointer 'next' that specifies what the next element should be in the path of packet treatment. An interface has ingress and egress directions and there can be different policies in each direction. As traffic enters an edge interface it can be classified, metered and marked. Traffic leaving the same interface might be re-marked according to the contract with the next network, queued to manage the bandwidth and so on. Functional data-path elements used on ingress and egress are of the same type, but may be structured in very different ways to implement relevant policies.

As an example, Figure 8.4 shows how elements are cascaded to form a typical EF edge policy configuration. The policy applies a single-rate, two-colour meter, dividing traffic into 'conforming' and 'excess' groups. The excess group is always dropped and the conforming group is served with high priority.

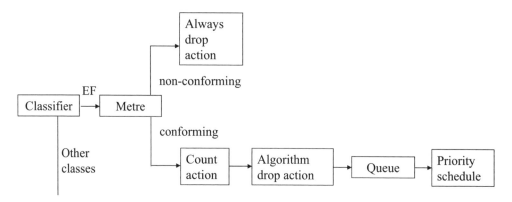

Figure 8.4   An example of DiffServ EF policy.

8.2.2   Provisioning QoS mapping in the network layer

As discussed at the beginning of this chapter, it is not enough to just provision QoS mechanisms at each network layer. Mappings between the layers must also be configured. The mapping functions that needs to be configured are in the network elements depicted in Figure 8.2 (Points 1–14 in the following list). In particular, the network administrator needs to configure and maintain all tables related to the following functions, which make it possible to implement end-to-end QoS as defined in Chapter 1:

- Point 1: a table that defines the mapping of service applications onto 3GPP QoS parameters (see Section 8.3.2).
- Points 2–7: a table related to the mapping of 3GPP QoS attributes onto IETF DiffServ, as described in Chapter 6.
- Point 8: a table describing the mapping of 3GPP QoS parameters onto Edge DiffServ QoS, as explained in Chapters 3 and 6.
- Points 9–14: a table reporting the mapping of IETF DiffServ onto a backbone class of service (CoS), as pointed out in Chapters 3 and 6.

The mapping between the 'authorised QoS' for the bearer and the UMTS QoS parameters (CN QoS) is specified in the standard (see Section 3.2.4.2).

8.3   Service and mobile terminal QoS provisioning

This section highlights the role and importance of the PDP context concept to manage end-to-end QoS differentiation and find simple ways to provision delay-critical service applications in GPRS, EGPRS and WCDMA networks.

8.3.1   Service QoS provisioning

Implementation of 3GPP QoS provides means to operators to change from 'best-effort' radio network planning, provisioning, optimisation and monitoring to a more service-oriented approach. This section briefly describes how a particular subset of QoS parameters (bearer service attributes) may be used by the network administrator to manage QoS differentiation when providing a wide range of value-added IP connectivity services to end-users.

The proposed methodology allows operators to logically divide their network into 'pipes' offering different service performance characteristics – for example, in terms of throughput. Individual service applications can then be mapped onto these different pipes according to corresponding performance requirements.

The subset of QoS attributes that allow an adequate number of priority 'pipes' is:

- UMTS traffic class (TC);
- traffic handling priority (THP);
- allocation/retention priority (ARP);
- maximum bit rate; and
- guaranteed bit rate.

As already mentioned, several functions at different layers affect QoS across the mobile network – each function having different capabilities for QoS control (see Figure 8.2). As the provided QoS must be in line with end-to-end bearer service characteristics, a management concept is needed to ensure consistent treatment and adequate quality for mobile services. This means consistently provisioning – that is, according to the defined priority 'pipes' – all QoS functions and mappings across all the layers shown in Figure 8.2. An example of service mapping onto distinct priority 'pipes' is depicted in Figure 8.5. Instead of one best-effort 'pipe', based on service requirements, ten priority pipes are offered using different combinations of the QoS attributes reported above.

Service QoS provisioning maps service applications onto these priority 'pipes'. As shown in Figure 8.5, within the same QoS class, further differentiation may be provided to diverse services using a different bit rate or ARP. For example, streaming video needs higher bit rate than streaming voice. Further, within the same service, different users may require different QoS depending on their service subscriptions. Using a different ARP, a VIP user may enjoy a higher bit rate and thus better video quality, for instance. Both service and user differentiation may be supported using distinct combinations of QoS attributes.

Mapping of services onto the above priority 'pipes' is based on service application characteristics and network administrator business strategy. Once mapping has evolved, its deployment depends on the inter-working of many network elements and mobile terminals. Proper configurations of them are hence necessary. As a result, there are many different provisioning approaches thereof. In the following, we report some examples.

8.3.1.1 Service provisioning in (E)GPRS and WCDMA networks

Almost all QoS differentiation related features and functionality in 3GPP standards is optional at both the terminal end and the network end:

- a UE and its individual application clients can either be a 'QoS-aware' or 'non-QoS-aware' terminal; and
- GPRS, EGPRS or WCDMA networks can be either 'best effort' or 'priority QoS' networks, or networks with 'priority QoS' and 'streaming/conversational class bit rate guarantees'.

In the standards, the QoS differentiation model is very tightly integrated to PDP contexts. In other words, QoS parameters are defined at the PDP context level and the mobile network thus treats all the traffic inside a single PDP context equally without any QoS differentiation between the applications inside a single PDP context (or 'priority pipe', as defined above, if a particular subset of QoS parameters is used to provision the bearer in question).

QoS differentiation for delay-critical applications like mobile terminal embedded streaming players and PoC/PTT clients in GPRS, EGPRS and WCDMA networks can be provided:

- by placing applications with delay-critical QoS requirements behind different APNs and setting the different QoS parameters per subscriber and APN from the HLR; or

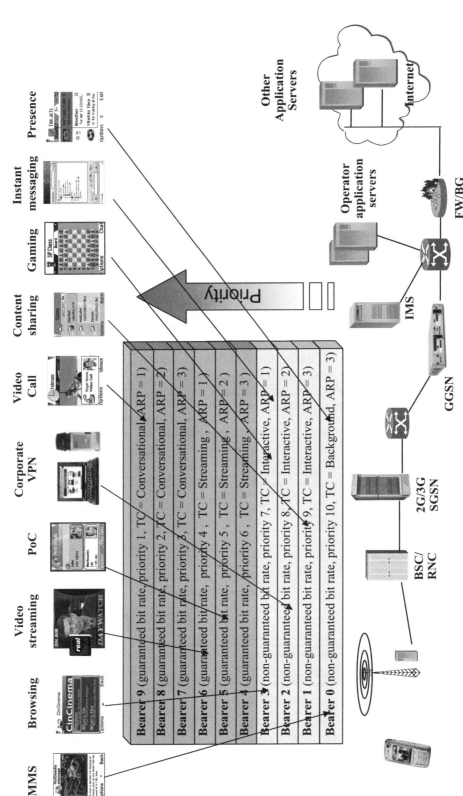

Figure 8.5 An example mapping of services onto priority pipes. The concept provides consistent treatment and adequate end-to-end quality for mobile services.

- for QoS-aware applications, by terminal clients activating parallel primary or secondary PDP contexts (with different QoS parameters) using the same APN. In this case the same QoS parameters stored in the HLR are used as the upper limit to QoS given by the network.

Priority QoS can be provided with these two alternatives for both QoS-aware and non-QoS-aware terminals and clients. However, according to 3GPP R99, bit rate guarantees are possible only for QoS-aware clients and terminals.

As described in Chapter 3, the introduction of SIP/IMS services brings new aspects to the QoS differentiation framework. 3GPP R5 defined service-based local policy (SBLP) implemented by means of the Go interface in-between the GGSN and IMS. The QoS working principle is illustrated in Figure 8.6. In addition to QoS authorisation, the Go interface can be used, say, for charging correlation between the IMS and GGSN. The policy decision function (PDF) implements the SBLP application in the IMS domain.

Usage of this functionality gives mobile operators the ability to ensure that, say, streaming class bit rate guarantees are given only to those PDP contexts that include the authorisation token, which prevents, for example, operator by-pass and usage of Internet-based 'hostile' servers. However, if the operator uses the 3GPP R5 Go/PDF mechanism for QoS control with a particular APN, then it is not possible to give, say, streaming class bit rate guarantees using the same APN to those QoS-aware mobiles that do not support Go usage. This is something that also needs to be taken into account in APN planning.

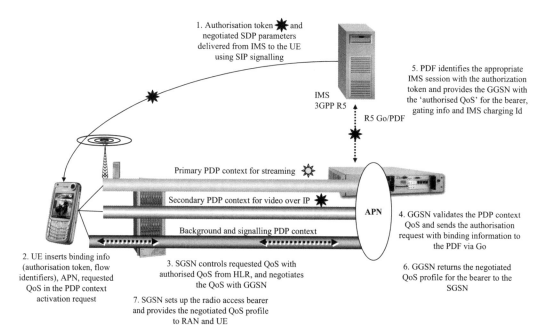

Figure 8.6 Working principle behind the Go interface for QoS authorisation in 3GPP R5-compliant networks.

The 3GPP standards also give an option to use dedicated PDP contexts for SIP signalling. This requires support from the terminals, the SGSN and GGSN.

8.3.2 Mobile terminal QoS provisioning

Once QoS is provisioned at the HLR and at various network elements, the next question is how an application on a mobile device knows which APN and hence associated QoS to use and, for an application on QoS-aware devices, what QoS to ask for. Applications are typically designed to handle a different bit rate or bandwidth. However, what QoS to request for a specific application depends partly on user subscription. It would be too great a hurdle to require end-users to manually configure QoS for different applications on mobile devices. The best way is to provision QoS over the air – for example, at service subscription time.

Though there are many proprietary ways to configure devices and applications over the air, the trend is toward Open Mobile Alliance (OMA) device management standardisation. They are described in the following sections.

8.3.2.1 OMA device management

'Device management' is the generic term used for technology that allows third parties to manage mobile devices on behalf of the end-user or customer. Third parties would typically be wireless operators, service providers or corporate information management departments. Through device management, a third party can remotely set parameters, conduct troubleshooting servicing of terminals, install or upgrade software.

In broad terms, device management consists of three parts:

- *Protocol and mechanism*: the protocol used between a management server and a mobile device.
- *Data model*: the data made available for remote manipulation.
- *Policy*: the policy decides who can manipulate a particular parameter, or update a particular object in the device.

Each part is briefly described in the following.

Protocol and mechanisms

The protocols used for device management are the OMA SyncML Representation Protocol [5] and the SyncML DM Protocol [6]. SyncML Representation Protocol is defined by a set of messages for synchronisation or device management. The messages are held in XML format. To reduce the data size, a binary coding of SyncML Representation based on the WAP Forum's WBXML is defined.

In the context of device management (DM), a package is a conceptual set of commands that could be spread over multiple messages. For example, a package from a server to a mobile device may include a GET command to retrieve the default APN used for web browsing.

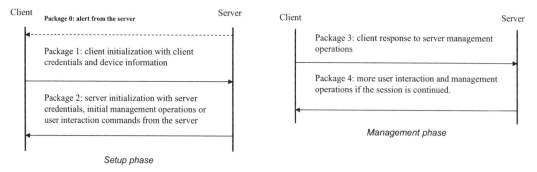

Figure 8.7   SyncML DM Protocol.

SyncML DM Protocol consists of two parts:

- setup phase (authentication and device information exchange); and
- management phase.

The management phase can be repeated as many times as the server wishes. Figure 8.7 depicts the two phases. Many devices cannot continuously listen for connections from a management server. Other devices simply do not wish to open a port (i.e., accept connections) for security reasons. However, most devices can receive unsolicited messages such as SMS, sometimes called 'notifications'. A management server can use this notification capability to cause the client to initiate a connection back to the management server. That is Package 0 in Figure 8.7.

The management phase consists of a number of protocol iterations. The content of the package sent from the server to the client determines whether the session must be continued or not. If the server sends management operations in a package that need responses (*Status* or *Results*) from the client, the management phase of the protocol continues with a new package from client to server containing the client's responses to those management operations. The response package from the client starts a new protocol iteration. The server can send a new management operation package and, therefore, initiate a new protocol iteration as many times as it wishes.

SyncML Representation and DM Protocols are transport-independent. Each SyncML package is completely self-contained and can be carried by any transport. OMA has specified bindings for HTTP, WSP and OBEX, but there is nothing to stop SyncML from being implemented using email or message queues, to list only two alternatives. Because SyncML messages are self-contained, multiple transports may be used without either the server or client devices having to be aware of the network topology.

Data model
The data model for device management is represented in a tree structure. In other words, each device that supports SyncML DM must contain a management tree [7]. The management tree organises all available management objects in the device as a hierarchical tree structure where all nodes can be uniquely addressed with a URI. Nodes are the entities that can be manipulated by management actions carried over SyncML DM

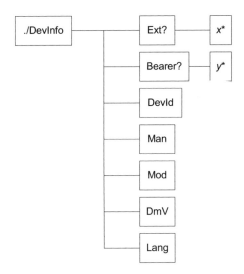

Figure 8.8   DevInfo management objects.

Protocol. The actions include ADD, COPY, GET, REPLACE, EXEC and DELETE. Nodes in the management tree can be either permanent or dynamic. Permanent nodes cannot be deleted whereas dynamic nodes can be created and deleted at runtime by management servers. A node can be either an interior node or a leaf node. An interior node may have child nodes, but cannot store any value. In contrast, a leaf node can store a value, but cannot have child nodes. The value of a leaf node can be as simple as an integer or as complex as a software package.

As an example, Figure 8.8 shows the DevInfo branch of objects standardised by OMA that stores information about a device [8]. A '?' in the figure indicates that the node may appear not at all or once whereas a '*' in the figure indicates that the node may have zero or more occurrences. The nodes have the following meanings:

- *Ext*. An optional, internal node, designating the only branch of the DevInfo subtree into which extensions can be added, permanently or dynamically.
- *Bearer*. An optional, internal node of the DevInfo subtree in which items related to the bearer (CDMA, etc.) are stored.
- *DevId*. A unique identifier for the device.
- *Man*. The manufacturer identifier.
- *Mod*. A model identifier (manufacturer-specified string).
- *DmV*. A SyncML device management client version identifier (manufacturer-specified string).
- *Lang*. The current language setting of the device.

Using this branch, a server can remotely retrieve the device identifier, manufacturer identifier and device model information through SyncML DM Protocol. It can also retrieve or update the current language setting of the device.

Policy

Each node on the management tree has properties (name, type, etc.) associated with it. The access control list (ACL) property defines which server can conduct what actions (e.g., ADD, GET, DELETE) on the node. Some parameters can only be configured and retrieved by the operator whereas parameters specific to an application may be set by the third party from whom the user purchased the application.

After the above introduction to OMA device management, provisioning QoS over the air to mobile devices should be quite straightforward. There must be a management subtree defined for the application with an APN and various QoS parameters as nodes on the tree. A management server can then remotely retrieve or set the information based on the defined tree.

8.4 QoS provisioning tools

In this section, the management system solutions for QoS configuration management and provisioning are introduced. It is possible to perform QoS provisioning using element manager functionality. The drawback to such an approach is the fact that the end-to-end view is not controlled since only one element at a time is configured. QoS consistency can be controlled only at a central point, like in NMSs.

8.4.1 *Configuration management in NMSs*

Configuration management (CM) has an important role in QoS provisioning. It provides the operator with the ability to ensure correct and effective operation of the networks as they evolve [9]. CM actions have the objective of controlling and monitoring the actual (QoS) configuration in NEs and network resources, and they may be initiated by the operator or by functions in operations systems (OSs) or NEs.

CM actions are needed as part of an implementation programme (such as additions and deletions), as part of an optimisation programme in terms of modifications and to maintain the overall performance and/or QoS. CM actions are initiated either as single actions on single NEs, or as part of a complex procedure involving actions on many network resources/objects (NRs) in one or several NEs.

The high-level requirements for network configuration management are listed in [2], where a distinction of three phases and states for a network and different degrees of stability is made. Once the first stage is over, the system will cycle between the second and the third phases. This is known as the network life-cycle and includes:

1. The network is installed and put into service.
2. The network reaches a certain stability and is only modified (dynamically) to satisfy short-term requirements; for example, by (dynamic) reconfiguration of resources or parameter modification. This stable state cannot be regarded as final because each item of equipment or SW modification will let the network progress to an unstable state and require optimisation actions again.

3. The network is being adjusted to meet the long-term requirements of the network operator and the customer (e.g., with regard to performance, capacity and customer satisfaction) through enhancement of the network or equipment upgrade.

During these phases, operators will require adequate management functions to perform the necessary tasks. The CM functions employed are:

1. Creation of NEs and/or NRs.
2. Deletion of NEs and/or NRs.
3. Modification of NEs and/or NRs.

The eTOM fulfilment process is responsible for configuring the network and services to meet QoS requirements. One of the challenges related to service configuration is that the setting in network elements must support the fulfilment and assurance of a service.

With network management systems, provisioning of configuration parameter values for network elements can be performed as a mass operation from a central location. This reduces the number of errors and is a fast method. Configuration management functionality provides operators with the basic functionality to interface planning tools, edit plans and provide plans to the network. CM tools should support fast 2G and 3G radio network roll-out and expansion and provide an efficient means for network configuration changes.

Part of the configuration management system is:

- Data repository containing the actual settings for network elements.
- Radio network plan management for provisioning/activating plans in the network.
- Support for rapid network roll-out by automating the most common roll-out-related parameter provisioning tasks like site creation and object reparenting.
- A rule system which comprises a set of configurable standard rules (value range checking rules or consistency rules that define relations between network objects) that define the technology and vendor-specific consistency requirements for configuration settings.
- Data exchange with external tools (requires an XML interface for the radio access planning data module).

For optimisation purposes, versioning of configuration settings should be supported. In cases where unexpected network behaviour results from provisioning a new configuration, a 'roll-back' functionality is needed in order to return to the original configuration.

Generally, optimisation actions are perceived to utilise measured data. More about optimisation can be found in Chapter 10.

8.4.2 Policy-based QoS management

The general policy management architecture consists of three layers of functionality (see Figure 8.2):

- *Policy management tool*: this provides the user interface to input policy rules; it validates the syntax and semantics of input data; and checks the consistency of high-level policies so that there are no conflicts with existing rules in the policy repository.

- *Policy decision point (PDP) or policy consumer (PC)*: this acquires policies from the policy repository, and deploys and translates policy rules into a form understandable by a policy target or PEP.
- *Policy enforcement point (PEP)*: this is a function of the NE that enforces (executes) policy decisions as network conditions dictate; it also monitors execution of the policies and reports them. This makes it possible to tune policies accordingly. In fact, for successful operation it is important to have feedback from the network in terms of performance statistics, usage of services, traffic volumes and QoS-related faults and alarms.

Specific protocols and interfaces for distributing rules from a centralized control point to network elements are also a part of this framework. COPS, SNMP and CLI are examples of distribution protocols between policy servers and policy targets.

A *policy* defines a definite goal, course or method of action to guide and determine present and future decisions [10]. A policy consists of a set of rules to administer, manage and control access to network resources. Each rule contains *conditions* and *actions* (e.g., [11]):

$$IF\ ARP == 1\ AND\ TC == Interactive\ (condition)$$
$$THEN\ limit\ the\ MinAllowedBitRate\ to\ 128\ kb/s\ (action)$$

where allocation retention priority (ARP) and traffic class (TC) are a part of the attributes the CN sends to the RNC when the bearer in question is set up; and *MinAllowedBitRate* is a management parameter of the RNC (see Section 5.2).

In policy-based management, a *role* is a type of property that is used to select one or more policies for a set of entities and/or components from (among) a much larger set of available policies. As a selector for policies, it determines the applicability of the policy to a particular managed element. Roles in this model are human-defined. For IP QoS provisioning, a set of traffic treatment templates are first defined, each template providing a particular behaviour – such as expedited forwarding (EF) traffic treatment. At provisioning time, a mapping between the templates and roles is specified. As a result, network elements with the same set of roles get the same set of policies.

8.4.3 Service configurator

In addition to network configuration, service configuration needs to be controlled. A service configurator (SC) is used when configuring customer-facing services [12]. While customer-facing services are deployed in the network, QoS/priority aspects are one of the so-called network-facing service components of the customer-facing service. This means that while deploying a new customer-facing service in the network, the SC will select what kind of treatment the corresponding traffic will experience while carried through the mobile network. The SC has to be aware of which QoS profiles/priorities are available and what are the QoS attributes of each of those priority pipes. Based on the needs of the newly created service the operator will select the most suitable pipe available in the network and assign the service to use that. Within pre-defined priority settings, the role of the SC is to ensure correct configuration settings in network elements. Service configuration settings need to be in accordance with the settings stored in the home location register (HLR).

8.4.4 Mobile terminal provisioning tools

A mobile terminal management server (MTMS) is a tool that makes it easy for mobile operators to install the right settings or configurations in mobile terminals. Device configurations are created with an MTMS and are then delivered to terminals over the air. Provisioning a large population of mobile devices becomes a relatively simple task. In addition, end-users are not being burdened with device configurations, which improves customer satisfaction.

The configuration manager module of the formation service manager suite is another example of terminal provisioning. It provides an over-the-air configuration solution for a wide range of wireless devices, enabling easy configuration and reconfiguration of settings over the device life-cycle [12].

8.5 Example of complete service management solution for NMS

In this section, we present a commercial solution for service management in IMS networks [13]. The complete solution is depicted in Figure 8.9. The operation support system (OSS) includes tools for centralised monitoring (network and service assurance), efficient service creation and deployment, and means for subscription and device management.

8.5.1 Centralised monitoring

The centralised monitoring solution includes tools that provide real time fault management of the IMS network and real time traffic monitoring all the way down to individual subscriber level. Advantages and problems related to media components, nodes and other resources, services and sessions can be visualised. The reporting tool efficiently collects, processes and visualises performance information from IMS network elements. Storing periods may range from days to years, depending on the operator's requirements. The reporting tool supports a variety of operator processes: daily operations that guarantee quality for end-users are covered, developing the network to cope with capacity and coverage demands, or carrying out long-term analysis of network behaviour. Using measurement administration, the operator can start and stop several measurements in several network elements in a single operation, receive change events from network elements and update the database automatically. More information on tools for service and network performance monitoring is given in Chapter 9.

8.5.2 Efficient service creation and deployment

This tool enables the operator to specify new services and deploy them in the underlying network and related systems, while maintaining an accurate service data structure

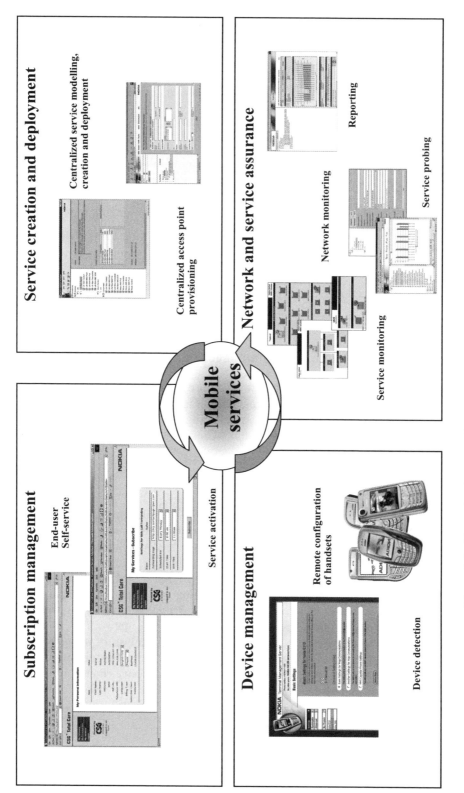

Figure 8.9 Example of commercial solution for IMS service management [13].

according to possibly standardised service modelling principles. The tool for service creation and deployment helps operators in:

- managing effectively a large number of services;
- generating new revenue from the currently untapped market by providing small-segment/short-term services;
- hiding the increasing complexity of the network;
- utilising the standardised service model in TMF;
- helping the operators and service/content providers to jointly build and manage services.

The tool supports a component-based service development model. New service components are normally developed only when necessary. For each service component type there is a specific user interface for creating/modifying/deleting service components. This user interface also includes functionality to define how the service component is meant to be used. In this way service components will build up a hierarchy [13].

8.5.3 *Centralised subscription management*

The solution includes a centralised subscription management system for GSM, GPRS and 3G networks with easy-access, user-friendly service activation and fast profile brokering capabilities in real time.

The tool provides one interface to all data related to a subscriber's subscriptions and services; it links the data to other user profile information such as the user's network location and presence, terminal capabilities, preferences and hobbies.

The main supported functionalities for the IMS are [13]:

- maintenance of a service catalogue;
- centralised service provisioning and management of subscription information;
- business rules for provisioning and subscription management;
- interfaces to other network elements like business support systems, terminal management systems, DNS/ENUM servers and application servers;
- an entry point for self-service management;
- monitoring of subscription-related activities and user profile information.

8.5.4 *Centralised device management*

The server for terminal management supports over-the-air (OTA) configuration of service settings, several ways of initiating remote configuration and updating the subscriber database with device capabilities. The tool also supports other manufacturers' devices by using open standards (e.g., OMA CP & SyncML DM), speeds up the roll-out of new services and reduces the number of calls to operators' helpdesks [13].

References

[1] TeleManagement Forum, *Enhanced Telecom Operations Map*, 2004, v. 3.0.
[2] 3GPP, R5, TS 32.101, 3G Telecom Management: Principles and High Level Requirements, v. 6.1.0.
[3] 3GPP, R5, TS 23.107, Quality of Service (QoS) Concept and Architecture, v. 5.13.0.
[4] IETF, RFC 3198, Terminology for Policy-based Management, November 2001.
[5] OMA, SyncML Representation Protocol, Device Management Usage, v. 1.1.2.
[6] OMA, SyncML Device Management Protocol, v. 1.1.2.
[7] OMA, SyncML Device Management Tree and Description, v. 1.1.2.
[8] OMA, SyncML Device Management Standardized Objects, v. 1.1.2.
[9] 3GPP, R5, TS 32.600, Telecommunication Management; Configuration Management (CM); Concept and High-level Requirements, v. 5.0.1.
[10] IETF, RFC 3198, Terminology for Policy-based Management, November, 2001.
[11] J. Laiho and D. Soldani, A policy-based quality of service management system for UMTS radio access networks, *WPMC, October 2003, Yokosuka, Japan*, pp. 298–302, Vol. 2.
[12] J. Laiho, A. Wacker and T. Novosad (eds), *Radio Network Planning and Optimisation for UMTS*, 2nd Edition, John Wiley & Sons, 2006, 630 pp.
[13] Nokia Operation Support Systems, *IMS Management Solution*. See *www.nokia.com*.

9

QoE and QoS Monitoring

David Soldani, Davide Chiavelli, Jaana Laiho, Man Li,
Noman Muhammad, Giovanni Giambiasi and Carolina Rodriquez

With the growth of mobile services, it has become very important for an operator to measure the QoS and QoE of its network and customers accurately. The measurements can be used to analyse problematic cases and improve the performance (optimise) of the network or services effectively and cost-efficiently. That will in turn help maintain both customer loyalty and competitive edge.

In this chapter, what the end-user perceives is mainly analysed in terms of the *integrity* of the service, which concerns throughput, delay (and delay variation or jitter) and data loss of a bearer service during communication. Service *accessibility*, which relates to availability, security, activation, access, coverage, blocking and setup time of a bearer service, and *retainability*, which in general characterises connection losses, are only partially covered in this book. Such performance metrics may be straightforwardly derived by signalling message flow charts available in the standards or by monitoring directly the messages exchanged between network domains through the corresponding interfaces.

Measuring QoS in the network is of vital importance for performance management and optimisation. The ability to measure QoE will help the operator to gather the contribution of network performance to the overall level of customer satisfaction.

In this chapter, an in-depth discussion is provided about different aspects of QoS and QoE assessments, including measures, frameworks, methods and tools used to achieve QoS and QoE targets.

9.1 QoE and QoS assurance concept

QoE and QoS monitoring in cellular networks consists of collecting/processing service and network performance statistics, usage data and QoS-related faults. In order to obtain end-to-end quality of service monitoring, network elements (NEs), the element management layer (EML) and network management layer (NML) must all be involved with the QoE and QoS monitoring process. Alarm and performance collection (counters and/or gauges) is performed at the network element layer and alarm/performance aggregation, report generation and analysis is done at element management and network management layers.

The QoE and QoS monitoring process consists of the following functions:

- Manage QoS fault conditions received from network elements.
- Retrieve QoE and QoS performance data from network elements, which includes mobile terminals.
- Collect and process usage data.
- Generate QoE and QoS reports.
- Undertake trend analysis of key QoE and QoS parameters.
- Audit/Analyse collected QoE and QoS parameters against expected values.

9.1.1 Conceptual architecture

The architecture of a QoE/QoS monitoring system is shown in Figure 9.1. The architectural components shown in the figure are described in the following sections.

The conceptual architecture for measurement results, aggregation, transfer and presentation may be designed as depicted in Figure 9.2. The element management layer and network management layer provide three distinct functions:

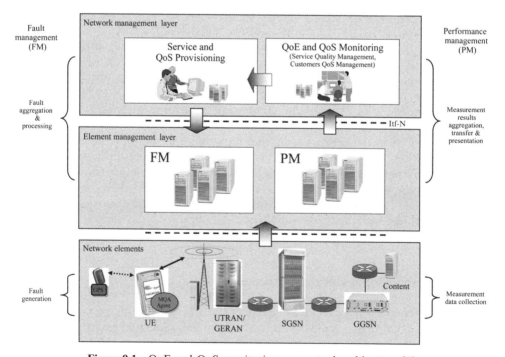

Figure 9.1   QoE and QoS monitoring conceptual architecture [1].

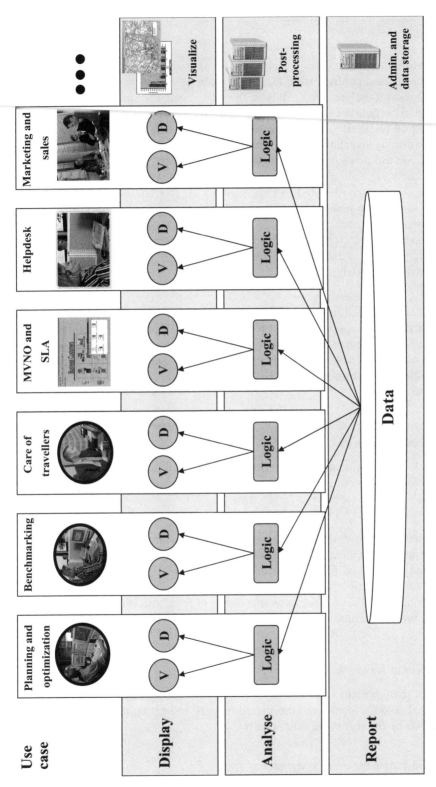

Figure 9.2 Conceptual architecture for data post-processing, visualisation (V) and drill-down (D) in particular locations/positions or in the raw performance metrics used in the analysis criteria.

- *Administration and data storage* (report layer) in a common database: this is where measurements and configuration files are collected.
- *Post-processing* (analyse layer): this is the logic used for intelligent data analyses according to particular use cases. Several analysis methodologies may be implemented depending on the desired level of accuracy and resolution (see Chapter 10).
- *Visualise* (display layer): this is where the performance results attained from intelligent data analyses are presented and visualised (denoted by V in Figure 9.2) according to network and service administrator needs. The results may be further displayed using a 'drill-down' approach (denoted by D in Figure 9.2). The 'zoom-in' of monitored data makes it possible to investigate, for example, what happened in awkward positions (street, area, etc.) or locations (cell, RNC, etc.) when a missing service or poor QoS performance was detected. The drill-down approach may also be used for identifying why certain criteria in data analyses were unsatisfactory, by displaying the combined (raw) metrics separately.

In the use cases depicted in Figure 9.2, the results of QoE and QoS performance data analyses are used for network (service) planning and optimisation; benchmarking (e.g., between different terminals and/or network/service providers); caring of corporate or other particular customers; monitoring of service level agreements; improving helpdesk efficiency; sales, marketing and product management; mobile virtual network operator (MVNO) cases, etc.

9.1.1.1   Network element

The network element (including the mobile terminal, UE) is responsible for collecting performance measures, usage data and generating alarms. It supports the following functions [1]:

- Collect performance data according to how measurements have been defined and return results to the element management layer.
- Collect usage data and forward the data to mediation.
- Perform fault detection, generation of alarms, clearing of alarms, alarm forwarding and filtering, storage and retrieval of alarms in/from the NE, fault recovery and configuration of alarms.

9.1.1.2   Element management layer

The element management layer (EML) is responsible for aggregating and transferring collected QoE and QoS performance measurements and generated alarms/events. The EML consists of the following functions [1]:

- performance management (PM); and
- fault management (FM).

Performance management

The PM function collects measurement data. Measurement types and measured network resources for QoE and QoS assurance are discussed in Sections 9.2 and 9.3, respectively. The related tools are described in Sections 9.6.1 and 9.6.2, respectively.

The resource(s) to which the measurement types are applied have to be specified; this includes the definition of a measurement recording period (periods of time at which the NE collects measurement data), measurement reporting criteria and measurement report file format. The measurement-related information to be reported has to be specified as part of the measurement. The frequency at which scheduled result reports are generated has to be defined.

Measurement results can be transferred from the NE to the EM according to the measurement parameters. They may also be stored locally in the NE and can be retrieved when required. Measurement results may also be stored in the EM for retrieval by the network manager (NM) when required.

Fault management

The main FM function is the management of alarm event reports (e.g., mapping of alarm and related state change event reports, real time forwarding of event reports, alarm clearing) and retrieval of alarm information (e.g., retrieval of current alarm information on NM request, logging and retrieval of alarm history information on NM request).

9.1.1.3 Network management layer

From the QoE and QoS monitoring perspective, the NML is responsible for the collection and processing of performance, fault and usage data. The NML QoE and QoS monitoring layer provides the following functions [1]:

- *Service quality management* (SQM): this is responsible for the overall quality of a service as it interacts with other functional areas to access monitored information, process that information to determine quality metrics and initiate corrective action when the quality level is considered unsatisfactory. Inputs to SQM include both performance and fault data.
- *Customer QoS management* (CQM): this includes monitoring, managing and reporting the QoS customers receive against what has been promised them in service level agreements and any other service-related documents. Inputs to CQM include data from SQM and usage data.

An example of a complete service assurance solution for the network management system (NMS) is presented in Section 9.7.

9.2 QoE monitoring framework

As discussed briefly in Chapter 1, there are two practical approaches to measuring QoE in mobile networks.

1. Service level approach using statistical samples.
2. Network management system approach using QoS parameters.

These two methods are not mutually exclusive. If these two approaches are used to complement each other, the QoE measurements are much more accurate and realistic from the end-user point of view. For example, the operator could identify the root causes of service application performance not being met, or relate QoE performance measures to other metrics measured in the network (such as throughput) in order to evaluate the spectral efficiency of the particular interface at which throughput is collected.

Tools used for QoE monitoring are introduced in Section 9.6.1. A complete solution for service assurance is presented in Section 9.7.

9.2.1 Service level approach using statistical samples

Key to the first approach is statistical sampling and taking the most relevant and accurate measurements according to that sample. In this approach, most measured performance indicators are at application level, providing the real end-user perspective.

The service level measurement approach relies on a statistical sample of overall network users to measure the QoE for all users in the network. If the size of the statistical sample (number of observations) is correctly selected, the results will get close to achieving 100% precision (repeatability of measures), much like the most reliable polls before elections. Section 9.4 provides some insight into various statistical methods and confidence intervals.

This process involves:

- *Determining the weighting of key service applications.* Many service applications are used by various network users. The right service mix has to be established and reflect the right kind of services mostly used in the network. At the same time, some applications are more popular and more frequently used than others. Also, some services are more important to end-users than others, even if their usage frequency is not very high. Based on some network statistics history, market understanding, trends and operator preferences, each application in the selected service mix needs to be given a weight percentage. The total weight of the service mix needs to be 100.
- *Identifying and weighting QoE KPIs.* Each application has unique key performance indicators (KPIs) that need to be identified. Various studies have been made on this subject and different standardisation bodies have defined these KPIs in their own way as general guidelines. These guidelines can be used as a basis to find KPIs for each application. On top of that, the weight of each KPI has to be defined, as it varies from application to application. For example, jitter could have almost no impact on perceived performance in the case of file downloading, but it can have significant impact on streaming and video sharing types of applications. The network point of measurement of each KPI also needs to be defined for proper reference.
- *Devising a proper statistical sample (geographic areas, traffic mix, time of day, etc.) and collecting QoE KPIs.* This is the most important step in this approach and is key to the accuracy of results. The help of a professional statistician is required to ensure proper sampling. A good sample would consider the proper representation of all kinds of

users, including various terminals, proper service mix and their weights, usage patterns like time of the day, various geographic distributions, etc.

- *Utilising mobile agents in handsets to make the results more accurate.* The mobile agents in the handsets can provide extra information on this process. These mobile agents can be installed in the handsets of selected users from a carefully selected sample. In practice, various strategies can be deployed. For example, incentives can be given to those users who agree to install mobile agents in their handsets. Information from these agents needs to be collected on a regular basis for incorporation into other network measurements.
- *Giving an overall QoE score (index) from KPI values for each separate service and a service mix.* Some kind of spreadsheet or application would be needed to calculate a final QoE index based on the inputs provided. The tool needs to be very flexible and adaptable to changing circumstances – like new applications, new usage patterns, etc. Regular revision of the above-mentioned steps would ensure accurate and realistic QoE index determination. Weak areas need to be improved ensuring better QoE.

This approach has many benefits, for example:

- The QoE for any network can be measured without access to operator products/ NMSs, etc.
- Operators can also measure QoE provided by its competitors.
- It is vendor-independent and, to a large extent, bearer-independent.
- Inexpensive tools can be used and the process is highly scalable.
- Good for benchmarking various services and applicable globally.

Despite being very useful, regular drive-through testing is not the only and definitely not the most practical way to implement this approach. Several combinations can be applied, including, for example, the mobile agents in UEs already mentioned above, the data collected from some performance measurement tools which use remote probes and stethoscopes, etc.

Frequent tests may improve the accuracy of results.

9.2.2   *Network management system approach using QoS parameters*

In the NMS approach, hard QoS performance metrics from various parts of the network are mapped onto user-perceptible QoE performance targets. These QoS measurements are made using a network management system, collecting KPI figures from network elements and comparing them with the target levels. The process involves:

- *Identifying the relationship between QoS KPIs and their effect on QoE.* This is the key step in this approach. This area has been under the spotlight for some time and several studies have looked into understanding the exact relationship between some QoS KPIs and end-user perception of performance (QoE). Although progress is being made, due to the many practical problems a comprehensive solution is not straightforward to achieve. Adding various bits and pieces would still make some kind of relationship model that can be used in this approach, though accuracy of QoE determination with this approach alone might be questionable.

- *Measuring QoS KPIs in the network.* Various network management systems and their supplements take lots of QoS measurements from the network. It would be practically impossible to measure all parameters for all users at all times, thus a good statistical sampling of users and services would be required in this method as well.
- *Rating QoE through measured QoS KPIs using some mapping metrics.* Mapping of various QoS KPIs onto user perception of performance will need to be made on a spreadsheet, for example. The inputs from hard QoS performance metrics would be fed into that sheet to calculate the QoE based on the identified mapping.

In principle, this approach could be ideal for operators, though it does depend on how well the mapping reflects real end-user experience. Identifying the proper relationship between network QoS KPIs and user experience (QoE) is an extremely challenging task, but is very unlikely to provide the highest level of accuracy. This approach is very much vendor-oriented and lacks flexibility and scalability.

The best and more realistic option would be to use both these methods in a complementary way. Together these two methods will provide the most accurate picture of user experience and give the network operator the possibility to assess the utilisation of radio and transport resources at a given QoE (percentage of satisfied users).

9.2.3 QoE metrics

Although QoE is subjective in nature, at the same time it is very important to have it measured. Waiting for end-users to vote with their money might turn out to be very expensive for stakeholders. As such, a strategy has to be devised to measure QoE as realistically as possible.

The top-down approach could be useful in this regard:

- The key is to understand the factors (metrics) contributing to user perception.
- Apply that knowledge to define the operating requirements (values).
- Devise a methodology to measure these factors constantly (tools, location, statistical sampling) and improve them as and when needed.

When identifying QoE metrics, we must first ask at a broad level, 'What does the end-user expect from any or all of the entities in the QoE value chain?' (see Figure 1.2). There will be as many different expectations as there are users, but most of these expectations can be grouped under two main categories: *reliability* and *quality*.

- Reliability, in this context, is the availability, accessibility and maintainability of the content, the service network, and/or the end-user device and application software.
- Quality, on the other hand, refers to the quality of the content, the bearer service, and/ or the end-user device and application software features.

Having established the two dimensions in which end-users will judge their QoE, we can now have a look at the performance metrics that describe these dimensions from the network viewpoint. The metrics are listed in Table 9.1. It is however important to evaluate these KPIs, first, on the basis of the four QoS classes (conversational, inter-

Table 9.1 QoE key performance indicators.

Reliability (service quality of accessibility and retainability)

QoE KPI	Most important measures	Explanation
Service availability (anywhere)	Ratio of territory under coverage to not under coverage (%)	Territory covered within a country. Since the word 'universal' is an important component of UMTS, it is quite likely that many users will find coverage to be an important issue in their judgement of quality. Operators should plan their network and also their roaming agreements in such a way that they satisfy every single segment of users. Global coverage using roaming partners can be included later if required.
Service accessibility (anytime)	Ratio of refused connections or ratio of PDP context failed to establish in first attempt (%)	Once a user knows that service is available in a particular area, it is important that the service is always up and accessible at all times. 'Please try again' messages are frustrating when one needs to download an important email.
Service access time (service setup time)	Average call or session setup time (s)	If service is accessible at any given time, it is important that the call setup or session establishment time to access the service is not too long. Long waiting times can frustrate the user.
Continuity of service connection (service retainability)	Service interruption ratio (%)	Nothing can be more exasperating than to be cut off from the service while accessing (using) it.

Quality (service quality of integrity)

QoE KPI	Most important measures	Explanation
Quality of session	Service application layer packet loss ratio (%)	This is a measure of the number of packets lost out of a thousand packets sent during a session. Users will always be more tolerant to packet loss in conversational and streaming applications, than in background and interactive applications. For these applications, TCP/IP retransmissions are measured as an indication of packet loss and hence delay, lowered throughput, high response time and network congestion.
Bit rate	Average bearer bit rate achieved as ratio of bit rate demanded by application (%)	This measure is especially important for conversational, streaming and some interactive service applications such as interactive gaming. The PDP context allocated bit rate can change during the session, as network conditions change or

Table 9.1 (*cont.*)

Quality (service quality of integrity)

QoE KPI	Most important measures	Explanation
		as a result of user mobility. Hence looking at the average bit rate over a session is important. If the bit rate achieved is lower than the encoded bit rate, the user will have a frustrating experience.
Bit rate variation	Bearer stability: bit rate variation around negotiated bit rate (%)	A user will always prefer a stable bit rate, especially for real time and streaming applications. This measure calculates the standard deviation around the average bit rate measured above.
Active session throughput	Average throughput towards mobile (kb/s)	It is normally assumed that an end-user who has a high throughput towards his/her mobile will always have a high QoE. This assumption holds true for most interactive and background class applications and hence it is important to include this measure in QoE calculations. However, as was pointed out above, in certain applications, as long as the average bearer bit rate achieved is higher than the bit rate demanded by an application, there will be no perceptible difference in the QoE experienced by the two users.
System responsiveness	Average response time (s)	In streaming and some interactive applications, it is important to look separately at the average response time. The average response time is the time it takes the first packets of information to arrive after a request has been sent (provided the PDP context has already been established). It does not apply to certain applications such as voice and videoconferencing.
End-to-end delay	Average end-to-end delay (ms or s)	End-to-end delay here refers to the time it takes a packet to traverse the network and get from host to destination or *vice versa*. Conversational, streaming and certain interactive class applications are very much delay-sensitive. A long delay will lead to very negative end-user experience.
Delay variation	Jitter (%)	Even if delay is minimised, variation in it could lead to poor QoE. Hence it is important to factor in the jitter which is the standard deviation experienced in the delay times of the packets during a particular session.

active, streaming and background) and, second, on the basis of the characteristics of some of the most popular service applications that fall under each of these traffic classes. In this section, we mostly focus on 3GPP QoS models; insight into the approaches adopted in other standardisation bodies is given in the sections that follow. More information on service applications and corresponding target performances can be found in Chapter 2.

The value of each of these metrics would translate to different level of impact on the actual QoE. Some will be totally irrelevant in one case while being the most important in another. It all depends on the type of service application being run by the user. Characteristics are different and as a result the requirements for all applications may not be the same.

In order to reduce the scope of performance analysis, it would be good to understand the different service applications in terms of the four traffic classes defined in 3GPP. As mentioned above, the four traffic classes are conversational, streaming, interactive and background.

Voice and videoconferencing calls are the two most popular service applications that need the conversational class traffic attribute specified in 3GPP R99. Both applications share certain characteristics that are common to services that fall under this traffic class:

- They preserve the time relation (variation) between information entities of the stream to minimise delay variation.
- A conversational pattern that is stringent and low delay.
- They are relatively insensitive to packet loss.
- Guaranteed resource allocation, no retransmissions.
- Real time traffic.

Audio and video streaming are the two most popular service applications that should use streaming class. For these service applications, we can state that:

- A short delay is important for the application.
- A low bit error rate is often requested.
- Important to minimise delay variation.
- Guaranteed resource allocation.
- Light retransmissions are used.

Web and WAP browsing, remote server access (Telnet) and interactive gaming are envisaged as the four most popular applications likely to use the QoS traffic attributes assigned for interactive class. The important characteristics of these service applications are:

- Request–response pattern of use.
- Preserve payload content (i.e., minimise bit error rate).
- Transmission delay must be acceptable for interactive use.
- Dynamic resource allocation.
- Retransmissions are used.
- Best-effort traffic (non-real time).

There are many service applications that fall under background class. These include email, messaging and FTP. The most important characteristics of background class are:

- Some amount of delay is tolerable by the application.
- Lower priority than interactive traffic.
- Responsiveness of interactive applications is ensured.
- Applications need reliable data, transmission, delay is not important.
- Dynamic resource allocation.
- Retransmissions are used.
- Best-effort traffic (non-real time).

9.2.3.1 ETSI view

ETSI has put a lot of effort into explaining QoS metrics from the end-user point of view (QoE) in [2]. All defined quality of service parameters and their computations are based on field measurements. That indicates that the measurements were made from the customer point of view (full end-to-end perspective, taking into account the needs of testing). By no means is it a complete specification, but it still contains enough information on this subject to warrant reading. The QoE metrics identified and grouped in the specifications are in line with the description made at the beginning of this section. Figure 9.3 illustrates this.

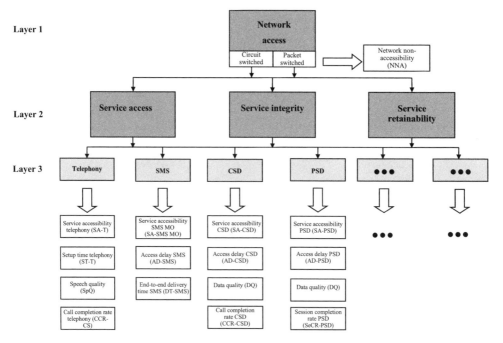

Figure 9.3 ETSI QoS parameters from end-user perspective [2].

9.2.3.2 ANSI view

ANSI T1.522-2000 specifies the classes of QoS sufficient to support business multimedia conferencing on IP networks, defined as being equivalent to legacy conference system performance. In this standard, two baseline conferencing systems are defined [3]:

- Tier 1: desktop PC systems.
- Tier 2: group conference room systems.
- The specification defines the perceptible performance level and the acceptable performance level. For some aspects (e.g., bit rate and loss) there are different levels specified for Tiers 1 and 2. The performance parameters of interest are summarised in Tables 9.2 and 9.3.

Table 9.2 ANSI-defined quality of service parameters – multimedia (MM) user interface [3].

Parameter	Quality criteria		
Communication function	Speed	Accuracy	Availability and reliability
Connection establishment	Setup time Transfer time	Mis-directed	Accessibility ratio (among media) Connection failure
User information	Delay (spontaneity) Delay variation Contention resolution	Media quality Media synchronisation	Dropped connection
Connection release	Take-down time		Release failure

Table 9.3 ANSI-defined quality of service parameters – end-to-end interface [3].

Parameter	Quality criteria		
Communication function	Speed	Accuracy	Availability and reliability
Connection establishment	Setup time Transfer time	Mis-directed	Accessibility Connection failure
User information (packet) transfer	Delay (network latency) Delay variation (within a single media stream and between streams) Information bit rate (committed bit rate and delivered bit rate)	Lost transport Packet Rate (combines IP packet Defects, such Error & Lost Packets)	Dropped connection (IP availability)
Connection release	Take-down time		Release failure

Error-tolerant	Conversational voice and video	Voice/video messaging	Streaming audio and video	Fax
Error-intolerant	Command/control (e.g. Telnet, interactive games)	Transactions (e.g. e-commerce, WWW browsing, email access)	Messaging, Downloads (e.g. FTP, still image)	Background (e.g. Usenet)
	Interactive (delay << 1 s)	Responsive (delay ~ 2 s)	Timely (delay ~10 s)	Non-critical (delay >>10 s)

Figure 9.4 ITU-T model for user-centric QoS categories [4].

9.2.3.3 ITU-T view

ITU-T Recommendation G.1010 defines a model for multimedia QoS categories from an end-user viewpoint [4]. By considering user expectations for a range of multimedia applications, eight distinct categories are identified, based on tolerance to information loss and delay. It is intended that these categories form the basis for defining realistic QoS classes for underlying transport networks and associated QoS control mechanisms. Figure 9.4 describes this model.

In an informative annex to ITU-T Recommendation G.1010 [4], indications of suitable performance targets for audio, video and data applications are given. These targets are reproduced in Tables 9.4 and 9.5.

9.3 QoS monitoring framework

This section presents means and methods of measuring the quality of distinct services across UMTS cellular networks. The approach is based on mapping service applications with different performance requirements onto distinct PDP contexts (QoS profiles). By means of a subset of bearer service attributes (bit rates, priorities and QoS classes), it is possible to formulate metrics to measure separately the performance of services within different network domains, in uplink and downlink directions, without any visibility of the content carried by upper layer protocols. Besides this, we present performance indicators to monitor the utilisation of interfaces between network domains and performance deterioration due to hardware limitations in network elements. The proposed measures collected by access and core network elements provide essential inputs to ensure quality compliance to service layer management commitments and optimisation solutions.

Table 9.4 ITU-T performance targets for audio and video applications [4].

Medium	Service application	Degree of symmetry	Typical data rates	Key performance parameters and target value			
				One-way delay	Delay variation	Informaton loss**	Other
Audio	Conversational voice (e.g., telephony)	Two-way	4–64 kb/s	<150 ms preferred* <400 ms limit*	<1 ms	<3% PLR	–
Audio	Voice messaging	Primarily one-way	4–32 kb/s	<1 s for playback <2 s for record	<1 ms	<3% PLR	–
Audio	High-quality streaming audio	Primarily one-way	16–128 kb/s ***	<10 s	≪1 ms	<1% PLR	–
Video	Videophone	Two-way	16–384 kb/s	<150 ms preferred**** <400 ms limit		<1% PLR	Lip-synch: <80 ms
Video	Broadcast	One-way	16–384 kb/s	<10 s		<1% PLR	

* Assumes adequate echo control.
** Exact values depend on specific codec, but assumes use of a packet loss concealment algorithm to minimise effect of packet loss.
*** Quality is very dependent on codec type and bit-rate.
**** These values are to be considered as long-term target values which may not be met by current technology.
PLR = Packet loss ratio.

Table 9.5 ITU-T performance targets for data applications [4].

Medium	Service application	Degree of symmetry	Typical amount of data	Key performance parameters and target values		
				One-way delay	Delay variation	Information loss
Data	Web browsing – HTML	Primarily one-way	~10 kB	Preferred <2 s/page Acceptable <4 s/page	N.A.	Zero
Data	Bulk data transfer/retrieval	Primarily one-way	10 kB–10 MB	Preferred <15 s Acceptable <60 s	N.A.	Zero
Data	Transaction services – high priority (e.g., e-commerce, ATM)	Two-way	<10 kB	Preferred <2 s Acceptable <4 s	N.A.	Zero
Data	Command/control	Two-way	~1 kB	<250 ms	N.A.	Zero
Data	Still image	One-way	<100 kB	Preferred <15 s Acceptable <60 s	N.A.	Zero
Data	Interactive games	Two-way	<1 kB	<200 ms	N.A.	Zero
Data	Telnet	Two-way (asymmetric)	<1 kB	<200 ms	N.A.	Zero
Data	Email (server access)	Primarily one-way	<10 kB	Preferred <2 s Acceptable <5 s	N.A.	Zero
Data	Email (server-to-server transfer)	Primarily one-way	<10 kB	Can be serveral minutes	N.A.	Zero
Data	Fax ('real time')	Primarily one-way	~10 kB	<30 s/page	N.A.	<10^{-6} BER
Data	Fax (store & forward)	Primarily one-way	~10 kB	Can be serveral minutes	N.A.	<10^{-6} BER
Data	Low-priority transactions	Primarily one-way	<10 kB	<30 s	N.A.	Zero
Data	Usenet	Primarily one-way	Can be 1 MB or more	Can be serveral minutes	N.A.	Zero

N.A. = Not available.

An introduction to tools used for QoS monitoring is given in Section 9.6.2 and an example of a complete solution for service assurance is reported in Section 9.7.

9.3.1 Performance monitoring based on bearer service attributes

When a PDP context is established, the attributes of the negotiated QoS profile for that particular bearer service are available in the UE/MS, RNC/BSC, SGSN and related GGSN. To assess service applications carried at different network elements (i.e., RNC/BSC, SGSN and GGSN), bearer service performance needs to be ascertained from a subset of attributes of the subscribed QoS profile – for example, traffic class (TC), traffic handling priority (THP) for interactive class, allocation retention priority (ARP) and bit rates (maximum and guaranteed) – which unambiguously relate to the offered service application (one-to-one mapping between subsets of PDP context attributes and upper layer protocols). (This concept was presented in Chapter 8 for consistent and differentiated QoS provisioning.) For inter-working between different releases, the following mapping (provisioning) rules between R97/98 and later 3GPP releases apply [5] (see also Section 3.2.2.11 and Chapter 8):

- Delay Class 1 corresponds to interactive THP1.
- Delay Class 2 corresponds to interactive THP2.
- Delay Class 3 corresponds to interactive THP3.
- Delay Class 4 corresponds to background TC.

Furthermore, for consistent traffic treatment through the mobile network, we need to set ARP = THP = Precedence class = Delay class. Background class (Delay Class 4) is mapped by the 2G SGSN onto ARP = Precedence Class 3. Hence, for (E)GPRS R97/98, without taking into account the bit rate attribute, four combinations can be defined, although in the BSS only two combinations (i.e., Precedence Class 3 and background class, which corresponds to Delay Class 4) undergo the same treatment (only three precedence class settings are available); whereas for 3G, again without considering the bit rate attribute, ten possible combinations of attributes (PS conversational ARP1–3, PS streaming ARP1–3, interactive THP1–3 and background) can be defined for legacy and QoS-aware terminals. The practical combination, definition and mapping of services onto these particular subsets of QoS attributes is for the operator to manage.

This means that counters in network elements should be classified with a granularity that allows the NMS to make statistics based on these particular combinations of attributes. In Figure 9.5, we illustrate an example of how different counters collected by 3G/2G SGSN and RNC/BSC can be classified to support PDP context based monitoring of throughput by the management layer. The classification of measurements by the RNC/BSC and 3G/2G SGSN includes attributes – such as traffic class and traffic handling priority (in the case of interactive class), or precedence class and delay class – that enable the NMS to filter out indicators and compute throughput per bearer service.

In the following sections, we present some metrics for differentiated QoS analyses. From these measured metrics, taking into account one-to-one mapping of applications onto distinct QoS profiles, it is possible for the operator to draw conclusions on the

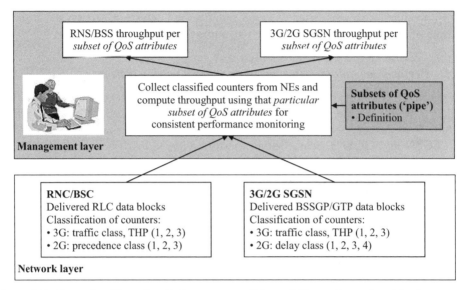

Figure 9.5 Example of how classified counters at different NEs can be retrieved by the management layer to compute the throughput of a particular bearer service ('priority pipe') consistently.

performance of each provisioned service application. However, if more services (or media components in the case of 3GPP R6 or later releases of IMS) are carried by the same PDP context, or placed behind the same access point name (APN) for legacy terminals (i.e., mapped onto the same subset of QoS attributes on which metrics are classified), QoS monitoring of services using this method is not possible. In this case, only the overall performance of the corresponding traffic aggregate (multiplexed IP flows) would be assessed, and the integrity, retainability and accessibility of different services can no longer be distinguished from each other.

9.3.2 QoS monitoring in BSS

The monitoring process is implemented in all network elements of the BSS. Bearer service performance needs to be assessed at each layer of the user plane protocol stack (see Figure 4.1). For each protocol, different performance metrics can be collected. For example, at the RLC level it is possible to measure RLC throughput and BLER (RLC SDU error ratio), if AM RLC is used. Also, at the BSSGP level, the BSC is able to measure essential metrics for assessing system performance (e.g., BSSGP buffer utilisation, BSSGP throughput, ratio of discarded BSSGP PDUs per cell or MS). Other key performance indicators that reflect the quality experienced can be derived from BSC/PCU statistics. For example, the paging success ratio relates to accessibility of the service, while PDTCH blocking deals with availability and the TBF establishment success rate gives an idea of the integrity of the service.

Troubleshooting activities lead to a second tier of performance indicators. This is needed for identifying the network conditions that deteriorate performance – for example, PS territory utilisation, MS multislot requested/allocated ratio, MCS (or CS) usage ratio.

In order to correctly interpret the values of measured metrics, it is important to understand the object/level aggregation of the statistics given by the network management system. Moreover, for troubleshooting purposes, values at the RLC level (within agreed quality criteria) may indicate good RF performance, but not necessarily satisfactory QoE: upper layers may be underperforming, for instance. Hence, complete analysis of performance at different protocol layers must be carried out. In this case, measuring LLC packet delivery success and the IP packet loss ratio would be essential inputs to QoE performance monitoring.

In the following sections, some examples of performance indicators are presented. Other measures that enable analysis of current network resources as well as the accessibility, availability and integrity of the session are also possible. Each vendor (or operator) has its specific means and methods of implementing a monitoring process, which includes counters at BSS level, formulas for computing performance indicators and tools for measurement results, aggregation, transfer and presentation in the NMS.

9.3.2.1 Classification of counters in the BSS

In the BSS, counters for differentiated performance monitoring need to be collected in both uplink and downlink directions – separately for GPRS and EDGE networks – and classified based on:

- IMSI.
- RLC transmission mode (acknowledged, unacknowledged).
- EDGE modulation and coding schemes (MCS-1 to MCS-9).
- GPRS coding schemes (CS-1 to CS-4).
- Precedence class or allocation retention priority (1–3).
- Cell identifier (cell ID) or BSS virtual connection identifier (BVCI);
- Attributes of BSS QoS profile [6]:
  - ○ peak bit rate, coded as the value part in the bucket leak rate – downlink and uplink;
  - ○ type of BSSGP SDU (signalling or data) – downlink;
  - ○ type of LLC frame (ACK, SACK or not) – downlink;
  - ○ precedence class used at radio access (1–4) – uplink.

This make it possible for the network administrator to collect measurements for consistent service performance monitoring across (E)GPRS network domains, as explained in Section 9.3.1, or for gathering performance metrics according to specific needs. In practice, only a subset of the above attributes may be required. For example, in the following sections only the precedence class attribute, denoted by p, is used.

9.3.2.2 Integrity monitoring

In this section, we present some metrics to assess the integrity of a service application carried on a particular precedence class (or ARP, as discussed in Section 9.3.1) in the BSS.

BLER monitoring

In AM RLC, BLER can be derived from the RLC SDU error ratio, which indicates the fraction of incorrectly received RLC blocks over the measurement period for a certain connection. Thus, the following ratio for EGPRS BLER computation can be used:

$$BLER_p = 1 - \frac{\sum_{i=1}^{MCS-9} N^p_{CorrectRxRLCblocks,i}}{\sum_{i=1}^{MCS-9} N^p_{TotalTxRLCblocks,i}} \tag{9.1}$$

where $N_{CorrectlyRxRLCblocks}$ is the number of RLC blocks correctly received by the MS or PCU (BSC), $N_{TotalTxRLCblocks}$ is the total number of RLC blocks sent for each of the modulation and coding schemes employed (MSC-1 to MSC-9) and p is the precedence class set in the QoS profile. The same formula can be used for calculating the SDU error ratio in the GPRS BSS, in this case MSC-1 to MSC-9 is replaced by CS-1 to CS-4.

Throughput computation

Let $B_i^{p,k}$ be the number of delivered RLC blocks (without retransmissions) of TBF i of duration d_i^p, where p is the precedence class, and k is a modulation and/or coding scheme that has 13 possible values (CS-1 to CS-4, MCS-1 to MCS-9). Total correctly delivered bits b^p and the related duration D^p in a cell of the (E)GPRS network can be calculated as:

$$b^p = \sum_{k=CS-1}^{MCS-9} \sum_{i=1}^{N} r_k B_i^{p,k}, \quad D^p = \sum_{i=1}^{N} d_i^p \tag{9.2}$$

where the radio block size in bits without RLC header r_k depends on the CS or MCS k of that particular TBF i (see Tables 4.1 and 4.2), and N is the total number of TBFs collected by the BSC in that particular cell during the measurement period S. The average throughput per user in the cell in question, from (9.2), is therefore:

$$t^p = \frac{b^p}{D^p} \tag{9.3}$$

Ultimately, for each precedence class p, mean cell throughput t^p_{cell} can be derived from (9.3) by summing up all correctly delivered GPRS and EGDE bits in the cell, over the monitored period S, and dividing the attained value by S; that is:

$$t^p_{cell} = \frac{b^p}{S} \tag{9.4}$$

An experimental validation of the above throughput analysis was presented in [7].

9.3.2.3 Accessibility monitoring

This section presents two examples of service accessibility measures for the BSS.

Ratio of NACK/rejected PS immediate assignment

Establishment of both uplink and downlink TBFs follows the 'PS-Immediate Assignment' procedure, unless the TBF is established via a PACCH, in which case this message is not used. By using this message the assigned TCH and TFI are communicated to the MS. An 'ACK' message is sent to the BSC to acknowledge the received AGCH messages, but only if the BSC explicitly requests it; whereas a 'PS-Immediate Assignment NACK' message is sent to the BSC for all AGCH messages that are deleted from TRX buffers due to 'buffer overflow', 'max. lead-time expiry' or 'start time expiry'. In this case, AGCH messages do not reach the MS and the attempt to establish a TBF is unsuccessful. When there are no physical data channels available for an MS, which has requested establishment of an uplink or downlink TBF, the message sent to the MS in question is a 'PS-Immediate Assignment Reject'.

Hence, a high blocking ratio, when accessing the PS territory by means of sending packet/channel request messages on a (P)RACH, can be a reason for deterioration in QoE when accessing the network to make use of certain service applications. The following formula indicates the ratio of allocation failure when an MS is accessing the network, due to the reasons already mentioned:

$$ImmAssNACKRejectRatio.Cause = \frac{AttPCReqAss.Cause - SuccPDTCHAssProc.Cause}{AttPCReqAss.Cause}$$

$$(9.5)$$

where $AttPCReqAss.Cause$ is the number of packet channel assignment requests per cause, and $SuccPDTCHAssProc.Cause$ is the number of successful packet channel assignment procedures per cause. A packet channel assignment is considered successful when either the 'Packet Uplink Assignment' message or the 'Immediate Assignment Command' message is sent to the terminal.

Number of packet pages discarded from the PPCH queue on PCCCH

The following measure provides the number of 'Packet Paging' messages that are discarded from the PPCH queue before they can be transmitted on a PCCCH:

$$NbrOfPSpagesDiscardedFromPPCHQueueOnPCCCH \qquad (9.6)$$

Page messages can be discarded from queues (assuming queuing is in operation) for a number of reasons – for example, access grant (AG) queue overflow, priority insertion in the queue causing an overflow or in queue timer expiry. More information on this metric can be found in [6].

9.3.2.4 Availability monitoring

This section presents some examples of service availability measures for the BSS.

PDCH availability

This metric yields the amount of time allocated to all available physical packet data channels. This metric is an indication of the congestion time in PS territory (see Sections 5.1 and 7.3). The measure starts when the last PDCH is assigned and ends when one of the allocated PDCHs is released (made free):

$$AvailablePDCHsAllocatedTime \qquad (9.7)$$

This measure indicates the time over the measuring period where no PDCH resources are available for TBF allocation. This period is perceived by the end-user as 'no-service time' or 'congestion time'. In a less extreme situation, it is the time where any new TBF is allocated with performance degradation of existing communications (TSL sharing perceived by the end-user as 'service degradation time'). The impact of the lack of PDCH availability has to be determined by considering existing services in the network and the minimum QoS (e.g., throughput, delay, etc.) required for a given service. More information on performance indicators can be found in [6].

Mean and maximum PDCH usage

Mean PDCH usage in a cell provides the average number of occupied PDCHs. This measurement is obtained by sampling at a pre-defined interval the number of PDCHs that are carrying packet traffic and then taking the arithmetic mean:

$$MeanNbrOfOccPDCHs \qquad (9.8)$$

The maximum number of PDCHs allocated in a cell gives an idea of how congested the PS territory in the cell is and may indicate a lack of PS resources and therefore QoE degradation. This measurement is obtained by collecting the maximum number of PDCHs that are carrying packet traffic while sampling them at a pre-defined interval.

$$MaxNbrOfOccPDCHs \qquad (9.9)$$

The above metrics give an idea of how the PS territory is used in a cell, giving information on possible resource shortages to carry all data traffic. PS resource redimensioning may be needed when *MeanNbrOfOccPDCHs* and *MaxNb OfOccPDCHs* values are similar and close to the maximum PS territory size in the cell, thus denoting that the cell is loaded to its limit and more time slots may be needed to cope with data traffic growth. See Section 7.3 for more information on (E)GPRS radio interface dimensioning.

9.3.3   QoS monitoring in RAN

This section presents theoretically how the QoS of differentiated PS and CS services can be assessed through counters collected and classified in the RNS. The analytical approach assumes that the network topology in which service performances are analysed is already defined (or selected within a wider *scope*) together with the measurement period (history) and user *satisfaction criteria*. The identified area may encompass radio network controllers (RNCs), base stations (BSs or Node Bs), cells and the interface between the base station and the radio network controller (Iub). Our aim is to define essential counters and performance indicators that need to be retrieved, and/or derived from measures in NEs, to ascertain a capacity and QoS status view in the RNS and/or its detailed performance analysis. In the latter case, for example, performance results may be compared directly with target values or, since only end-user perceived service quality matters, expressed in terms of satisfied users. The network administrator may then compare the number of satisfied users with the related target thresholds defined *a priori*. Average and deviation values of the proposed performance indicators may be computed as explained in Section 9.4. More information on the described performance measures and data analysis methodology is available in [8].

9.3.3.1 Classification of counters in the RNS

Cumulative counters or gauges (when data being measured can vary up or down during the measurement period) presented in the following sections are collected in the RNC. For differentiated performance monitoring, measurements, wherever possible, should be collected in uplink and downlink directions and classified based on the following attributes (for more information see Chapter 3 and [8]):

- CS conversational:
  - call type: speech or transparent (T) data;
  - guaranteed bit rate;
  - transport channel type (e.g., DCH).
- CS streaming:
  - non-transparent (NT) data;
  - guaranteed bit rate;
  - transport channel type (e.g., DCH).
- PS conversational:
  - guaranteed bit rate;
  - allocation retention priority (ARP);
  - transport channel type (e.g., DCH).
- PS streaming:
  - guaranteed bit rate;
  - allocation retention priority (ARP);
  - transport channel type (e.g., HS-DSCH, DCH).
- PS interactive:
  - maximum bit rate;
  - traffic handling priority (THP);
  - allocation retention priority (ARP);
  - transport channel type (e.g., HS-DSCH, DCH, E-DCH, FACH, RACH).
- PS background:
  - maximum bit rate;
  - allocation retention priority (ARP);
  - transport channel type (e.g., HS-DSCH, DCH, E-DCH, FACH, RACH).
- Cell, RNC, URA, RA and LA identifiers.

This makes it possible for the operator to filter out measurements for consistent service performance monitoring across network domains, as pointed out in Section 9.3.1, or for gathering performance metrics according to specific needs. In practice, only a subset of the above attributes may be required. In the following sections, such a combination of attributes (particular subset of QoS parameters) is denoted by m.

9.3.3.2   Integrity monitoring

This section presents how some useful performance metrics for QoS integrity monitoring – such as throughput, delay (and delay variation or jitter) and data loss of bearer services – can be derived.

Uplink E_b/N_0, BLER and BER derivation
This section enables determination of the BLER, BER and E_b/N_0 associated with a selected TrCH multiplexed with more TrCHs onto a DPCH based on the available target signal-to-interference ratio (SIR) in the RNC.

The desired E_b/N_0 for each transport channel can be expressed as a function of determined SIR according to the following equation [8]:

$$\frac{E_{b,DCH}}{N_0} = SIR_{DPCCH} - 20\log(r) - 20\log\left(\frac{R_{DCH}}{R_{DPDCH}(N + r^2)}\right)$$

$$- 10\log\left(\frac{\sum_{DCH \in RL} RM_{DCH}N_{DCH}^C}{RM_{DCH}N_{DCH}^C}\right) - 10\log\left(\frac{SF_{DPCCH}}{SF_{DPDCH}}\right) \quad (9.10)$$

where SIR_{DPCCH} is the SIR target per symbol on the DPCCH estimated in the RNC or the actual SIR_{DPCCH} value measured at the Node B [9], r is the uplink DPCCH overhead in terms of power and may be expressed as a function of 3GPP-specified gain factors (amplitude offsets) β_c and β_d; that is [10]:

$$\frac{RSCP_{DPCCH}}{RSCP_{DPDCH}} = \left(\frac{\beta_{c,TFC_{Max}}}{\beta_{d,TFC_{Max}}}\right)^2 = r^2 \quad (9.11)$$

R_{DCH} is the bit rate prior to CRC attachment, R_{DPDCH} is the DPDCH symbol rate, N is the number of codes employed in uplink transmission, RM_{DCH} is the rate matching attribute for a specific dedicated channel belonging to the radio link (RL) in question [11], N_{DCH}^C is the number of bits per radio frame subsequent to radio frame equalisation (i.e., prior to rate matching) and SF is the spreading factor. SF_{DPCCH} is always 256 in the uplink direction. Equation (9.10) defines the calculation of the E_b/N_0 requirement for a specific transport channel. The composite E_b/N_0 figure for all transport channels belonging to a physical channel (denoted as the code composite transport channel [11], CCTrCH) can be obtained by summing individual E_b/N_0 values.

In the case of HSDPA, (9.10) needs to take into account the overhead introduced by the uplink HS-DPCCH. Let Δ be the power offset between the uplink-associated DPCCH and the HS-DPCCH [10]. The energy per bit for a specific dedicated channel can be expressed as:

$$E_{b,HSDPA} = E_{b,DCH}\left(1 + \frac{r^2}{(N + r^2)}\Delta\right) \quad (9.12)$$

Hence, (9.10) for HSDPA becomes:

$$\frac{E_{b,HSDPA}}{N_0} = \frac{E_{b,DCH}}{N_0} + 10\log\left(1 + \frac{r^2}{N + r^2}\Delta\right) \quad (9.13)$$

The RNC is thus able to determine an average E_b/N_0 for this TrCH as follows:

$$\left(\frac{\overline{E_b}}{N_0}\right)_{DCH, Active} = \frac{\sum_{SW_1}\left(\frac{\sum_{RP} 10^{\frac{E_b}{N_0}}}{RP}\right)_{DCH}}{SW_1} \qquad (9.14)$$

where SW_1 denotes the sliding window size for the average computation and RP the reporting period of the counter in question, which should be classified based on the bearer attributes of the user data transported (see Section 9.3.3.1). Further, a BLER can be determined for the transport channel by:

$$\overline{BLER}_{DCH} = \frac{\sum_{SW_2}\frac{\sum_{RP} CRC\_NOK}{\sum_{RP}(CRC\_OK + CRC\_NOK)}}{SW_2} \qquad (9.15)$$

where SW_2 denotes the BLER sliding window size, and CRC_NOK, CRC_OK is the result of the CRC the RNC receives from the BS together with the received transport block. This counter should also be classified based on the bearer attributes of the user data transported (see Section 9.3.3.1). In addition or alternatively, a BER can also be determined for the monitored DCH by:

$$BER_{DCH} = \frac{\sum_{SW_3}\frac{\sum_{RP} QE}{RP}}{SW_3} \qquad (9.16)$$

where SW_3 denotes the BER sliding window size and QE is an estimation of the average BER of the DPDCH data of a radio link set at the Node B [9], which is reported to the RNC after the end of each TTI of the TrCH. BER computation is only possible, however, when turbo-coding is used. This counter should also be classified as stated in Section 9.3.3.1. Furthermore, sliding window content for quality computations needs to be reset when the target SIR is changed and thus sent to a WCDMA Node B.

Connection-based counters should be updated cell by cell, since the RNC is aware of the cell participating in diversity handover (DHO), and based on the bearer attributes of the user data transported (see Section 9.3.3.1). If measurements are needed for an online – and/or trace of a – specific radio connection, counters should be delivered together with the actual connection frame number (CFN) of each counter update period.

Downlink BLER computation

The downlink transport channel block error rate (BLER) is based on evaluating the CRC of each transport block associated with the measured transport channel after RL combination. The BLER is computed over the measurement period as the ratio between the number of received transport blocks with CRC errors and the number of received transport blocks. The mobile when explicitly ordered by the network may report such a measurement periodically in an event-based manner [12]. In the case of AM

RLC, the BLER can be estimated in the RNC, reducing the signalling load in the cell, as follows:

$$BLER^m = \frac{\sum\limits_{i=1}^{N} \overline{B}_i^m}{\sum\limits_{i=1}^{N} (B_i^m + \overline{B}_i^m)} \tag{9.17}$$

where $\overline{B}_i^m$ and B_i^m denote, respectively, the number of RLC blocks unsuccessfully and successfully transmitted during sampling period s, and N is the total number of samples collected by the RNC for the particular subset of QoS attributes m during measurement period S.

Throughput computation

Throughput relates only to the correctly received bits during a pre-defined measurement period (observation time), denoted by S in the following, where RLC buffers are not empty. In the literature the metric is presented as 'active session throughput' or 'circuit-switched equivalent bit rate', which is an essential indicator for assessing QoE and how spectral-efficient the provisioned QoS is [8]. In the RNC, accurate measurements are only possible in the downlink direction, since the exact status of uplink buffers is known only at the UE. (Throughput measurements in the uplink may be made as described for the downlink direction by considering the correctly received RLC blocks instead of the transmitted ones by the corresponding protocol entities.)

Let $B_i^{m,k}$ be the number of delivered RLC blocks (PDUs without retransmissions in the case of AM RLC) by RLC entity i, where m denotes the particular subset of QoS attributes upon which counters are classified and k the related PDU size. Let d_i be the related transmission duration during the sampling period s, in number of transmission time intervals (TTIs). Throughput for a particular m in a cell of the WCDMA network (SAP 6 in Figure 4.6) can be calculated as:

$$t^m = \frac{\sum\limits_{i=1}^{N} \sum\limits_{k=1}^{C} r_k B_i^{m,k}}{\sum\limits_{i=1}^{N} d_i^m} \tag{9.18}$$

where the RLC PDU size in bits without header r_k depends on the amount of data that can be transferred to MAC during each TTI, C denotes all possible combinations of dynamic parts of the set of transport formats (TFs) associated with the transport channel [13] and N is the total number of samples collected by the RNC for that particular measurement type during measurement period S. (*Note*: for conversational and streaming traffic classes there is no need to measure the bit rate per connection, since services carried on those QoS classes are offered with a guaranteed bit rate. However, (9.18) may be used to compute throughput per cell as follows.)

Ultimately, for each particular subset of QoS attributes m, mean cell throughput T^m can be derived by summing all correctly delivered bits in the cell over the monitored period S, and dividing the attained value by S; that is:

$$T^m = \frac{\sum_{i=1}^{N} \sum_{k=1}^{C} r_k B_i^{m,k}}{S} \qquad (9.19)$$

Total cell throughput can be obtained from (9.19) by summing all bearer contributions. Despite the high computational time, the advantage of measuring at the RLC is that the obtained measures are independent of the transport channel employed for user data transmission. Statistics based on transport channels may be derived by filtering collected data based on m, as explained in Section 9.3.3.1.

RLC retransmission rate

This indicator relates to the number of retransmissions required to deliver an RLC PDU when the first transmission through the radio interface was unsuccessful. The metric can be used to set the maximum number of allowed link layer retransmissions without compromising the load of the cell for the sake of global quality of service requirements [14].

Let F_i^m be the number of RLC blocks (PDUs) to be transmitted (or received in the uplink case) for the first time after segmentation by AM RLC entity i, where m denotes the particular subset of QoS attributes. Let P_i^m be the total number of transmitted (or received in the uplink) blocks (including retransmissions) during sampling period s. The RLC retransmission rate U for m in a cell of the WCDMA network (SAP 6 in Figure 4.6) can be calculated as:

$$U^m = 1 - \frac{\sum_{i=1}^{N} F_i^m}{\sum_{i=1}^{N} P_i^m} \qquad (9.20)$$

where N is the total number of samples collected by the RNC for the particular subset of QoS attributes m during measurement period S.

Service data unit error ratio

The SDU error ratio is defined as the fraction of SDUs lost or detected as erroneous [5]. This metric is an essential indicator for assessing the performance of protocols, algorithm configurations and error detection schemes.

Let Y_i^m be the total number of SDUs obtained from the upper layer (PDCP or RRC) by AM RLC entity i for transmission, where m denotes a particular subset of QoS attributes and X_i^m is the number of SDUs detected as erroneous or discarded (e.g., due to error detection, too many retransmissions or timer discard expiry [15]), during sampling period s. The SDU error ratio for m in a cell of the WCDMA network (SAP 6 in

Figure 4.6) can be calculated as:

$$Q^m = \frac{\sum\limits_{i=1}^{N} X_i^m}{\sum\limits_{i=1}^{N} Y_i^m} \tag{9.21}$$

where N is the total number of samples collected by the RNC for a particular subset of QoS attributes m during measurement period S.

Downlink transfer delay computation

'Transfer delay' is defined as the maximum delay for the 95th percentile of the distribution of delay for all delivered SDUs during the lifetime of a bearer service, where 'delay for an SDU' is defined as the time from a request to transfer an SDU at one SAP to its delivery at another SAP [5]. Based on this definition, assuming the distribution of transfer delay is normal, the statistical value of the transfer delay TD for a particular subset of QoS attributes m may be estimated as follows [8]. The method is valid for AM RLC in the downlink direction:

$$\overline{TD}^m = \frac{E(D_1^m) + \cdots + E(D_{N_B^m}^m)}{N_B^m} + n \cdot \frac{\sigma(D_1^m) + \cdots + \sigma(D_{N_B^m}^m)}{N_B^m} \tag{9.22}$$

where D_i is the time interval from a request to transfer an SDU at SAP 6 to its delivery at SAP 4 for bearer service i (see Figure 4.6), $E(D_i)$ and $\sigma(D_i)$ are the expectation and standard deviation of D_i taken over the number of transferred SDUs for bearer service i, N_B is the number of different bearer services of class m during measurement interval S and n relates to the level of confidence, which is a parameter for the operator to set. If $n = 1.95$, (9.22) yields the transfer delay of bearer service i as defined in 3GPP (i.e., the maximum delay for the 95th percentile of the distribution of delay for all delivered SDUs during the lifetime of bearer service i.

Downlink jitter computation

The jitter of a specific bearer service is defined as the difference between the one-way delays of the selected packet pair (e.g., consecutive packets). This section proposes a method for assessing delay variation as the difference between maximum and minimum one-way delay for a pre-defined percentile of the distribution of delay for all delivered SDUs during the lifetime of the bearer in question. Assuming that the distribution of delay is normal, this is exactly the confidence interval corresponding to a given probability. Hence, for bearer service i the delay variation for the Xth percentile of the distribution of delay for all delivered SDUs during the lifetime of bearer service i is:

$$J_i = 2 \cdot n \cdot \sigma(D_i) \tag{9.23}$$

where n relates to the level of confidence chosen by the operator.

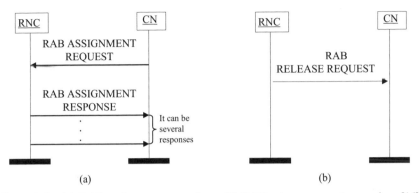

Figure 9.6   (a) RAB assignment procedure; (b) RAB release request procedure [16].

The statistical value of delay variation J for a particular m is the average jitter value on all bearer services N_B of type m during measurement interval S; that is [8]:

$$\bar{J}^m = 2 \cdot n \cdot \frac{\sigma(D_1^m) + \cdots + \sigma(D_{N_B^m}^m)}{N_B^m} \qquad (9.24)$$

Therefore, the statistical value of the delay variation of a particular subset of QoS attributes m can be expressed as $2n$ times the average value of the standard deviation of SDU delay, where n relates to the level of confidence set by the operator.

9.3.3.3   Accessibility and retainability monitoring

Accessibility and retainability measurements are based on the success/failure of the procedures needed to set up, modify or maintain a certain bearer service or signalling connection [16]. Hence, the proposed measurements are attached either to the successful or the unsuccessful issue of a procedure for RAB or signalling connection management. The procedures of interest are depicted in Figures 9.6 and 9.7; other relevant procedures, such as radio link and handover management, for a more detailed performance analysis, may be found in [16].

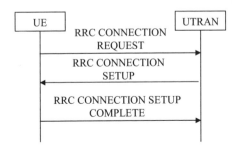

Figure 9.7   RRC connection setup procedure [16].

RAB management

Five measurement types may be defined for CS and PS domains. Measurements need to be split into sub-counters according to the classification of Section 9.3.3.1. The subset of relevant attributes is denoted by m in the extension of the following performance indicators.

- *Number of RAB assignment attempts*: on receipt by the RNC of a 'RANAP RAB Assignment Request' message by the CN, each RAB assignment request is added to the *RAB.AttEstab.m* counter.
- *Number of successfully established RABs*: on transmission by the RNC of a 'RANAP RAB Assignment Response' message to the CN, each successfully established RAB is added to the *RAB.SuccEstab.m* counter.
- *Number of RAB establishment failures*: on transmission by the RNC of a RANAP RAB ASSIGNMENT RESPONSE message to the CN, each RAB that failed to establish is added to the *RAB.FailEstab.Cause.m* counter according to the failure cause [17].
- *RAB connection setup time (mean)*: this measurement is obtained by accumulating the time intervals *RAB.SuccEstabSetupTimeMean.m* for each successful RAB establishment, which are then divided by the number of successfully established RABs observed in the granularity period to give the arithmetic mean.
- *RAB connection setup time (maximum)*: this measurement may be obtained by the high-tide mark *RAB.SuccEstabSetupTimeMax.m* of the monitored time intervals for each successful RAB establishment.
- *Number of RAB releases*: on transmission by the RNC of a 'RANAP RAB Release Request' message, each RAB requested to be released is added to the relevant per-cause measurement *RAB.Rel.Cause.m*. Possible causes are included in [17].

From the above measurements the following key performance indicators may be derived. Bearer service accessibility refers to the percentage of attempts that have been followed by assignment of a bearer service:

$$RAB.SetupSuccRatio.m = \frac{\sum RAB.SuccEstab.m}{\sum RAB.AttEstab.m} \tag{9.25}$$

The reliability of the bearer service relates to the number of RABs that have been terminated by the UE (as a result of UE-generated signalling connection release) and the number of RABs that have been started correctly:

$$RAB.RelSuccRatio.m = \frac{\sum RAB.Rel.UE.m}{\sum RAB.SuccEstab.m} \tag{9.26}$$

Both the accessibility and reliability of the bearer service can be assessed by considering the product of the previous performance indicators, which reduces to the RAB success ratio:

$$RAB.SuccRatio.m = \frac{\sum RAB.Rel.UE.m}{\sum RAB.AttEstab.m} \tag{9.27}$$

Signalling connection management

In order to assess the establishment and release of a signalling connection the following measurements can be made [16].

- *Attempted signalling connection establishments*: this measurement provides the number of attempts *SIG.AttConnEstab.m* made by the RNC to establish an Iu control plane connection with the CN. In this case, *m* may simply denote the PS or CS domain. The trigger point is the transmission of a 'RANAP Initial UE' message by the RNC to the CN, which is sent by the RNC on receipt of an 'RRC Initial Direct Transfer' message from the UE.
- *Attempted RRC connection establishments*: this measurement provides the number of RRC connection establishment attempts for each establishment cause. On receipt of an 'RRC Connection Request' message by the RNC from the UE, each received 'RRC Connection Request' message is added to the relevant per-cause measurement *RRC.AttConnEstab.Cause*. Possible causes are included in [12].
- *Failed RRC connection establishments*: this measurement provides the number of RRC establishment failures for each rejection cause. On transmission of an 'RRC Connection Reject' message by the RNC to the UE, or an expected 'RRC Connection Setup Complete' message not received by the RNC, each 'RRC Connection Reject' message received is added to the relevant per-cause measurement *RRC.FailConnEstab.Cause*. The possible causes are included in [12].
- *Successful RRC connection establishments*: this measurement provides the number of successful RRC establishments for each establishment cause. On receipt by the RNC of an 'RRC Connection Setup Complete' message following an RRC establishment attempt, each 'RRC Connection Setup Complete' message received is added to the relevant per-cause measurement *RRC.SuccConnEstab.Cause*. Possible causes are included in [12].
- *RRC connection setup time (mean)*: this measurement is obtained by accumulating the time intervals for every successful RRC connection establishment per establishment cause between receipt by the RNC from the UE of an 'RRC Connection Request' and the corresponding 'RRC Connection Setup Complete' message over a granularity period. The end value of this time, denoted as *RRC.AttConnEstabTimeMean.Cause*, is then divided by the number of successful RRC connections observed in the granularity period to give the arithmetic mean. The measurement is split into sub-counters per establishment cause [12].
- *RRC connection setup time (max.)*: this measurement is obtained by monitoring the time intervals for each successful RRC connection establishment per establishment cause between receipt by the RNC from the UE of an 'RRC Connection Request' and the corresponding 'RRC Connection Setup Complete' message. The high-tide mark of this time, *RRC.AttConnEstabTimeMax.Cause*, is the collected value. Possible causes are included in [12].
- *Attempted RRC connection releases*: this measurement provides the number of RRC connection release attempts per release cause sent from the UTRAN to the UE. On transmission of an 'RRC Connection Release' message by the RNC to the UE, each 'RRC Connection Release' message sent is added to the relevant per-cause measurement *RRC.AttConnRel.Cause*. Possible causes are included in [12].

From the above counters the following key performance indicator may be derived for each cause of RRC connection establishment:

$$RRC.SetupAccessCompleteRatio.Cause = \frac{\sum RRC.SuccConnEstab.Cause}{\sum RRC.AttConnEstab.Cause} \quad (9.28)$$

9.3.4 QoS monitoring in packet core and backbone networks

In this section we present QoS monitoring in the packet core and backbone networks. Network elements that should be considered when performing this task are:

- serving GPRS support node (SGSN);
- gateway GRPS support node (GGSN);
- switches and routers.

In the definition of QoS metrics we also have to consider that the protocol stacks between 2G SGSN and BSS in (E)GPRS networks and between 3G SGSN and UTRAN in WCDMA networks are different. For this reason the service access points where QoS performance is measured are different, depending on the radio access technology.

9.3.4.1 2G serving GPRS support node (2G SGSN)

Several performance metrics for a BSS packet flow context (PCF) or PDP context can be monitored by the SGSN. For example, throughput per PDP context can be measured at SNDCP SAPIs, as shown in Figure 4.1 (SAP 3). At this level, upper layer IP packets can be considered in delivered packet computation. Throughput per PFC can be monitored by the logical link entities (LLEs) at different LLC SAPIs (see Figure 4.1, SAP 5). In the case of AM LLC, both gross and net throughput (without LLC retransmissions) may be computed. Throughput per BVC and, within the BVC per MS, can be measured at BSSGP SAPIs (see Figure 4.1, SAP 9). Measurements are essential for SGSN flow control. Such measurements may be normalised with respect to the corresponding bucket leak rate (R). Also, the buffering delay of BSSGP PDUs can be monitored for each BVC (cell) or MS.

In the 2G SGSN, measurements may be classified on the basis of:

- IMSI or IMEI, for tracing either subscribers or terminals.
- Radio Priority – uplink.
- PDP type.
- APN.
- GGSN address in use.
- Release 97 negotiated QoS profile:
  - delay class (1–4);
  - reliability class (1–5);
  - peak throughput class (1–9);
  - precedence class (1–3).

- Release 99 negotiated QoS profile:
  - traffic class;
  - traffic handling priority (THP 1–3);
  - max. bit rate;
  - guaranteed bit rate;
  - allocation retention priority (1–3).
- Packet flow context.
- Aggregate BSS QoS profile.
- PDP context charging characteristics.
- Cell ID from BVCI.
- Routing area identifier.

Based on the previous classification, mean throughput can be measured at each interface (i.e., Gb and Gn) and for each connection type (particular combination of the above attributes) as follows:

$$MeanThroughput(i) = \frac{s}{S} \sum_{j=1}^{\frac{S}{s}} \frac{\sum_{i=1}^{ActiveConnections(j)} Throughput(i,j)}{ActiveConnections(j)} \qquad (9.29)$$

where S is the measurement period, s is the sampling period, $ActiveConnections(j)$ represents the number of connections active during sampling interval j, $Throughput(i,j)$ represents the throughput of connection i measured during sampling interval j.

Throughput measurements can be performed when the corresponding buffers are not empty, the PDP context is 'active' and the MM state is 'ready'.

LLC SDU transfer delay – that is, the maximum delay of the 95th percentile of the distribution of delay for all delivered LLC SDUs during the lifetime of LLC connections – can be measured only in the case of AM LLC. The delay for an LLC SDU is defined as the time from a request to transfer an SDU at one end to its delivery at the other end.

The service accessibility (blocking) and latency (delays) introduced by signalling can also be assessed in the SGSN during the PDP context activation procedure. The PDP context activation procedure for 2G SGSN is illustrated in Figure 3.5.

From that figure we can define how PDP context setup time can be measured by the SGSN. This can be done by considering the time difference between the 'Activate PDP Context Request' message sent from the terminal and receipt of the 'Activate PDP Context Accept' message sent by the SGSN. In the CN, these two messages are mapped onto the 'Create PDP Context Request' and 'Create PDP Context Response' messages, respectively, which can be used to measure PDP context activation delay.

The PDP context activation rejection ratio can be also measured in the packet core network.

Information from 2G SGSN counters

2G SGSN counters can give information about the actual occupancy status of different queues and the number of dropped packets. Such information is important because if the queue grows too much (average and peak utilisation) delay in the 2G SGSN grows and the round trip time in the network can be adversely affected, causing degradation in QoE.

Measurement data from 2G SGSN can be retrieved at different layers and aggregation levels, namely:

- GTP layer (AM LLC):
  ○ average utilisation of the GTP buffer;
  ○ peak utilisation of the GTP buffer.
- BSSGP layer:
  ○ average utilisation of the BSSGP buffer for each of the traffic class and traffic handling priorities implemented in the 2G SGSN;
  ○ peak utilisation of the BSSGP buffer for each of the traffic class and traffic handling priorities implemented in the 2G SGSN;
  ○ data lost at the BSSGP layer due to buffer overflow for each of the traffic class and traffic handling priorities implemented in the 2G SGSN.
- Network service–virtual channel (NS-VC) layer:
  ○ NS-VC-discarded data in packets for each of the traffic class and traffic handling priorities implemented in the 2G SGSN, defined as the number of data packets discarded by NS-VC CIR flow control when the packet lifetime expires;
  ○ NS-VC-passed data in packets for each of the traffic class and traffic handling priorities implemented in the 2G SGSN, defined as the number of data packets passing through NS-VC CIR flow control.

From the previous counters it is possible to build the following performance indicator, defined as the ratio of discarded NS-VC data packets at the NS-VC layer:

$$NS\text{-}VC\_discard_j = \frac{NS\text{-}VC\_discarded_j}{NS\text{-}VC\_passed_j} \tag{9.30}$$

where $NS\text{-}VC\_discarded_j$ represents the number of data packets in priority class j discarded by NS-VC CIR flow control when the packet lifetime expires; $NS\text{-}VC\_passed_j$ represents the number of NS-VC-passed data packets in priority class j; and j indicates the specific priority class in the 2G SGSN (i.e., refers directly to the traffic class and traffic handling priority under which the counters are classified).

9.3.4.2 3G serving GPRS support node (3G-SGSN)

The user plane and control plane protocol stack for UMTS are illustrated in Figures 4.6 and 4.7, respectively. From Figure 4.6 we can see that a tunnel is established both between the 3G SGSN and RNC and between the 3G SGSN and GGSN. The protocol involved in tunnel establishment is the same in both cases (GTP). For this reason the measurement of throughput can be similar at both Iu-PS and Gn interfaces. SAP 7 in Figure 4.6 can be used to measure throughput at the Iu interface, while SAP 5 is considered while evaluating throughput at the Gn interface. At this level, upper layer IP packets can be considered in the delivered packet computation.

In the 3G SGSN, measurements may be classified on the basis of:

- IMSI or IMEI, for tracing either subscribers or terminals.
- PDP type.

- APN.
- GGSN address in use.
- Release 99 negotiated QoS profile:
  - traffic class;
  - traffic handling priority (THP 1–3);
  - max. bit rate;
  - guaranteed bit rate;
  - allocation retention priority (1–3).
- PDP context charging characteristics.
- Routing area identifier.

Mean throughput can be measured for each interface (i.e., Gn and IuPS) and for each connection type based on the above classification. Throughput computation can be performed as defined in (9.29) for the 2G SGSN.

Service accessibility (blocking) and latency (delays) introduced by signalling can also be assessed in the WCDMA during the PDP context activation procedure, using the approach described in Section 9.3.4.1 for the 2G SGSN. The PDP context activation procedure for the 3G SGSN is presented in Figure 3.9.

Information from 3G SGSN counters

The following performance indicators could be used to assess the traffic related to active PDP contexts based on the traffic classes and traffic handling priority attributes on which the counters are classified.

From the information gathered from total bytes handled in the downlink per traffic class and dropped bytes in the downlink per traffic class, the following performance indicator can be built:

$$DroppedBytesRatio_j = \frac{DroppedBytesDL_j}{BytesDL_j} \tag{9.31}$$

where *Dropped Bytes DL_j* and *Bytes DL_j* are the number of bytes dropped and sent in the downlink direction by the 3G SGSN, respectively, and j represents the particular subset of QoS attributes in which the counters are classified (i.e., the traffic class in this case).

9.3.4.3   Gateway GPRS support node (GGSN)

At the GGSN site we may have maximum visibility of ongoing application services. Considering the user plane definition for GSM and UMTS we can see that in both cases the GGSN protocol stack (both 2G and 3G) is the same. For this reason we refer to only one user plane: the GSM user plane presented in Section 4.1.1.

Different tunnels are created for different PDP contexts (see Figures 4.1 and 4.6). In addition, if several services are carried (multiplexed) using the same PDP context and the IP payload is not encrypted (e.g., a VPN tunnel using IPsec, which allows confidentiality, integrity, authentication and cryptography), it is possible to separate the ongoing connection by considering the distinct sockets (micro-flows) visible in IP packets and measuring IP SAP (see Figure 4.6, SAP 5) throughput per micro-flow and per PDP context.

Furthermore, if higher layer reliable protocols (e.g., TCP) are used for transporting application services, unpacking the IP payload, it is possible to monitor upper-layer SDU transfer delay and error ratio.

Transfer delay may be computed as the maximum delay of the 95th percentile of the distribution of delay for all delivered SDUs during the lifetime of upper-layer connections, where the delay for an SDU is defined as the time from a request to transfer an SDU from SAP 5 of Figure 4.6 to SAP 4 of the same figure.

The upper layer SDU error ratio can be derived from the SDU discarded or detected as erroneous by the MS.

Mean throughput may be measured for each interface (i.e., Gi and Gn) using (9.29) and for each connection type based on the following classification:

- IMSI for tracing subscribers.
- PDP type.
- APN.
- Release 99 QoS profile negotiated:
  ○ traffic class;
  ○ traffic handling priority (THP 1–3);
  ○ max. bit rate;
  ○ guaranteed bit rate;
  ○ allocation retention priority (1–3).
- Charging characteristics.
- Routing area from the SGSN IP address.
- Micro-flow, to monitor application services multiplexed onto one PDP context:
  ○ source and destination addresses;
  ○ source and destination port numbers;
  ○ protocol type (ID).

Information from GGSN counters
The counters collected by the GGSN may be used for assessing the performance of the procedures related to PDP context activation/deactivation and throughput per user and APN. In this case the measured object in the GGSN is the APN, and the average throughput in both uplink and downlink directions per user and per APN can be computed as:

$$AveULthroughput.APN = \frac{GTPbytesSent.APN}{K \cdot PDPcontextActive.APN} \qquad (9.32)$$

where *GTPbytesSent.APN* counts the number of GTP bytes (overhead included) sent at the APN level to another network element; *PDPcontextActive.APN* counts the average number of PDP contexts active per APN; *K* is a constant that allows the conversion to bytes/s. If the performance indicator is calculated hourly *K* will be equal to 3600, if the sampling interval is 5 minutes *K* will be 600:

$$AveDLthroughput.APN = \frac{GTPbytesReceived.APN}{K \cdot PDPcontextActive.APN} \qquad (9.33)$$

where *GTPbytesReceived.APN* is the number of GTP bytes (overhead included) received at the APN level from another network element; *PDPcontextActive.APN* counts the average number of PDP contexts active per APN; and *K* is as defined above.

This last indicator should not be confused with real throughput perceived by the user, which is difficult to monitor by means of counters only. The reliability of this indicator is related to the application that is considered and to the measurement and sampling interval.

The PDP context activation failure ratio per APN may be derived as follows:

$$PDPactFail.APN = \frac{PDPactFail.APN}{PDPactAtt.APN} \tag{9.34}$$

where *PDPactFail.APN* counts the number of failed PDP contexts related to a specific APN and *PDPactAtt.APN* counts the total number of PDP context activation requests received by the GGSN, both 2G and 3G.

Furthermore, performance indicators can be used to assess the traffic related to the active PDP context, not based on APN but on traffic classes, on any particular subset of bearer service attributes used in counter classification.

The relationship between traffic classes and traffic handling priorities (i.e., the subset of the QoS attributes on which measures are classified) depends on how the APNs are provisioned by the operator (e.g., service differentiation). The object measured at this stage is the whole GGSN, and the data aggregation level can be different depending on the particular need for QoS monitoring.

In this case we can discriminate between the 2G GGSN and 3G GGSN. The traffic classes that can be monitored (both uplink and downlink) for the 2G GGSN are background and interactive, in the case of an R98-compliant QoS profile, and background, interactive and streaming in the case of an R99-compliant QoS profile. The traffic classes that can be monitored (both uplink and downlink) for the 3G GGSN are background, interactive, streaming and conversational. Analysis of the interactive class can be further specified according to the THP parameter.

In the GGSN, we can get information about downlink average throughput per traffic class and per subscriber. The performance indicator is calculated as:

$$AveDLuserThroughput_j = \frac{1}{Samples_j} \sum_{i=1}^{Samples_j} \frac{(DLtrafficBytes_j)_i}{(ActivePDPcontext_j)_i} \tag{9.35}$$

where *DLtrafficBytes_j* represents the number of bytes transmitted in the downlink direction belonging to traffic class *j* – the amount of traffic is measured over a certain sampling interval *s* (e.g., 5, 15, 30 or 60 minutes); and *ActivePDPcontext_j* represents the number of active PDP contexts belonging to traffic class *j*.

This performance indicator is dependent on the sampling interval. The information given by this performance indicator could give input into how the service agreement signed by the user is respected by the network. Anyway, it should be noted that the average throughput measured by means of internal GGSN resources hides subscriber behaviour – that is, it should only be considered as a performance indication for services such as WAP and web browsing, while it is much more realistic for streaming and FTP services. Obviously, the information given by (9.35) is more accurate when the sampling interval is decreased. A trade-off between monitoring accuracy and network element load should also be considered.

The same indicator could be built for uplink traffic, as:

$$AveULuserThroughput_j = \frac{1}{Samples_j} \sum_{i=1}^{Samples_j} \frac{(ULtrafficBytes_j)_i}{(ActivePDPcontext_j)_i} \tag{9.36}$$

where $ULtrafficBytes_j$ represents the number of bytes transmitted in the downlink direction belonging to traffic class j – the amount of traffic is evaluated over sampling interval s (e.g., 5, 15, 30 or 60 min); and $ActivePDPcontext_j$ represents the number of active PDP contexts belonging to traffic class j.

9.3.4.4   GPRS backbone network

Considering the GPRS backbone, different transport solutions can be implemented. The most common are (see Chapters 3 and 6 for more information):

- IP over ATM; and
- IP over MPLS.

Both solutions use the DiffServ approach in order to provide differentiated treatment to the traffic. Based on specific DiffServ code points the characteristic per-hop behaviour is implemented in the GPRS backbone. The inter-working between 3GPP QoS implementation and DiffServ-based QoS is performed at the edges of the GPRS backbone – namely, in the SGSN and GGSN – where marking of the transport IP layer is performed. The DiffServ Code Point (DSCP) field of the transport IP layer is marked according to bearer service attributes.

The DSCP field is then used in different ways, depending on the GPRS backbone solution, in order to ensure different treatment of packets:

- In the IP over ATM network, ATM VCs can be selected based on DSCP values (different VCs for real time and non-real time traffic).
- In the MPLS backbone, the MPLS label path can be selected based on DSCP.

On this basis, the main indicators that can be monitored in the different network elements in the GPRS backbone (i.e., routers and switches) are:

- Link occupancy.
- Dropped and passed packets per Diffserv code point.
- Dropped and passed bytes per DiffServ code point.

Based on the previous information it is possible to build the following performance metrics:

$$DroppedPks_{DSCP_j} = \frac{DroppedPks_{DSCP_j}}{PassedPks_{DSCP_j} + DroppedPks_{DSCP_j}} \tag{9.37}$$

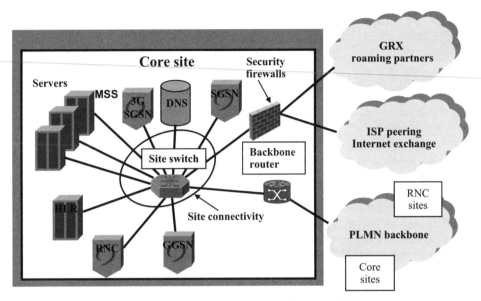

Figure 9.8 Packet core site and site connectivity.

and

$$DroppedBytes_{DSCP_j} = \frac{DroppedBytes_{DSCP_j}}{PassedBytes_{DSCP_j} + DroppedBytes_{DSCP_j}} \qquad (9.38)$$

where *Pks* is an abbreviation used for packets.

After computation of the previous indicators, the correlation between them and the occupancy level of the different links could be also evaluated.

When monitoring the QoS in the GPRS backbone, the specific QoS features of the network element part of the backbone solution have to be considered. As illustrated in Figure 9.8 the site switch is a fundamental element of the packet core connectivity solution. Some features give the option of overriding the DSCP field within the IP transport layer: operators having huge amounts of available bandwidth in the backbone could choose this implementation. This means that the DSCP marking performed by the GGSN in the downlink and by the SGSN in the uplink could be changed by the site switch to the DSCP corresponding to best-effort traffic (i.e., 0x00, which is the default value for best-effort traffic). In this way the different treatment of packets belonging to different priority 'pipes' cannot be implemented in the GPRS backbone. In this situation, QoS monitoring in the GPRS backbone gives more generic information without any link to the priority 'pipe'. QoS monitoring can only be based on:

- Link occupancy.
- Dropped and passed packets.
- Dropped and passed bytes.

Based on this information, performance indicators similar to those defined in (9.37) and in (9.38) can be built, but without any differentiation per DSCP.

9.3.5   QoS service level agreement

A service level agreement (SLA) is a formal, negotiated contract between two parties. The two parties may be a *customer* and an operator, or two operators where one takes the customer role by buying services from another service provider. The purpose of an SLA is to create a common understanding about services, priorities, responsibilities, etc. An SLA serves as a tool to ensure that customers get what they paid for. If an operator fails to deliver the promised service quality, the SLA also includes procedures on how customers will be compensated. An SLA also serves as a tool to ensure that customers do not impose unlimited traffic load on the network. If a customer fails to comply with the agreement, his applications may experience performance degradation.

There are important differences between mobile networks and fixed IP networks that should be taken into considerations in designing mobile SLA. Fixed IP networks are connectionless networks whereas UMTS networks are connection-oriented. To begin a packet data communication, a UE must first establish a data connection or a 'pipe' through the UMTS network. In addition, fixed IP networks are fairly reliable with redundant routing and fault tolerance technologies. In contrast, mobile network air interfaces are not as reliable – connections may be dropped occasionally. Furthermore, with fibre-optic technologies, fixed IP networks can achieve a very low transmission error ratio whereas mobile network air interface quality varies and may produce a much higher error ratio.

A typical SLA QoS for either fixed or mobile network contains the following elements:

- *Network scope* specifies where the SLA applies. For example, the scope may be between a group of mobile phones and the Gi interface, when mobile phones are within network coverage.
- *Traffic flows* contain parameters that identify customer traffic flows. Each cuatomer traffic flow is further associated with QoS performance metrics, traffic profile and non-conformance actions, which are subcategories of traffic flows:
  - *QoS performance parameters* specify the QoS performance the flow will experience.
  - *A traffic profile* describes the traffic characteristics of the flow.
  - *Non-conformance actions* specify the treatment of traffic that violates the traffic profile. For example, non-conformance actions can drop or delay packets.
- A *service schedule* specifies the periods during which the SLA contract is valid. It may include daily, weekly or long-term service schedules.
- *Actions of SLA violation* specify the penalties that apply to the operator for violating the SLA contract. This may be a reduction in service fees or early termination of the contract.

The QoS performance parameters, traffic profiles and non-conformance actions are three integral parts associated with a customer traffic flow. While the QoS performance parameters convey an operator's promises, the traffic profile and non-conformance actions protect the operator's network from being overloaded with user traffic. In other words, they help the operator to keep the network in good shape to deliver on its promises.

What sets a 3G network SLA apart from that for fixed networks – such as fixed IP networks – is the information contained in customer traffic flows, QoS performance parameters and traffic profiles, each discussed in the following subsections, respectively.

9.3.5.1 Traffic flows

A generic traffic flow consists of a stream of data packets that share common character-istics – for example, have the same source and destination IP addresses and are carried by the same PDP context. A traffic flow is uniquely identified by a set of attributes called 'flow identifiers'. In IP networks, the attributes of a micro-flow include source and destination IP addresses, protocol and port numbers. In a 3G mobile network, however, the IP addresses of mobile phones are often dynamically assigned. Hence, additional attributes are needed to identify *customer traffic flows*. We propose the following flow identifiers for 3G mobile networks:

- An *IMSI* combined with an *MSISDN* can uniquely identify a subscriber of a mobile network. A range of IMSIs or MSISDNs can be used to identify a group of mobile users. (The SIP address may also be considered for IMS.)
- *Traffic class, allocation/retention priority* and *traffic handling priority* that are asso-ciated with a QoS profile.
- *Access point name* (APN), a logical name referring to a service or a network a user wants to access.
- *Source and destination IP addresses, protocol, port numbers and higher layer protocol information* of user IP packets carried by a PDP context. If static IP addresses or a pool of IP addresses (e.g., for a corporation) are pre-assigned to mobile phones, IP addresses can also be used to identify mobile users. Protocol and port numbers can identify an application. They may not be known at the time of signing an SLA. In such a case, high-level application names (e.g., FTP, video gaming, etc.) can be used instead. Network administrators may map these high-level names onto protocol and port numbers at a later time.

Not all the attributes must be present in order to identify a customer traffic flow. For example, streaming traffic to and from a corporation network named XYZ can be identified simply as:

- Traffic class = Streaming.
- APN = corporate_XYZ.com.

The above customer flow consists of an aggregate of traffic flows carrying streaming traffic. Packets belonging to the flow may be to and from different mobile phones and hence do not necessarily follow the same path.

9.3.5.2 QoS performance parameters

Once a customer traffic flow is identified, an SLA further lays out the QoS performance that will be experienced by the flow. Typical QoS performance parameters for a fixed IP network include packet delay, jitter, loss, throughput and bearer availability. Since a 3G mobile network is a connection-oriented network, we envision two types of QoS performance parameters for specification: PDP context session performance parameters that indicate the ability of a network to establish and maintain PDP context sessions, and packet data performance parameters that specify the delivery quality of the user IP packets carried by PDP contexts. The statistics apply to an identified customer traffic flow over a pre-defined period of time (e.g., a month).

PDP context session performance parameters

These parameters measure the ability of the network to establish and maintain PDP context sessions. Let N be the total number of activation attempts, S the number of activations that are completed successfully and F the number of PDP context activations that cannot be established due to network problems. We have:

$$N = S + F + a \qquad (9.39)$$

where a is the number of failures that are not caused by network problems. For example, failures caused by incorrect settings on mobile devices include:

- *PDP context session blocking ratio*: the ratio of PDP context activations that cannot be established due to network problems over the total number of PDP context activations. It is defined as:

$$F/N \qquad (9.40)$$

A mobile operator should only be responsible for failed attempts that are due to network problems such as network congestion. The operator should not be penalised for failures due to, for example, authentication failure or incorrect mobile phone settings. Hence, for every failed attempt, it is worth recording failure reasons.
- *PDP context session retain ability*: the ratio of PDP context activations that are completed successfully using a normal deactivation procedure over the total number of PDP context activations. This is defined as:

$$S/N \qquad (9.41)$$

This parameter measures how well a network can maintain PDP context sessions, as some established PDP context sessions might be dropped, for example, due to bad air interface quality.
- *PDP context session access time*: the 95th percentile of the distribution of delay in establishing a PDP context. The delay for establishing a PDP context is defined as the time between a mobile terminal sending an 'Activate PDP Context Request' and when it receives an 'Activate PDP Context Accept'.

Packet data performance

The following parameters measure the delivery quality of user IP packets carried by PDP contexts. A service access point (SAP) is the logical point between the UMTS bearer layer and the upper user IP layer (see SAP 4–5 in Figure 4.6). A service data unit (SDU) refers to the data carried by a UMTS bearer. It is typically a user IP packet. The parameters are:

- *Packet transfer delay*: the 95th percentile of the distribution of delay for all delivered SDUs of the traffic flow, where delay for an SDU is defined as the time from a request to transfer an SDU at one SAP to its delivery at another SAP.
- *SDU error ratio*: the fraction of erroneous SDUs. This includes those SDUs that are detected as erroneous and are dropped by the lower layer protocol. An indication to the upper layer is provided when the lower layer drops an erroneous SDU.
- *SDU loss ratio*: the fraction of lost SDUs. This excludes those erroneous SDUs that are dropped by the lower layer protocol.
- *Throughput*: the number of bits delivered by UMTS at an SAP within a period of time divided by the duration of the period. When a customer traffic flow contains an aggregate of PDP contexts, it may be desirable to specify the throughput for each PDP context as well. In which case, it shall be specifically pointed out that the throughput target is for the individual PDP context.

As an example, the performance targets for video streaming traffic from a corporate network may be:

- PDP context session blocking ratio $\leq 1\%$,
- PDP context session retainability $\geq 98\%$
- PDP context access time $= 500\,\text{ms}$
- Packet transfer delay $= 250\,\text{ms}$
- SDU error ratio $\leq 1\%$
- SDU loss ratio $\leq 1\%$
- Downstream throughput at the Gi $= 10\,\text{Mb/s}$
- Downstream throughput for each PDP context $= 500\,\text{kb/s}$

All performance metrics may be measured as presented in Section 9.3.3 over a period of 1 month.

9.3.5.3   Traffic profile

Traffic profiles are typically defined for and enforced at the entry points of a network. The attributes contained in a customer traffic profile shall include:

- *Maximum bit rate* (kb/s): the upper limit at which a user or application can send data to the network. The traffic is conformant with the maximum bit rate as long as it follows a token bucket algorithm where 'token rate' equals 'maximum bit rate' and 'bucket size' equals 'burst size'.
- *Guaranteed bit rate* (kb/s): sustained bit rate at which a user or application can send data to the network. The traffic is conformant with the guaranteed bit rate as long as it follows a token bucket algorithm where 'token rate' equals 'guaranteed bit rate' and 'bucket size' equals 'burst size'.
- *Burst size* (octets): the size of a burst that is allowed in the network. It is used with the maximum bit rate or guaranteed bit rate in the token bucket algorithm for conformance tests.

At times when a customer traffic flow consists of an aggregate of PDP contexts, it may be desirable to specify traffic profiles for both aggregate and individual PDP contexts.

9.4 Post-processing and statistical methods

Data collected from various measurements need to be post-processed by intelligent means to draw various conclusions according to given conditions. The aim of post-processing and data analyses is to discover pre-defined patterns in a set of measurements, so as to allow us to draw conclusions on the underlying process. These measurements may constitute a sample of a certain population of data under test, so that the eventual match with a given probabilistic distribution allows us to interpret possible irregularities in the population itself.

This section focuses on the procedures to be used for statistical calculations in the field of QoE/QoS measurement of mobile communications networks. It does not aim at giving detailed mathematical definitions, but just to present an overview of the main methods and various alternatives that should be used for correct data analysis, as, for example, presented in [18].

9.4.1 Data types

Data analysis methods are heavily dependent on the type of data to be analysed as well as on the scope of the analysis. Therefore, before analysis methods are introduced, the different types of data will be briefly reviewed. Four general categories of measurement results are expected when measurements are performed in mobile communication networks:

1. *Data with binary values*: that is, only two outcomes are possible (e.g., for voice calls every successfully completed call leads to the positive result 'call completed', every unsuccessfully ended call is noted as a 'dropped call', which represents a negative outcome).
2. *Data from time interval measurements*: that is, where the result is the time span between two timestamps marking the starting and end point of the time periods of interest (e.g., end-to-end delivery time for the MMS service).
3. *Measurement of data throughput*: that is, values which describe the ratio of transmitted data volume to required portion of time (e.g., 1 Mb transmitted in 60 s equals roughly 16.66 kb/s).
4. *Data concerning quality measures*: for example, evaluations of speech quality measured on a scale.

Measurements related to audio-visual quality can be done *objectively* by algorithms or *subjectively* by human listeners. The outcome of audio-visual quality evaluation is related to a scaled value that is called a 'mean opinion score' (MOS) for subjective testing.

Table 9.6 reports different kinds of QoE-related measurements, typical outcomes and examples.

Table 9.6 QoS-related measurements, typical outcomes and examples [18].

Category	Relevant measurement types	Examples
Binary values	Service accessibility, service availability	Service accessibility, telephony, service non-availability, SMS
	Service retainability, service continuity	Call completion rate, call drop rate
	Error ratios, error probabilities	Call setup error rate
Duration values	Duration of session or call	Mean call duration
	Service access delay	Service access delay WAP
	Round trip time, end-to-end delay	ICMP Ping round trip time
	Blocking times, system downtimes	Blocking time telephony, SGSN downtime
Throughput values	Throughput	Mean data rate GPRS
		Peak data rate UMTS
Content quality values	Audio-visual quality	MOS scores from subjective testing

9.4.2 Probability model and key parameters

If specific distributions are assumed, it is known from the mathematical literature (see, e.g., [19]) that they can be specified through parameters like the mean and variance of the dataset. Any variation in the dataset will generally result in a change of those distributional parameters, so that they constitute a useful way to 'compress' the information of the dataset in a suitable statistical distribution. It is important, though, to explicitly state any assumptions behind a given probabilistic process, as they only hold under precise mathematical constraints, and not knowing about them can lead to wrong usage of statistical applications and calculations. Therefore, to correctly interpret any statistical measurement, we need to know the hypotheses behind it and how well the data match these assumptions.

The probability model of a random process, which matches a certain distribution, can generally define a random variable. This is commonly denoted by:

$$X \sim Distribution(parameters) \tag{9.42}$$

where X is the random variable.

Once the conditions for a distribution are verified, we can proceed with calculation of their specific parameters – like mean, variance and quantiles. Data can be thought to be the physical 'materialisation' of a certain random process and the variable that describes it. This means that data analysis can be thought to be a sort of 'pattern-matching' search, in which data should be fitted with an appropriate distribution model (and its defining assumptions) so that conclusions and predictions can be drawn from it

The data types described in Section 9.4.1 differ substantially in their statistical behaviour, and their differences can be singled out in terms of continuous and discrete distributions.

Data with binary values can be described by discrete random processes, since the probability of getting certain results only refers to a fixed number of possible values. The same holds for quality measurements where subjective evaluations are normally spread on a scale with a limited number of possible values – that is, marks 1 to 6 or similar (e.g., when marking the quality of a service from 'poor' to 'excellent').

Time interval measurements, on the other hand, together with objective quality measurements and their quantitative variables, may be distributed over an infinite range of possible outcomes. In theory, since the possible results can assume an infinite number of values, the probability that a single exact value happens is 0. Probabilities greater than 0 are only meaningful for intervals with positive width.

During any measurement campaign, it will only be possible to collect a certain number of discrete measurements (spread over a wide interval of possible outcomes), because all the tools have a limited resolution and accuracy. This notwithstanding, all the measurements collected from these devices (within reasonable resolution limits) can be treated as being continuous. So, this leads us to assume that, despite the physical limits of measurement systems, continuous distributions can be assumed wherever appropriate. Abstract definitions for continuous and discrete distributions are normally given by their probability density functions and will be described hereafter.

Probability density functions (PDFs) define probabilities either for single results (*discrete* distributions) or for intervals of possible outcomes (*continuous* distributions).

A PDF is defined as a function f: [Set of real numbers] $\rightarrow [0, +\infty)$ satisfying these requirements [18]:

$$\begin{cases} f(x) \geq 0 \text{ for all } x \in S, \text{ where } S \text{ is the set of real numbers} \\ \int_S f(x)\,dx = 1 \text{ for continuous or } \sum_S f(x) = 1 \text{ for discrete distributions} \end{cases} \quad (9.43)$$

In other words, the values of a PDF are always non-negative (i.e., negative probabilities cannot be assigned to values nor intervals) and summation or integration over the PDF is always 1 (= 100%) – that is, all values will be assumed during the process.

Sometimes a cumulative distribution function (CDF) is used instead of a PDF, and the former can be calculated by summing (for discrete distributions) or integrating (for continuous distributions) over the probability values of the latter up to the current point.

Important quantities to define discrete and continuous distributions are *moments* and *quantiles*. The most important moments are (see, e.g., [20] for their mathematical definition):

- *Expected value* or *mean* μ (first central moment), which gives the position of distribution in the probability space. A sample-based estimator of the expectation of the underlying distribution is given by the empirical mean $\bar{x}$, which is defined as:

$$\bar{x} = \frac{1}{n} \sum_{i=1}^{n} x_i \quad (9.44)$$

where $x_i = 1, \ldots, n$ are the observations of the population sample of size n.

- *Variance σ^2* (second central moment), which gives the dispersion around the expected value. The variance of a distribution is commonly estimated by:

$$s = \frac{1}{n-1} \sum_{i=1}^{n} (x_i - \bar{x})^2 \tag{9.45}$$

- *Skewness* (third central moment), which expresses how symmetric the distribution is.

Whenever data distribution is asymmetric (i.e., skewness $\neq 0$), the usage of quantiles instead of moments becomes more appropriate. The α-*quantile* can be thought of as dividing the distribution in parts such that one part $\alpha * 100\%$ is smaller than an α-quantile and the other $(1 - \alpha) * 100\%$ is bigger than an α-quantile.

For example, to describe the time interval measurements gathered during a packet data session – for example, round trip time (RTT) or session setup time – we might at first think to derive them as sample mean and variance – or its square root (i.e., standard deviation). But, in practice, it often happens that a connection observes one or two RTT values that are way higher than the rest. These abnormal values greatly influence the sample mean and variance, so much so that the aggregated values no longer reflect 'typical' behaviour. Robust statistics have been developed to address these issues (see, e.g., [21]), providing us with statistics that are the least influenced by the presence of extremes, or 'outliers.'

One way to smooth away the effect of such outliers is to use the median, or 50th-quantile, for summarising the central location of a distribution, instead of the mean. Conversely from the mean, the median is virtually unbiased by any abnormal values or outliers. Therefore, use of the median is recommended as a robust estimate of a central location. Another statistic, which is robust enough for measuring variations around the mean (i.e., variance or standard deviation), is the inter-quartile range (IQR) [20]. The IQR is defined as the difference between the 75th quartile and the 25th quartile of a distribution. Thus, it properly defines a distribution's 'central variation'. It is likewise virtually unaltered by the presence of outliers (25% of the data have to be outliers to affect IQR), since these by definition fall outside the value range used to compute the IQR: hence, the IQR is robust. One could use the IQR instead of variance or standard deviation.

Network performance data often come in samples of the overall measurement population – that is, we can only collect a limited set of instrumental values because of space and time limitations. This means that the previously mentioned moments and quantiles must be estimated (i.e., evaluated over the limited amount of available data instead of over the whole population). Several empirical estimators have been devised to do this (see, e.g., [21]).

9.4.3 Distribution types

The most important continuous distributions (such as those for time-interval measurements) should be considered when making QoS measurements or traffic models (see [8] and [10]):

- *Normal* (or Gaussian or bell-shaped) distribution, commonly used when there is a symmetric deviation from the mean (e.g., for certain types of time measurements).
- *Log-normal* distribution, when the logarithm of the random variable is normally distributed (e.g., radio signals subject to slow (shadowing) fading).
- *Exponential* distribution, often used to model arrival processes (e.g., inter-arrival of call attempts to a mobile network).
- *Weibüll* distribution, describing rare but non-negligible processes due to their weight (e.g., fatigue in materials).
- *Pareto* distribution, useful also for heavy-tailed random processes (e.g., HTTP requests and FTP downloads).
- *Extreme* distribution (Fisher–Tippett distribution), for extremely rare events which have a big impact on the data (e.g., a single 100-GB download while the average traffic is around some tens of megabytes).

When it comes to measurement values only of integer type, the most important distributions are the:

- *Bernoulli* distribution, for a single test with a dual outcome (e.g., drawing a black/white ball once from a box).
- *Binomial* distribution, for multiple tests with a dual outcome (e.g., multiple draws, like the previous one).
- *Geometric* distribution, which models the probability of the first success after n consecutive failures (e.g., first successful streaming call after n tries).
- *Poisson* distribution, describing the number of events happening in a certain time period provided they happen at a constant arrival rate λ (e.g., call arrivals at a telephone exchange).

Examples of normal and binomial distributions are depicted in Figure 9.9.

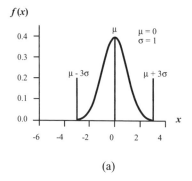
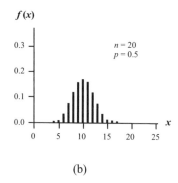

(a)   (b)

Figure 9.9 Examples of (a) normal and (b) binomial distributions. In the binomial distribution, n is the number of tests, m the number of successful test outcomes and $p(= m/n)$ the observed probability of successful outcomes.

9.4.4 Calculating the confidence interval

In order to decide whether a dataset realises one or another distribution, several test procedures have been devised (e.g., the chi-square or the Kolmogorov–Smirnov test). By running these tests on the available dataset, we can establish how closely measured values match a certain distribution [18].

Once the distribution is known, it is possible to calculate all the distributional parameters within certain confidence intervals, which describe an interval that covers the true parameter value with a certain probability. Confidence intervals are related to the statistical confidence in measurements or simulation results [8]. A percentage *confidence interval* – that is, $100(1 - \alpha)$, where $1 - \alpha$ is the *confidence coefficient* – correlates with the reliability (or repeatability) of performance results. In other words, the higher the number of collected samples, the more precise and trustworthy the results are.

Confidence intervals are heavily linked to the distribution type, so that it is extremely important to run a statistical test correctly in order to choose the most appropriate distribution for the dataset.

Two of the most common distributions for discrete and continuous processes are, respectively, the binomial and normal distributions. We now describe how to compute a confidence interval, denoted by $[p_1; p_2]$, up to level $1 - \alpha$ for the p value (proportion) of a binomial distribution and for the mean value μ of a normal distribution.

In the case of a binomial distribution, if the condition $n * p * q \geq 9$ holds – where n is the number of tests, m the number of successful test outcomes, p the observed probability of successful outcomes (i.e., $p = m/n$) and q the probability of unsuccessful outcomes (i.e., $q = 1 - p$) – then the distribution can be approximated by a Gaussian, which makes calculation of the confidence interval simpler. Hence, the values for p_1 and p_2 can be expressed as:

$$p_1 = \frac{2m + u_{1-\frac{\alpha}{2}}^2 - u_{1-\frac{\alpha}{2}} \cdot \sqrt{u_{1-\frac{\alpha}{2}}^2 + 4m\left(1 - \frac{m}{n}\right)}}{2\left(n + u_{1-\frac{\alpha}{2}}^2\right)}$$

$$p_2 = \frac{2m + u_{1-\frac{\alpha}{2}}^2 + u_{1-\frac{\alpha}{2}} \cdot \sqrt{u_{1-\frac{\alpha}{2}}^2 + 4m\left(1 - \frac{m}{n}\right)}}{2\left(n + u_{1-\frac{\alpha}{2}}^2\right)}$$

(9.46)

where $u_{1-\alpha/2}$ represents the $1 - \alpha/2$ quantile of the standard normal distribution $N(0, 1)$. (Note that $u_{1-\alpha/2} = u_{\alpha/2} = z_{\alpha/2}$, where $z_{\alpha/2}$ is the upper $\alpha/2$ percentage point of the $N(0, 1)$.) For example, if the confidence level is $1 - \alpha = 0.95$, then quantile $u_{1-\alpha/2} = u_{0.975} = 1.96$.

If the above condition is not valid, then the confidence interval has to be calculated from the binomial distribution itself. In this case, the calculation is much more complex and is based on the relationship between the binomial and the F distribution, giving as a result the so-called 'Pearson–Clopper limits' for the values of p_1 and p_2 (see any advanced statistical text, such as [20], [21]).

In the case of a normal distribution, the formula for calculating the confidence interval $[p_1; p_2]$ of the mean μ (assuming the standard deviation σ is known) is:

$$[\bar{x} - u_{1-\alpha/2} \cdot \sigma/\sqrt{n}; \bar{x} + u_{1-\alpha/2} \cdot \sigma/\sqrt{n}] \qquad (9.47)$$

where $\bar{x}$ is the estimated empirical value mean using (9.44), $u_{1-\alpha/2}$ the $1 - \alpha/2$ quantile of $N(0, 1)$ and n the sample size.

When σ is unknown it must be estimated using (9.45) and, in this case, the Student t-distribution rather than the normal distribution should be used. The formula for computing the confidence interval for μ when σ is estimated is:

$$[\bar{x} - t_{n-1,1-\alpha/2} \cdot s/\sqrt{n}; \bar{x} + t_{n-1,1-\alpha/2} \cdot s/\sqrt{n}] \qquad (9.48)$$

when $t_{n-1,1-\alpha/2}$ is the quantile $1 - \alpha/2$ of the Student t-distribution, it depends on the degrees of freedom $n - 1$ and the requisite level of confidence $1 - \alpha$. Moreover, $t_{n,\alpha}$: $t_{n,\alpha} = -t_{n,1-\alpha}$. As the values of a Student t-distribution are larger than those of a Gaussian distribution, the confidence interval when σ is estimated are wider than confidence intervals when σ is known.

Once the distribution and all its parameters are known, we will be able to properly visualise and analyse the data in terms of, say, the respect of certain thresholds or the trend over a certain time span. Different spatial or temporal data aggregation techniques can also be used to single out weird phenomena and raise specific performance alarms against an expected reference level.

9.4.5 Statistical confidence on measured data

A common issue in performing measurements (e.g., using a mobile QoS agent, see Section 9.6.1.4) is determination of the minimal sample size for a particular confidence interval on the measured parameter. The following sections present two methods for deriving the number of observations needed to reach a certain statistical confidence (precision) and accuracy on a measured proportion and time interval. Besides this, we present a criterion for setting the sampling rate to estimate the local average power of a mobile radio signal.

9.4.5.1 Statistical confidence on a proportion

In estimating a proportion p (e.g., unsuccessful call ratio), in situations where the sample size n (e.g., number of calls to be observed) can be selected, we may choose n to be $100(1 - \alpha)$ percent confident that the relative error $\Delta p/p$ is less than some specified value E. The appropriate sample size is [22]:

$$n = \frac{\sigma(\alpha)^2}{(E)^2} \frac{(1 - p)}{p} \qquad (9.49)$$

where $\sigma(\alpha)$ is the $(1 - \alpha/2) * 100$ percentile of the $N(0, 1)$, p is the expected unsuccessful call ratio and $E(= \Delta p/p)$ denotes the required relative accuracy for p.

Table 9.7(1) shows an example calculation of the required number of call attempts and needed mobile agents using (9.49). In this case, assuming an unsuccessful call ratio p of

Table 9.7 Examples of (1) needed mobile agents for a desired accuracy and statistical confidence on a proportion and (2) on a measure of time [22]; (3) is an example minimal sampling rate calculation for estimating the local average power of a mobile radio signal [23].

(1) *Relationship between the accuracy of the estimator of a proportion and the number of calls to be observed*

Confidence interval	**95 %**		
p	**5 %**		Proportion (e.g., *expected unsuccessful call ratio*)
$\Delta p/p$	**10 %**		Error (*required accuracy* for p)
Measurement interval	**3600** s	A	
Mean arrival rate per UE	**600** s	B	
α	0.050	C	
$\sigma(\alpha)^2$	3.841	D	
$(1-p)/p$	19.000	E	
$(\Delta p/p)^2$	0.010	F	
n	**7299**	G = D*E/F	*Required number of call attempts*
UE	**1217**	H = G/A*B	*Needed mobile agents*

(2) *Method of calculating the number of observations required for measures of time*

Confidence interval	**95 %**		
s	**0.2** s		Expected *standard deviation* of the call setup time (calculated from former measurements using (9.45))
$\bar{x}$	**3.5** s		Expected *mean value* of the call setup time (calculated from former measurements using (9.44))
a	**2 %**		Relative accuracy
Measurement interval	**3600** s	A	
Mean arrival rate per UE	**600** s	B	
α	0.050	C	
$\sigma(\alpha)^2$	3.841	D	
a^2	0.000	E	
$(s/\bar{x})^2$	0.003	F	
n	**32**	G = D/E*F	*Required number of observations*
UE	**6**	H = G/A*B	*Needed mobile agents*

(3) *Estimate of local average power of a mobile radio signal*

v	**3** km/h	A		UE speed
f	**2150** MHz	B		Transmission frequency
λ	**0.14** m	C		
Miminal sampling period	**134** ms	D = 1000*0.8*C*3.6/A		

5% with a relative accuracy E of 10%, at a confidence level of $100 * (1 - \alpha) = 95\%$, the required number of call attempts n to be observed is about 7300. Then, if a user makes one call every 600 s and the measurement period is 1 hour, the needed mobile agents are about 1217.

9.4.5.2 Statistical confidence on measurement of time

The number of observations n for quantitative variables (e.g., call setup time) depends on the variability of measurements. n can be estimated using the following formula [22]:

$$n = \frac{\sigma(\alpha)^2}{(a)^2}\left(\frac{s}{\bar{x}}\right)^2 \tag{9.50}$$

where $\sigma(\alpha)$ is the $(1 - \alpha/2) * 100$ percentile of $N(0,1)$, s is the expected standard deviation of the call setup time (calculated from former measures using (9.44)), $\bar{x}$ is the expected mean value of the call setup time (calculated from former measures using (9.45)) and a is the relative accuracy.

An example calculation of the required number of observations and needed mobile agents using (9.50) is illustrated in Table 9.7(2). In this case, the expected standard deviation and mean value of the call setup time are 0.2 s and 3.5 s, respectively, with a relative accuracy of 2%. Hence, at a confidence level of $100 * (1 - \alpha) = 95\%$, the required number of observations is about 32, and if a user makes one call every 600 s and the measurement period is 1 hour the needed mobile agents are about 6.

9.4.5.3 Estimate of local average power of a mobile radio signal

Another issue may arise when setting up (in the mobile terminal) the *sampling period* for power measurement of the radio signal. In order to minimise the variance of the estimate of the local mean of the field strength, we need to operate on samples of a route whose length is between 20 and 40 wavelengths. Should the estimate not deviate more than 1 dB from its real value with a degree of confidence of 90%, the minimum number of independent observations must be at least 50 [23].

On these grounds, let v be the mobile speed (in m/s), f the downlink transmission frequency of the signal (in hertz) and λ the corresponding wavelength in the vacuum (in metres); the minimal sampling rate r (in seconds) is thus given by:

$$r = 0.8\frac{\lambda}{v} \tag{9.51}$$

In the example given in Table 9.7(3), assuming the terminal moving at 3 km/h and the transmission frequency of the downlink signal of 2150 MHz, the minimal sampling period is 134 ms.

9.5 Mapping between QoE and QoS performance

In order to define the quality of end-user experience on a cellular network, we should conduct proper psychological studies into human perception for every mobile application. Only then would it be possible to frame the experiences that drive the satisfaction of an end-user, defining certain QoE indicators for each service (like the ones described in Chapter 2: MMS, WAP, etc.). For these indicators, we can thereby distinguish two types of quality measurement: *subjective* and *objective* measurements. If we identify quantitative measures which are highly correlated to the quality of interest, this will simplify the analysis. However, if this is not possible, some kind of evaluation on a standardised scale by qualified experts is needed. The result may therefore be given either as the measurement result or as a mark on a pre-defined subjective scale.

Luckily, at least some of these evaluation studies are already available for applications on other networks like the Internet, whereby factors affecting end-user satisfaction have been studied through subjective performance tests (see, e.g., [24] and [25]). The important point here is that while these psychological satisfaction factors (e.g., the expectation of first page rendering time) remain more or less constant to keep users happy, marked differences in throughput and delay between the Internet and a mobile network must be taken into account properly to see how user perception of network quality changes.

One of the ways that could be used to measure the overall satisfaction of a user for an application is to map the previously mentioned QoE indicators onto easily measurable QoS network performance indicators. In this way, every operator could simply measure the (average) end-user perception of the network from the network management system. The complexity of this approach lies in one-to-many mapping between the QoE and QoS and in the aggregation levels of performance counters for the different parts of the network. The first is due to the fact that one QoE indicator is normally related to many QoS parameters, and it is not easy to establish a mathematically unique relationship between them. Some authors have tried to do that – for example, [26], where the satisfaction of the user (= QoE), represented as 'cost', is mapped onto a number of application level parameters by means of a standard mathematical regularisation technique for the related cost functional. The second reason is that although it is relatively easy to get cell-based performance counters from any element in the access network, it is not at all easy to get them from the core part of the network because any SGSN, GGSN or MSC normally collects these data only at a higher aggregation level than cells on their own (e.g., access point nodes).

A good example of this approach was presented in [27] where the authors try to correlate quantities like objective network conditions (e.g., bandwidth and latency) to the subjective perception of the Internet service by a human user. In this study, QoE is defined as 'cancellation rate' – that is, the number of cancelled HTTP requests over the total number of HTTP requests (if network quality is not good, the end-user gets impatient and cancels the download of a specific Internet page). QoS, in contrast, gets defined through the following network quantities:

- Response time – that is, time to first byte from the HTTP request.
- Delivery bandwidth – over sufficiently large web objects for accuracy (e.g., $>8\,\text{kB}$).
- Object delivery time – that is, time to last byte (from the HTTP request) for web objects.

By mapping, say, the cancellation rate (= QoE), as measured by tracing the corresponding HTTP messages, onto response time (= QoS), we can build an empirical relationship of how QoE as defined in this case depends on QoS. The resulting graph is depicted in Figure 9.10.

As the authors [27] point out, it does not make sense to try to improve the response time in the 50–500-ms range, as any additional improvement here will not result in a significantly better cancellation rate. A similar methodology can help operators to

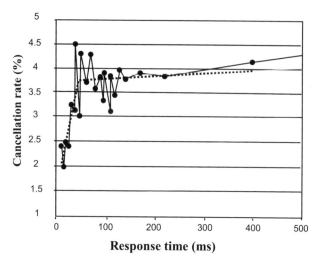

Figure 9.10 QoE vs. QoS mapping. Cancellation rate ($=$ QoE) dependence on response time (QoS).
Taken from [26].

correctly gauge their end-to-end network optimisation efforts ($=$ QoS) vs. the right levels of QoE for different applications.

In fact, following a similar approach we could, say, measure WAP characteristics on a mobile network and build a similar mapping between end-user perceived quality and those network quantities which influence the performance of the service itself. An example of WAP/xHTML browsing QoE–QoS mapping is shown in Table 9.8.

A precise and full characterisation of the end-user perceptual domain for each application must await further developments in psychophysics, though.

9.6 QoE and QoS monitoring tools

This section introduces the tools used for QoE and QoS monitoring in UMTS cellular networks. More information on the topics presented may be found on the webpages

Table 9.8 Example of a WAP/xHTML browsing QoE–QoS mapping.

QoS KPIs	QoE-subjective scale				
	Excellent	Very good	Average	Fair	Poor
End-to-end delay (median)	$\leq 2\,s$	$\leq 4\,s$	$\leq 8\,s$	$\leq 15\,s$	$\geq 15\,s$
Packet loss ratio	$\leq 0\%$	$\leq 0.1\%$	$\leq 1\%$	$\leq 5\%$	$\geq 5\%$
Mean throughput	$\geq 200\,kb/s$	$\geq 120\,kb/s$	$\geq 60\,kb/s$	$\geq 20\,kb/s$	$\leq 20\,kb/s$

reachable thorough the shortcuts reported on the list of references at the end of the chapter.

9.6.1 Introduction to QoE monitoring tools

QoE monitoring tools provide mobile operators with the ability to measure end-user perceived service quality. The tools generally fall into different categories. The following subsections explain the general categories and provide some examples of tools. See also Section 9.7, where a complete service assurance solution for an NMS is described.

9.6.1.1 Application layer tools

These tools are capable of generating a different kind of traffic and have the ability to run various service applications from the user end. The tools in this category distinguish themselves for their ability to handle a variety of traffic types, automation, ease of use, logging, reporting, etc. Examples of supported measurements include Ping for round trip delay monitoring, FTP, HTTP browsing, MMS, SIP, WAP, etc. Measurements can be shown in real time or stored for post-analysis.

In many cases, these tools not only handle application layer data but also record other vital information about radio conditions, signal quality, etc. In its simplest form, this function can be performed by using the tools and accessories available in normal operating systems (e.g., browsers, media players, etc.). But logging and automation would be a challenge in this case.

In a practical measurement scenario no single tool would be sufficient for the whole range of available applications. It would be unrealistic to assume a single tool at this stage capable of running HTTP and WAP browsing, FTP uplink and downlink, email with different protocols, audio and video streaming, PoC, MMS, etc. Multiple tools are required to handle these functions appropriately.

There are several tools for application layer testing available in the market. An example of a lightweight Windows-based tool used for automated application measurements may be found in [28]. Various scripts can be written as simple text files and the tool will act accordingly. Supported measurements include Ping for RTT with and without an interval, FTP downlink and uplink, and HTTP browsing. Measurement activity is shown in real time graphs at the application and RLC layer. These measurements are also recorded in comma-separated value (CSV) files for detailed post-processing and replay. Another example of a tool for active measurements is described in Section 9.7.2.

9.6.1.2 Field measurement tools

Field measurement tools (FMTs) are tools capable of monitoring the air interface, giving information not only on the session established but also on air interface conditions, such as coverage level, interference, radio channels assigned to the terminal, etc.

FMTs offer the possibility to monitor certain service QoE by properly configuring the script that generates traffic (e.g., Internet browsing, FTP down/upload, etc.). By monitoring the air interface, the QoE experienced during each session can be correlated to the air interface performance to observe whether the QoS offered (e.g., cell reselection delays) has any adverse impact on service QoE. An FMT also gives the possibility to monitor network resources, channel allocation problems, (E)GPRS or WCDMA availability, high interference level in certain areas that may prevent expected QoS to be attained, etc.

There are many tools available on the market with the above capabilities. Most provide cross-technology scalability and allow early verification, troubleshooting, optimisation and maintenance of network deployments for 2G, 2.5G and 3G wireless networks. An example of a field measurement tool can be found in [29].

9.6.1.3 Protocol analysers

Protocol analysers are vital for QoS and QoE measurements in the network. They have the ability to capture traffic non-intrusively at high speed at various network interfaces. Protocol analysers are generally categorised in two groups: for TCP/IP traffic and for telecom protocols. Protocol analysers in the first category are widely available and have the ability to capture and decode user protocols from the physical layer all the way up to the application layer. They are normally less expensive, and the ease of use, decoding ability for various protocols, reporting capability, etc. are the main differentiating factors.

Examples of tools in this category include Ethereal [30] and TCPdump [31]. To get an in-depth understanding of the captured traffic, tools such as TCPtrace [32] can be used for further analysis of TCPdump files.

For telecom interfaces and protocols, special protocol analysers are available that can connect to a specific interface and decode all the protocols used in that interface. Extra features such as post-processing ability, deciphering functionality and usability, etc. are the distinguishing factors.

9.6.1.4 Mobile QoS agent

A mobile QoS agent is a piece of software running on a standard mobile terminal for the purpose of measuring service quality. (In practice, Layer 1–7 protocols in the control and user plane stack implemented in the UE may be measured. This includes signalling and radio parameters for troubleshooting, throughput, delay, jitter, and packets lost or detected as erroneous for integrity monitoring.) It conducts active probing and/or passive monitoring according to a configurable monitoring profile installed on the phone. Because of their proximity to end-users, mobile agents are able to accurately capture end-user service experience. Mobile agents can be installed on a selected group of end-user mobile devices. This can effectively turns thousands of phones into service quality probing stations.

In addition to being valuable tools for mobile operators in monitoring multimedia service quality, mobile QoS agents are also important tools for emerging mobile virtual network operators. MVNOs typically do not own any network infrastructure but buy network capacity from network operators to offer their own branded mobile subscriptions and value-added services. Typically, there are service level agreements between an MVNO and the mobile operator from whom the MVNO purchases network capacity. Since MVNOs do not have access to the network infrastructure, mobile QoS agents become an effective means for them to monitor service quality and to monitor whether the mobile operators adhere to service level agreements.

The responsibilities of a mobile QoS agent include measuring mobile multimedia service quality, and producing and reporting performance statistics to central management servers. A monitoring agent can conduct both active and passive measurements. For active measurements, the agent actively initiates services and records the performance of each service instance. For passive measurements, the agent observes the services initiated by an end-user and records the service performance.

A QoS agent periodically submits reports to a central server. Since mobile devices move around networks when recording measurements, an agent also attaches time and location information to each service instance monitored.

A QoS agent is also configured with a set of thresholds for each service. When the statistics for a service exceed the thresholds, a monitoring agent will warn the server.

A central management server derives key performance indicators from the reports provided by QoS agents; it also manages QoS agents by dynamically dispatching, installing, activating or deactivating them on the phones.

Key performance indicators are performance statistics that are crucial for evaluating the QoS provided over a network. The statistics computed by each monitoring agent are not considered to be KPIs because they reflect the service experience of a single user only. However, a central management server can derive KPIs from reports made by multiple agents.

These KPIs show the level of service quality within a given time and space. When KPIs for different areas are computed, a 'service weather map' can be effectively constructed with different colours indicating the performance in each area. Such a map would be a very convenient tool for monitoring and displaying overall service quality.

Figure 9.11 shows the key functional elements of a mobile QoS agent. The agent uses the SyncML DM Protocol [33] as its interface to external servers.

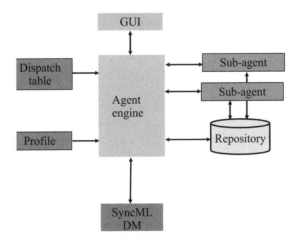

Figure 9.11 MQoS agent functional elements.

The agent engine is the core element that co-ordinates activities. It parses commands received from SyncML DM and takes corresponding actions. It composes XML messages for reporting measurements or configurations to the external server as scheduled or as requested by the server.

For active measurements, the agent engine schedules and activates sub-agents according to the monitoring profile and makes sure that no two sub-agents are monitoring at the same time. This is to ensure that the monitored performance is not skewed by competing for resources on the phone. A sub-agent notifies the agent engine after a measurement instance is complete. The agent engine waits for at least a pre-defined interval (e.g., 5 s) before activating another sub-agent. This is to allow any background tasks associated with the first sub-agent to run down. The agent engine also starts a timer after instructing a sub-agent to take a measurement. If the sub-agent does not report back before the timer runs out, the agent engine can delete the sub-agent.

At power-up, the agent engine reads the dispatch table from the file. It also queries the repository to see which active monitoring profile was used before the power was turned off. It then reads the profile into memory. If the repository query returns an empty result or the returned profile name does not exist as a file, the agent reads in the default monitoring profile and enters into the repository the default profile as the active profile. The agent engine also queries the repository for the list of services that was being monitored before power-off and their next reporting times. It then resumes monitoring by rescheduling the monitoring of different services according to the active monitoring profile.

When the server sends a new monitoring profile, the agent stores it as a local file. When the server instructs replacement or deletion of a monitoring profile, the agent engine will act accordingly.

The agent engine also instructs the GUI to display the status or progress of measurements. On mobile devices dedicated to drive-through tests, the agent engine also receives commands from the GUI and acts accordingly. All commands that are sent from the server can also be input from the GUI.

A sub-agent takes care of the details of monitoring a specific service, storing, retrieving and deleting measurements from the repository, and aggregating measurements to derive performance indicators for the specific service.

The dispatch table is a table that maps service names to sub-agent names. When the agent receives a command that contains a service name (e.g., 'start browsing'), it consults the dispatch table to determine which sub-agent it shall activate or pass the command to.

The dispatch table is stored in a file that comes with installation of the agent. At power-up, the agent will read the file and store the dispatch table in memory. The dispatch table file is updated whenever a new sub-agent is added for monitoring a new service.

The use of a dispatch table and sub-agents makes the agent architecture scalable. To add the function of monitoring a new service, we only need to add a new sub-agent and a new row in the dispatch table, leaving the rest of the codes unchanged.

The repository stores measurements and persistent data such as agent ID, the name of the active profile, the list of services that are being monitored, etc. The agent engine and sub-agents both have access rights to the repository.

The GUI displays measurement-related information. On mobile devices dedicated to drive-through tests, the GUI is used to display the status of the agent. Examples of

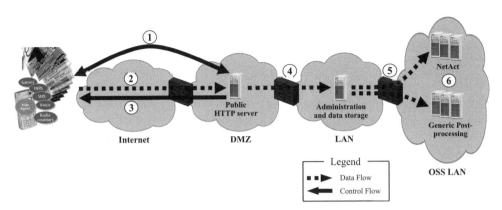

Figure 9.12 Mobile quality analyser system architecture. ① Self-registration, MQA agent and configuration download. ② Secure measurement reporting via HTTPS post. ③ Download and update of MQA configuration if it is available. ④ Measurement collection in database. ⑤ Measurement forwarding to reporting system (e.g., Nokia NetAct™, third-party, etc.). ⑥ Data analysis and report generation.

displayed status are 'ready to make measurements', 'accessing *www.yahoo.com*', 'no response from server'. On end-user devices, whenever an active measurement is in progress, the GUI displays a message 'testing in progress, please wait for 30 seconds at least before starting any application'. For passive measurement on end-user devices, nothing is displayed since all measurements are conducted silently in the background.

Another function of the GUI is to allow user input. On mobile devices dedicated to drive-through tests, the GUI allows someone to conduct a test at any place and any time. The test results are also viewable from the GUI. In addition, the user shall be able to locally issue all the commands that can be remotely issued by the server.

The profile contains all instructions on how to monitor each type of service.

In addition to the MQoS agent described above, there are also simple local MQoS management functions (not shown in Figure 9.11) that install, upgrade, start and stop the agent.

An example of a commercial tool is available in [34]. The system architecture thereof is illustrated in Figure 9.12.

9.6.2 Introduction to QoS monitoring tools

In Section 9.6.1 we introduced the tools for QoE monitoring in order to obtain information on end-user perceived service quality. Equally important for mobile network operators is the ability to monitor QoS performance, in order to ensure that the network is able to provide satisfactory services. To assess the performance of the network, two different approaches can be used:

- Passive approach – that is, a non-intrusive way of collecting and analysing measured data.

- Active approach – that is, a monitoring process that implies the generation of traffic in a controlled manner and the analysis of the involved network elements performance.

The advantage of the passive approach is the small influence of the monitoring process on system performance: The network is seen as it is. However, for the statistical reliability of performance results, a longer monitoring period is required resulting in a huge amount of data. In the passive monitoring approach, the metrics presented in Section 9.3.3 and in Section 9.3.4 may be implemented to measure QoS performance in the BSS/RNS and core/backbone networks, respectively. These tools – mostly suitable for passive measurements – are useful for general performance monitoring and troubleshooting activities, since they offer an overview of actual network behaviour and data can be collected automatically, usually over a long period of time. They mainly comprise the general operation and management tools that allow network operation and implementation of certain types of measurements in the network elements, which offer a certain degree of monitoring.

The active approach is more statistical in nature and allows end-to-end analysis of the network and emulation of QoE when accessing the available services in the network. Some of the tools used in active tests are capable of generating traffic – e.g., on an interface – or recreating end-user sessions to analyse the end-to-end performance of the network. The first type may be used to create situations of high load and monitor the network/interface response. The second type can simulate user behaviour in certain locations of the network and collect bearer service statistics, such as service access time, throughput, session failure rate, etc.

Passive and active monitoring approaches complement each other, and selection of a particular tool depends on the use case. The same tools can be used for both active and passive measurements. A comprehensive QoE and QoS measurement tool setup is illustrated in Figure 9.13 and an example complete service assurance solution for an NMS is presented in Section 9.7.

9.7 Example of complete service assurance solution for NMS

In this section we present an example centralised performance management solution for an NMS. As part of this framework, tools for active service monitoring and a tool for service quality management are also described.

9.7.1 Centralised performance management

An example architecture for a centralised PM solution is shown in Figure 9.14. The data are collected from the network element layer, stored in performance data databases, from where they are pre-processed, summarised and prepared for visualisation and textual reporting. As already pointed out in Section 9.1.1, it is necessary for a NE to retain the measurement data it has produced until they have been sent to, or retrieved by, destination management systems. The storage capacity and duration for which the data need to be retained at the NE are operator- and implementation-dependent. If measurement

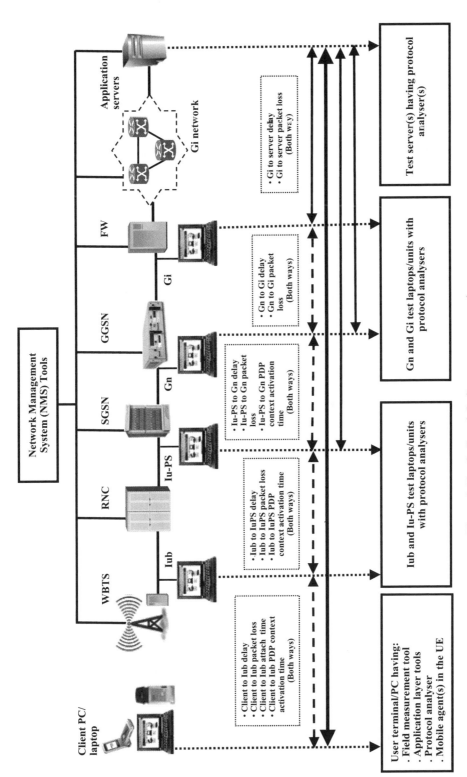

Figure 9.13 QoE and QoS measurement tool setup.

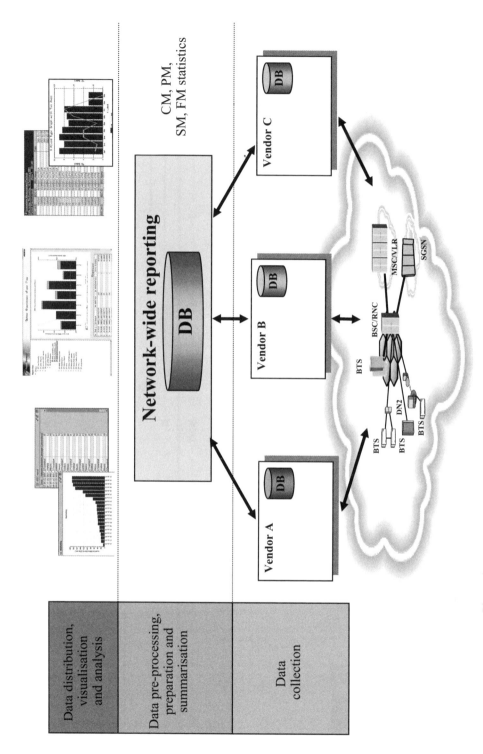

Figure 9.14 Centralised performance monitoring solution for a multivendor environment.

results are routed to an NM via the EM, then it is necessary for the EM to retain the data at least until they have been successfully transferred to the NM.

Typically, the measurement results produced by network elements are transferred to an external NMS for storage, post processing and presentation to the system operator for further evaluation.

Generally, the differentiating factor between an EM and NM function is related to the amount of data. The reporting functions at the NMS level contain information from a multitude of network elements network-wide. The requirement of being network-wide can limit the usability of certain data sources, because there are numerous data sources that do not provide naturally network-wide data (e.g., field measurement tools).

In addition to network-wide highly summarised data, an NMS system supports drill-down functionalities, in order to zoom in to the needed details (e.g., in troubleshooting cases). Summarisation can be done in terms of elements/objects, time or the measurement content. Element level summarisation provides averaged reports on network elements, such as RNC level KPI reporting. Time axis summarisation provides, for example, daily average or busy hour KPIs, instead of collecting values individually from NEs. Measurement content summarisation is related to the grouping of measurements. In high-level reports, for example, the handover performance figure is presented as one value. In the case of poor handover performance, support is offered to zoom in to the details of different handover types.

The power of an NMS lies in the fact that large amounts of data are available, and different views of such data can be provided. If the scope is the whole network (high numbers of objects), then it is obvious that the detail level of data in the report is small. Users should be supported so that they can easily find the relevant area and focus. In high-detail views the number of elements to be handled is typically lower, but different kinds of data are available, as are time series or distributions.

If the level of detail is very high (high number of counters, KPIs, time series, etc.) then the number of objects should be really low. The reasons are (a) the user cannot absorb too much information at a time and (b) data interfaces have a limited performance.

In addition to the collection and storage of PM data from the entire mobile network, other relevant data – such as configuration, alarm and service-relevant data – are stored in the solution. Performance management solution benefits from including configuration-specific or alarm data in the analysis. This aids manual troubleshooting and optimisation cases, one example being the combination of the configured theoretical maximum and measured throughput for a service. By comparing the CM/PM data from a network it can be concluded whether the current configured network capacity is sufficient when compared with current network performance data.

The actual KPI content makes the reporting solution QoS-specific. An example reporting content (KPIs) can be found in Section 9.3. An example of textual NMS level reporting application output is given in Table 9.9. Details on how to administer the measurement process can be found in [16].

9.7.2 Active, service monitoring tools

Traditional network management is based on monitoring individual network elements (NEs). This gives a view of the network and service status, but does not always guarantee

Table 9.9 An example report generated by an NMS application. Downlink DCH throughput per traffic class – RSRAN011 (2004.12.06–2004.12.19). Time aggregate: whole period. Object type: RNC, object(s): 45306021. Object aggregation: WCELL (rows 1–14/14).

Date	RNC name	WBTS name	WCELL code	WCELL name	WCELL ID	Allocated DLDCH capacity for CS voice	Allocated DL dedicated channel capacity for CS voice	CS conversational	CS conversational (Erlangs)	CS conversational (min)	Allocated DL dedicated channel capacity for CS streaming	Allocated DL dedicated channel capacity for data calls	Allocated DL dedicated channel capacity for PS streaming	Allocated DL dedicated channel capacity for PS interactive	Allocated DL dedicated channel capacity for PS background
Total	RNC4	Hippos	45321021	WHippos-1	1	0.00	0.00	0.00	0.00	0.00	0.00	0.00	0.00	0.00	0.00
Total	RNC4	Hippos	45325021	WHippos-5	5	6.47	6.39	7.54	0.12	105.97	0.00	3.81	0.00	0.00	0.10
Total	RNC4	Makkyla	45318021	WMakkyla-1	1	0.00	0.00	0.00	0.00	0.00	0.00	0.00	0.00	0.00	0.00
Total	RNC4	Makkyla	45319021	WMakkyla-3	3	0.00	0.00	0.00	0.00	0.00	0.00	0.00	0.00	0.00	0.09
Total	RNC4	Perkkaa	45329021	WPerkkaa-3	3	0.14	0.12	0.08	0.00	0.58	0.00	0.04	0.00	0.00	0.00
Total	RNC4	Sateri	45312021	WSateri-1	1	8603.56	8603.56	772.88	12.08	31156.78	0.00	56876.86	90366.18	35510.27	35526.31
Total	RNC4	Sateri	45313021	WSateri-2	2	2093.65	2093.65	4985.16	77.89	140207.74	0.00	58480.38	0.00	30450.46	62156.83
Total	RNC4	Sateri	45314021	WSateri-3	3	3476.66	3476.66	707.15	11.05	31158.69	0.00	126449.82	0.00	15040.34	136299.80
Total	RNC4	Sateri	45315021	WSateri-4	4	2120.55	2120.55	1146.02	17.91	124629.36	0.00	27236.25	44037.79	13186.06	22669.86
Total	RNC4	Sateri	45316021	WSateri-5	5	3019.20	3019.20	0.00	0.00	0.00	0.00	48969.89	8404.63	9304.65	51825.75
Total	RNC4	Sateri	45317021	WSateri-6	6	7832.44	7832.44	0.01	0.00	0.61	0.00	54506.60	114091.13	30334.46	41651.77
Total	RNC4	Sello	45330021	WSello-1	1	1463.63	1463.63	0.00	0.00	0.00	0.00	0.00	0.00	0.00	0.00
Total	RNC4	Sello	45331021	WSello-2	2	5428.94	5428.94	0.00	0.00	0.00	0.00	0.00	0.00	0.00	0.00
Total	RNC4	Sello	45332021	WSello-3	3	1.32	1.32	3.92	0.06	22.02	0.00	1.96	0.00	0.00	0.00

WCELL = WCDMA CELL.
WBTS = WCDMA BS (Node B).

that the end-user service itself is working effectively. Typically, these passive measurements, like counters and gauges from network elements, provide a statistical view of a component in the whole end-to-end chain. The network load situation can be analysed for business and resource (HW/SW) planning, and problems in the network may be found which currently have no impact on the service, but will have if not fixed in time.

Active measurements are implemented by probes performing regular tests at scheduled intervals. Service usage is emulated in the same way as if an end-user is using the service. Probes are distributed around the network to gather results from all its parts. With consistent test transactions performed at regular intervals, fault situations are immediately detected. Comprehensive statistical data are obtained as well, offering a reliable analysis of historical trends. Utilising active probing also allows proactive fault detection.

Active measurements generate additional traffic in the network. However, with a well-planned configuration of QoS data collection, traffic can be minimised, and in normal conditions the volume of test traffic is likely to be very small compared with actual end-user traffic. Due to the concept it is an easy way to monitor multivendor networks, as well as parts of the network that do not belong to the operator such as transmission lines, application server and IP backbone.

Passive network and active service monitoring should not be seen in opposition but as complementary methods to get the most out of the network. Probes can be centrally managed by an administration tool and can be distributed around the network. Measurements can be combined with other NMS level performance management applications. Active measurements can provide a comprehensive solution for end-user service monitoring in mobile operator and service provider environments.

Probe(s) collect QoS data from the services and forward the resulting data to the NMS, where monitoring, analysis and reporting products utilise the data. The probe verifies the service at regular intervals by simulating end-user behaviour. The service can be assessed end to end, meaning from the mobile to the application server (e.g., a WWW or MMS server). Another possible scenario includes verifying the service, starting at the Gb, Gn or Gi interface and ending at the application server. In the case of Gb or Gn the BSS (respectively, the SGSN) is emulated. This allows an isolated insight into how the radio access network, packet core network or service platform behave for a certain service. Example use cases are given in Figure 9.15.

Probes will produce PM and FM data for each service verification. Alarms are generated based on the number of configurable thresholds per probe. This information is sent to the NMS where several applications can use these data to come to a conclusion about both network and service performance. An example commercial tool for active measurements can be found in [35].

9.7.3 *Service quality manager*

As discussed in Chapter 8 the static business process model of TOM/eTOM [36] identifies the basic operational functions within telecom operations. These functions are the building blocks of different processes (e.g., those related to service assurance). The TOM model (see [37] and [38]) also points out the undisputable connection between

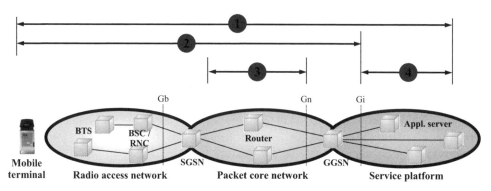

Figure 9.15 Example use cases for active measures. ① Verify QoE end to end via an MT connected to a stethoscope. ② Verify the mobile network only (GPRS or 3G). ③ Verify the IP backbone between core sites. ④ Verify the service platform (services offered via an AP).

the network layer and the service layer. Thus, the service quality manager (SQM) concept contains measures and elements from both layers.

The SQM concept must support the TOM service problem resolution process. This includes the functionality to identify and isolate the root cause of service-affecting failures. This is achieved by combining information related to alarms, KPIs and key quality indicators (KQIs) that has caused or can indicate service-related problems. In addition, the SQM concept can be linked to the trouble ticketing machinery of the network operations department.

The TOM service quality management process is engaged in monitoring service or product quality on a service class basis. The basic concept of SQM effectively supports modelling of and monitoring service classes. This enables observation of the service levels and general problems of different service classes for life-cycle management of the service product portfolio. Service-specific monitoring is determined by different rule sets. These rules can contain measures from, say, network elements, application servers, user equipment, mobile QoS agents, etc. Each rule set is refined to provide optimal monitoring capabilities for each service class (e.g., value-added services), comprising those elements most crucial in reaching a conclusion about service performance from the end-user point of view.

The SQM concept supports the service assurance process. All service-relevant information that is available in the operator environment can be collected. The information forwarded to the service quality manager is used to determine the current status of defined services. The current service level is calculated by service-specific correlation rules. There are pre-defined correlation rules for different types of services (e.g., MMS, GSM, WAP services) and the operator or a third party can develop new rules.

The general concept of SQM is illustrated in Figure 9.16.

Passive data provide information about the alarm situation and performance in terms of KPIs and KQIs within individual network elements. Depending on network element capabilities and implementation in the network element, passive measures can contain service class (priority class) related information as well.

RT traffic data related to charging and billing records provides additional information, which can be used to get a very detailed view of specific services.

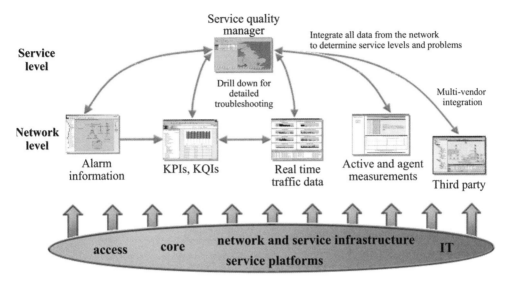

Service level

Network level

Alarm information

KPIs, KQIs

Real time traffic data

Active and agent measurements

Third party

Service quality manager

Integrate all data from the network to determine service levels and problems

Drill down for detailed troubleshooting

Multi-vendor integration

access   core   **network and service infrastructure**   IT

service platforms

Figure 9.16   The service quality manager and examples of possible data sources.

Active measurements (probing) and mobile QoS data (see Section 9.6.1.4) complement well previous data sources, providing a view on service usage and performance from the customer perspective.

All these different data sources can be combined in SQM. SQM correlates the data in order to provide a global view from the end-user perspective. The combined data can thus be used for several purposes. Some use cases were presented in Section 9.1.1.

In addition, the SQM concept can contain drill-down functionality to all underlying systems in order to allow troubleshooting and root cause analysis to aid the service problem resolution process. This is depicted in Figure 9.17.

The inference engine in SQM correlates all data in near-real time and determines the current service status of defined services. As soon as service degradations are detected, SQM will raise up any corresponding service problems (e.g., in terms of trouble ticketing or visual representation). Thus, the SQM concept can be used in SLA monitoring (see Section 9.3.5). SLAs that have been agreed upon with customers or within the organisation are setting cornerstones to determine service quality objectives. Measurable items derived from the SLAs could be used as thresholds to detect current service level violations. In Figure 9.18 a conceptual view of service level management using SQM is shown.

A description of a commercial product for service quality management in compliance with the above characteristics and functionalities can be found in [39].

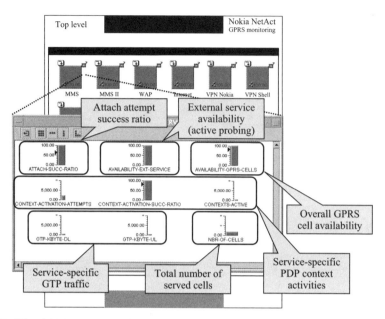

Figure 9.17 The SQM concept. The high-level view indicates service quality violation with 'discretised' severity. The concept allows zooming in to individual attribute levels for troubleshooting purposes. Ultimately, it is possible to follow the time series of an individual measurement to isolate the problem related to a service.

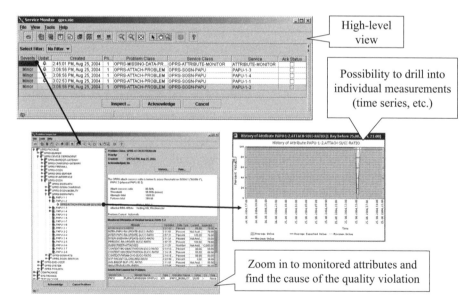

Figure 9.18 An example of service level monitoring. High-level view contains individual services like MMS, VPN, etc. Lower level view zooms in to individual measurements that could be part of a service level agreement.

References

[1] 3GPP, R6, TS 32.101, Telecommunication Management; Principles and High Level Requirements, v. 6.10.0.

[2] ETSI, TS 102250, Speech Processing, Transmission and Quality Aspects (STQ); QoS Aspects for Popular Services in GSM and 3G Networks, v. 1.2.1.

[3] ANSI, T1.522-2000, Quality of Service for Business Multimedia Conferencing.

[4] ITU-T, G.1010, End-User Multimedia QoS Categories.

[5] 3GPP, R5, TS 23.107, QoS Concept and Architecture, v. 5.13.0.

[6] 3GPP, R5, TS 48.018, BSS GPRS Protocol (BSSGP), v. 5.12.0.

[7] D. Soldani, N. Lokuge and A. Kuurne, Service performance monitoring for GPRS/EDGE network based on treatment classes, *IEEE, 12th International Workshop on QoS, iwQoS04, June, 2004*, pp. 121–128.

[8] D. Soldani, QoS management in UMTS terrestrial radio access FDD networks, dissertation for the degree of Doctor of Science in Technology (Doctor of Philosophy), Helsinki University of Technology, October, 2005. See *http://lib.tkk.fi/Diss/2005/isbn9512278340/*

[9] 3GPP, R5, TS 25.215, Physical Layer – Measurements (FDD), v. 5.7.0.

[10] 3GPP, R5, TS 25.214, Physical Layer Procedures (FDD), v. 5.11.0.

[11] 3GPP, R5, TS 25.212, Multiplexing and Channel Coding (FDD), v. 5.10.0.

[12] 3GPP, R5, TS 25.331, Radio Resource Control (RRC) Protocol Specification, v. 5.15.0.

[13] 3GPP, R5, TS 25.302, Services Provided by the Physical Layer, v. 5.9.0.

[14] H. Holma and A. Toskala (eds), *WCDMA for UMTS*, John Wiley & Sons, 3rd Edition, 2004, 450 pp.

[15] 3GPP, R5, TS 25.322, Radio Link Control Protocol Specification, v. 5.12.0.

[16] 3GPP, R5, TS 32.403, Telecommunication Management; Performance Management (PM); Performance Measurements – UMTS and Combined UMTS/GSM (Release 5), v. 5.12.0.

[17] 3GPP, R5, TS 25.413, UTRAN Iu Interface RANAP Signalling, v. 5.12.0.

[18] ETSI, TS 102 250-6, Speech Processing Transmission and Quality Aspects (STQ), QoS Aspects for Popular Services in GSM and 3G Networks, Part 6: Post Processing and Statistical Methods, v. 1.2.1.

[19] A. M. Mood, F. A. Graybill and D. C. Boes, *Introduction to the Theory of Statistics*, McGraw-Hill Statistics Series, 1974.

[20] J. R. Rice, *Mathematical Statistics and Data Analysis*, Duxbury Press, 2nd Edition, 1995.

[21] V. Barnet and L. Toby, *Outliers in Statistical Data*, John Wiley & Sons, 3rd Edition, 1994.

[22] ETSI, EG 201769, Speech Processing, Transmission & Quality Aspects (STQ); QoS Parameter Definitions and Measurements; Parameters for Voice Telephony Service Required under the ONP Voice Telephony Directive 98/10/EC, v. 1.1.2.

[23] W. C. Y. Lee, Estimate of local average power of a mobile radio signal, *IEEE Trans. Veh. Technol.*, Feb. 1985, **VT-34**(1), 22–27.

[24] J. Nielsen, *Designing Web Usability: The Practice of Simplicity*, New Riders Publishing, 2000.

[25] N. Maraganore and A. Shepard, *Driving Traffic to Your Web Site*, Forrester Research. See *http://www.forrester.com*, January, 1999.

[26] A. Richards, G. Rogers, M. Antoniades and V. Witana, Mapping user level QoS from a single parameter, *Proc. International Conference on Multimedia Networks and Services (MMNS '98)*, Nov. 1998. See *http://citeseer.nj.nec.com/richards98mapping.html*

[27] S. Khirman and P. Henriksen, Relationship between quality-of-service and quality-of-experience for public Internet service, *Proc. of the 3rd Workshop on Passive and Active Measurement*, March 2002.

[28] Nokia Apptester, Tool for service application layer testing. See *http://www.nokia.com*

[29] Nemo Outdoor™, Portable engineering tool designed for measuring and monitoring the air interface of wireless networks. See *http://www.nemotechnologies.com*

[30] Ethereal, The world's most popular network protocol analyzer. See *http://www.ethereal.com*

[31] TCPdump/libpcap. See *http://www.tcpdump.org*

[32] TCPtrace, Tool for analysis of TCP dump files, written by Shawn Ostermann at Ohio University. See *http://www.tcptrace.org*

[33] Open Mobile Alliance™, SyncML Device Management Protocol, v. 1.1.2.

[34] Nokia Mobile Quality Analyzer, Mobile QoS agent using passive measures. See *http://www.nokia.com*.

[35] Nokia NetAct™ InSPector, Monitoring the service quality of ISP networks. See *http://www.nokia.com*.

[36] TeleManagement Forum, Enhanced Telecom Operations Map, 2004, v. 3.0.

[37] Telemanagement Forum, *SLA Management Handbook: Concepts and Principles*, GB917-2, April 2004, Vol. 2.

[38] J. Laiho, A. Wacker and T. Novosad (eds), *Radio Network Planning and Optimisation for UMTS*, John Wiley & Sons, 2nd Edition, 2006, 630 pp.

[39] Nokia NetAct™ Service Quality Manager. See *http://www.nokia.com*

10

Optimisation

David Soldani, Giovanni Giambiasi, Kimmo Valkealahti,
Mikko Kylväjä, Massimo Barazzetta, Mariagrazia Squeo,
Jaroslav Uher, Luca Allegri and Jaana Laiho

UMTS cellular networks provide an opportunity for operators to offer new services to potential and existing subscribers. The negotiated or contracted QoS needs to be sustained without overprovisioning network resources. Operators need to have means and methods to improve the performance of the network within a number of constraints and wishes – for example, expenditure, QoE, changes in traffic and/or service portfolio, revenue, low complexity, effectiveness, efficiency and network resource utilisation.

This chapter discusses the objectives, concepts, processes, algorithms and means for QoS and QoE optimisation across UMTS cellular networks. In particular, Section 10.1 presents the conceptual breakdown of QoS management for mobile networks and processes for service performance optimisation. The importance of intelligence and automation is pointed out and the optimisation process using OS tools is introduced. Simple approaches to QoS optimisation for UTRA and GERA networks are described in Sections 10.2 and 10.3, respectively. In these sections we show how the parameter settings of the packet scheduling algorithms presented in Chapter 5 can be tuned to get better spectral efficiency, at a given QoE. Several methods to improve system performance are described. Performance results are attained by means of simulations. Some insights into the problem of QoS optimisation in the core and backbone networks are given in Section 10.4, which further discusses parameters and configuration options for improving the performance of traffic management functions and signalling procedures that may adversely affect the user perceived experience of what is being presented by the communication service. Some remarks on how the traffic path between core network elements could be optimised are also given. Performance results are from measurements made in real networks. Besides this, we present a process for troubleshooting that can be embedded in the end-to-end QoS optimisation procedure presented in Section 10.5, which concludes the chapter with several use cases on service application optimisation.

QoS and QoE Management in UMTS Cellular Systems
Edited by David Soldani, Man Li and Renaud Cuny © 2006 John Wiley & Sons, Ltd

10.1   Service optimisation concept and architecture

The optimisation of mobile services not only faces several challenges, but also provides operators the opportunity of improving revenue by exploiting capacity investments more effectively and differentiating in the market with QoS. Moreover, without a continuous optimisation process, operators would not be capable of adapting their cellular networks to the evolution of mobile service applications.

10.1.1   Conceptual breakdown of service and QoS management

QoS management in cellular networks primarily consists of three functional areas: QoS provisioning, QoE and QoS monitoring and optimisation. QoS provisioning is the process of configuring and maintaining selected network elements based upon customer service level agreements (SLAs) and observed quality performance (see Chapters 8 and 9). QoE and QoS monitoring is the process of collecting QoE and QoS performance statistics, faults and warnings; these data are then used for generating analysis reports for making changes/upgrades to the network (see Chapter 9). QoE and QoS optimisation is the process responsible for accessing monitored information, processing the data to determine service and network quality metrics, and initiating corrective actions when any of the quality levels is considered unsatisfactory.

   A conceptual breakdown of the QoS management framework is shown in Figure 10.1. The network management layer is responsible for collecting and processing of QoE and

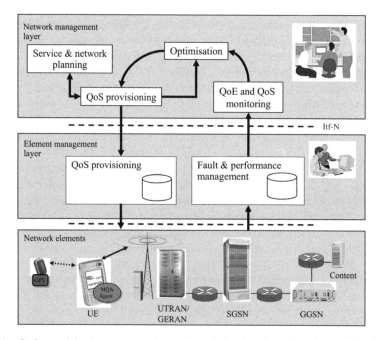

Figure 10.1   QoS provisioning, monitoring and optimisation functions located in the network and element management layers [1].

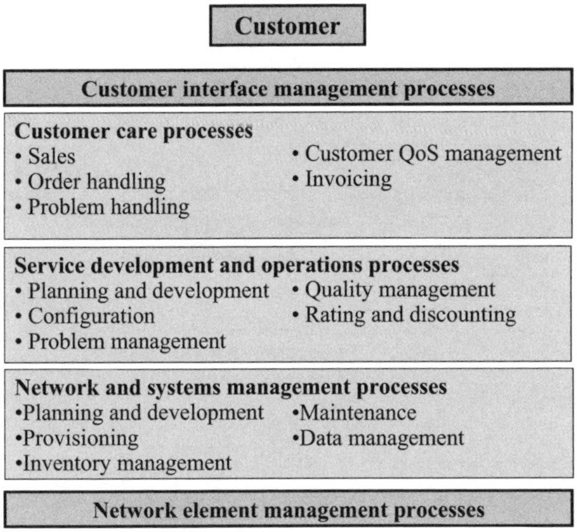

Figure 10.2 Operator processes needed to provide satisfactory services through network elements [2].

QoS performances, faults and data usage. The element management layer is accountable for aggregating and transferring the collected performance measurements and generated warnings/events to the above layer [1].

The operator processes for public land mobile network (PLMN) management can be represented as illustrated in Figure 10.2. Customer care processes may give rise to optimisation processes. Customer complaints, and changes in sales or marketing are typical triggers for optimisation. From the QoE and QoS perspective, the network optimisation process must take into account packet domain subscription data (QoS profiles subscribed in PDP context subscription records in the HLR), customer service level agreements, and observed service applications and network performances.

10.1.2 Service optimisation framework and process

As illustrated in Figure 10.3, service performance improvement (QoE optimisation), as well as network performance improvement (QoS optimisation), can be abstracted to a control loop, which takes into account the factors influencing the processes. Compared with other processes – for example, those adopted in manufacturing industries – in telecommunications, there are a number of factors that cannot be controlled by the process owner – for example, operators.

The optimisation loop can be divided into four distinct parts, each influenced by its own external factors. Service and network configuration aim at setting up the network elements and their parameterisations to support operator business functions. Service and network configurations cannot be separated completely from each other, since the network is the platform for QoS provisioning.

Service and network configurations are the control points for the operator to deploy the strategy to support its business. However, there are external factors that restrict or

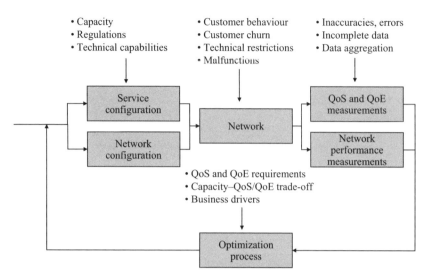

Figure 10.3 Service optimisation loop and external factors influencing the different phases.

direct possible solutions. The capacity owned by operators is constrained by, for example, radio bandwidth and number of communication elements. The limited number of sites for positioning radio base stations, and country-by-country legislation thereof, reduce the possibilities to provide good performance to all mobile users in the network. The deployed technology also has several restrictions for service and network configurations. Also, making changes in the network is not always cost-free. Normally, it requires engineers to visit the sites or network elements to be rebooted with temporary service interruption.

Mobile cellular networks have several uncontrollable factors that further challenge optimisation. For instance, subscriber behaviour cannot be controlled nor fully predicted. Some customer behaviour phenomena can be foreseen, like big sport events or festivals. In general, subscriber behaviour can only be monitored statistically, but the fluctuation of the traffic generated is an inevitable effect that needs to be taken into account. The same oscillation also applies to the different services provided to customers. For example, if by chance there are several mobile streaming users using the service in the area of one radioelement, this will definitely adversely affect the quality they experience. Another crucial, uncontrollable aspect influencing user satisfaction is how the provided content (e.g., webpages) is designed.

As mobile networks have a large number of elements and connections, the number of malfunctions per day may be massive. Severe malfunctions are rare and minor ones occur more often, but all affect the optimisation control loop by generating inaccuracies or breakdown of QoS performance measures.

Being able to make frequent and accurate service and network performance measures, as well as to detect and report faulty situations, is one of the cornerstones of network optimisation. What cannot be measured cannot be optimised! There are a number of metrics providing different information about network behaviour that can be used to improve service and network performance:

- *KPIs collected and stored in the NMS* are the most relevant source of information for *statistical optimisation*. The indicators typically cover the whole network, but the amount of detail is low. These measures are often averages calculated over one or more days or during busy hours, which can result in patchy information. Another drawback is that these measurements do not typically capture (assess) QoE properly.
- *Detailed logging of protocol performance at different interfaces* provides information on a part of the network. Standard communication protocols enable the operator to monitor different interfaces and collect specific performance data. This information is commonly used for optimisation, but the challenges are in filtering (the data amount is usually gigabits per hour), scalability (usually only a small part of the network can be assessed at once) and pre-processing (frequently requires a third-party tool).
- *Detailed logging of elements* gives rise to files of gigabyte size inside the network element. This is possible and reasonable only for selected elements – for example, base station or radio network controllers. A case in point might be, for example, monitoring one RNC by detailed signalling logging and, based on the data captured, handover, interference, coverage or QoS parameters are tuned. This method for collecting data is proprietary, and usually only vendors' own personnel can utilise it. The other drawbacks are similar to detailed logging of interfaces: data amount, coverage and pre-processing.
- *Drive/walk tests* provide very accurate and actual information not only on the radio conditions in the mobile station, but also application layer QoS can be captured. Drive and walk tests are very useful for troubleshooting the network. This information can also be used for QoS optimisation. However, drive and walk tests are expensive and can only capture the performance of a small part of the network during that particular measurement period. Moreover, drive and walk tests can only be used to monitor roads and public places; offices and private houses cannot be accessed. Thus, forming statistical relevancy with drive or walk tests is quite difficult.
- *Mobile QoS agent (MQA) technology* is rather new but extremely promising to collect QoS data for optimisation purposes. MQA technologies can overcome most of the challenges mentioned above for collecting and reporting L1–L7 measurements made more precise by GPS positions (e.g., latitude and longitude) and network locations (cell, URA, LA, RA, etc.). Two of the challenges are, for example, persuading subscribers to install this particular application and minimising the inconvenience caused to them.

The above-measured metrics are related to several phases of the optimisation process. Data may be used for deciding whether there is a need for optimisation, for selecting elements with poor performance, for making decisions inside the optimisation process (e.g., automated parameter optimisation based on performance data) and verifying the results of the optimisation process.

The targets of optimisation processes are to exploit actual capacity maximally, at a given QoE, and to delay forthcoming investment. Optimisation process can be initiated for a number of reasons, the most typical of which are:

- *New technologies or elements are taken into use.* In this case, the initial parameters for elements are selected in the network planning process, which is dependent on simulations, models and other assumptions. When network elements are installed or upgraded in the real environment, QoS or service performance optimisation can be executed based on real measurements and also in a real traffic scenario.

- *External edge conditions have changed.* There are different edge conditions for optimisation that may change. For example, if the strategy on how to utilise network resources has changed, or a business decision to support a new type of service has been made, such occurrences will bring about QoS or service performance optimisation.
- *Detection of decreased QoS performance in a particular network area.* Some external source indicates QoS or service performance degradation to the personnel responsible for optimisation. The source can be a subscriber, service provider or any interested party.
- *As a part of daily network operation process.* QoS performance is constantly monitored and unsatisfactory quality is reacted to by the optimisation process. There are two common ways to select the focus of daily performance routines. First, the worst performing network elements (typically base stations or controllers) can be selected for optimisation. Second, the elements that do not satisfy certain quality criteria can be chosen.

10.1.3 Benefits of intelligent and automated optimisation process

In Section 10.1.2 the main reasons to initiate an optimisation process were pointed out. In the following the motivation and benefits of *intelligent* and *automated* optimisation are discussed. Figure 10.4 shows the different types of benefit.

For *intelligence* and *automation* a rough division of benefits and features can be made:

- Intelligent optimisation algorithms aim at modelling expert decision-making and reasoning (knowledge modelling).
- Automation tries to reduce human-made repetitive and routine tasks.

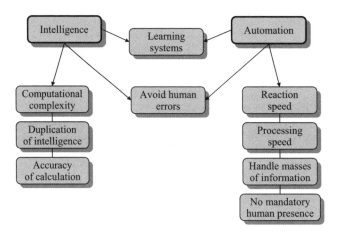

Figure 10.4 Classification of benefits and targets of automated optimisation processes. The benefits of intelligence and automation can be roughly separated. Different solutions target different types of advantages.

In many cases, the complexity of calculation and data analysis is so great that human reasoning itself is incapable of handling it properly. The reason computation is so complex is often the number of different types of evidence needed for decision-making. Problems require solutions that are based on multivariate analysis. For this purpose there are several different types of analysis algorithms available in the literature – for example, clustering algorithms, neural networks and statistical processing.

The precision of calculation made by an automated algorithm is greater than a human-made calculation. Automated algorithms can maintain the accuracy even when several decisions need to be taken during a day.

When a model of expert reasoning and an automated decision algorithm are available, that information can be reused in other parts of the network for the same purpose. For example, if an expert creates a model for automatic analysis of QoS deterioration in the base station, the same model likely can be reused for other base stations in the network. However, fine-tuning of the model might be needed to adapt the solution to different types of base stations or other network elements.

There are several ways to create models for automated decision-making. One of them is the *learning machine*. In an optimised area, the autonomic system learns from decisions made by several experts. After the learning period, the algorithm is able to take similar decisions autonomously. An example is utilisation of an artificial neural network for QoS monitoring. The neural network is capable of separating differently behaving elements from each other. Additionally, if the neural network is taught what kind of behaviour is acceptable and what is not, it can automatically detect poorly performing elements.

Well-configured automated and intelligent solutions are not as error-prone as humans simply because humans are not capable of handling the data as accurately as algorithms. Algorithms can also handle huge masses of data, without increasing the number of errors. A good example is detection of anomalies from large datasets, a task that a proper algorithm can do fluently at a low error rate. Moreover, the processing speed (computational power) of computers is far higher than any human-based effort.

One important aspect in automation is the speed of reaction. Networks generate performance data 24 hours per day. Computer-aided solutions can react to the required situations fast and no human presence is required. In reality, most of the processes are still fully controlled by engineers, especially to verify telecom network parameter changes. Fully automated optimisation loops controlled only by algorithms are still rare. However, certain tasks, such as data transfer and provisioning of new parameter settings, can be autonomously run at night when there is no traffic in the network (impact on business is low).

10.1.4   Optimisation using OS tools

During the last couple of years the optimisation scope has changed from network optimisation towards service and application monitoring and optimisation. Operators need to introduce new services in mobile networks more quickly and with less manual intervention, and the quality of service must be ensured in order to stay competitive and attractive. Network vendors have realised the need to hide the complex infrastructure of several radio access technologies. There is a clear analogy with the tendency in the process industry in the 1980s and early 1990s to install, for example, intelligent valves in process plants when OPEX became a major issue. Efficient management system support is needed not only for OPEX savings but also for faster service deployment. Key objectives are fast service creation, introduction and provisioning, and improved quality of service at a lower cost. These objectives can be achieved only through proper workflow support and automation of customer care and operational support processes, and a strong automated linkage between the management of customer service offerings and the underlying network [3].

As already pointed out, optimisation activity combines configuration management, performance management and fault management aspects. All this information is needed in other to reach a conclusions about optimisation actions in the network or services. Network and service configuration management and automated parameter tuning can be triggered by the monitoring or performance management process. There are many parameters in the network that can be tuned automatically without any manual intervention. Ideally, as much as possible of this kind of tuning should happen in the network in real time. However, there are also processes that need more statistical data or overall knowledge of the network, and therefore the tuning functionality is justified in a network management system.

Performance management for next-generation networks consists of two components. The first is a set of functions that evaluates and reports on the behaviour of telecommunication system equipment and the effectiveness of the network or network element. The second is a set of various subfunctions that includes gathering statistical information, maintaining and examining historical logs, determining system and service performance under natural and artificial conditions, and altering system modes of operation. Service providers must be able to monitor and manage traffic levels and concurrent network congestion to optimise network performance. They must analyse the data to correlate end-to-end service performance and take action based on a complete understanding of network behaviour. The ability to direct network performance effectively can be achieved with an application that allows service providers to control the operation.

Fault management tools should detect and log network problems and, wherever possible, fix them automatically in order to keep the network running effectively.

One strong driver for OS level optimisation and automation is operational efficiency. Network management systems with workflow support and the availability of actual data from the network target OPEX-efficient optimisation solutions. As discussed in [3], automation in connection with statistical optimisation includes automatic access to performance, fault and configuration data, functions to support network performance analysis and reporting, and automated workflow support in order to move from one process phase to another. Further, algorithms for automated optimisation can be provided in addition to manual optimisation actions. Figure 10.5 depicts network-wide optimisation workflow.

When it comes to the optimisation of services, it is important to note that there is a distinct difference between the user/service QoS requirements defined in service level agreements, network performance (NP) and QoS enabling mechanisms. The target of service optimisation activity is to ensure SLAs are not violated and, on the other hand, guarantee end-user satisfaction by providing a positive service experience. This is achieved, first, by optimisation of the network layer and, second, by tuning QoS enabling mechanisms in network elements and devices.

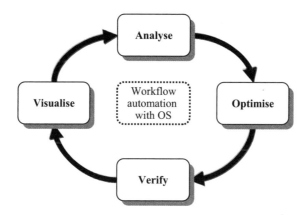

Figure 10.5 Network-wide optimisation workflow [3].

10.2 QoS optimisation in GERA networks

This section describes some aspects of QoS optimisation for GERA networks. The scheduling algorithms for BSS were presented in Section 5.1. More information on the results reported in the following is available in [4] and [5].

10.2.1 QoS optimisation in GPRS radio access networks

In this section we introduce a simple approach to QoS optimisation for GRPS radio access networks using the scheduling algorithm presented in Section 5.1. As a part of this framework, we present a simple model to estimate quickly the impact of several traffic mixes on the gain achieved with different queue weights in the adopted weighted fair queuing (WFQ) method. The proposed method makes it possible to adapt WRR queue weights to the offered traffic mix for optimal service performance.

10.2.1.1 Model description and assumptions

Assume that radio conditions are equal for all terminals. Given the number of TBFs sharing the same time slot and their associated SSS values, Equation (5.1) (p. 146) can be used for calculating the individual throughput for each TBF *on that particular time slot*. Generally, however, each cell has several time slots available for GPRS usage. A terminal might be able to use, for example, at most three time slots simultaneously. For instance, suppose there are three terminals, each using three time slots in the downlink and a cell with five slots available for GPRS. The total throughput of five time slots R_{tot} is known if we know the (equal and stable) radio conditions of the terminals. However, three phones using three slots each cannot be divided by the total five slots of the cell such that each TBF is in an equal position. In this example, one TBF necessarily gets allocated two time slots that are shared with some other TBF. The other two TBFs get one slot for their own use. Figure 10.6 illustrates the fairest allocation. In this case, (5.1) needs to be applied individually for each time slot. Cell throughput is given by the sum of all contributions

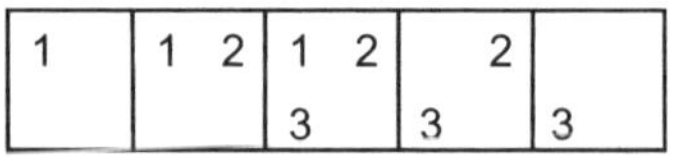

Figure 10.6 Example of TBF 1, TBF 2 and TBF 3 allocation to five time slots [5].
Reproduced by permission of © IEEE 2004.

over the allocated slots. In a real network implementation, the time slot chosen generally depends on several issues. Also, reallocations can be made periodically or in an event-based manner. In the model this issue is omitted and the error in throughput is accepted. In (5.1) R_{tot} is simply considered as total cell throughput. Under these hypotheses, computing the throughput of each TBF in a cell is very fast but inaccurate.

Traffic and user satisfaction criteria

The number of terminals is given as an input to the model. Each terminal in the network may use a streaming media (S), browsing (B) or messaging (M) service. The encoding rate for media streaming is assumed to be a constant 20 kb/s. The browsing service is supposed to be web surfing type of service. The user downloads content from the Internet, but due to the snapshot nature (time is not modelled) of the model there is no need to take into account file size or reading time length. The browsing service is simply assumed to require a certain throughput to avoid the user getting angry over delays in page download. Finally, the messaging service is one where users send messages or emails where time is not a critical factor.

The three services have different required throughputs in order to achieve satisfactory QoE. For the streaming service the required throughput is 20 kb/s based on the media encoding rate. It is also the maximum throughput taken by one streaming TBF regardless of the amount of available capacity. For the browsing service several throughput values are used (see Table 10.1). The messaging service is assumed to be such that throughput does not significantly impact QoE. Therefore, the minimum throughput is set to 2 kb/s, as this is assumed to be sufficient, for example, to avoid problems with retransmissions on higher layer protocols, such as TCP. Maximum throughput for both the browsing and messaging service is set at 30 kb/s. This is because mobile terminals are assumed to be capable of using at most three time slots in the downlink direction, and one slot is assumed to provide a throughput of 10 kb/s.

At the network level it is required that, within each of the three service classes, at least 90% of users have to meet the throughput criteria. Otherwise, it is considered that network level quality is unacceptable and there are too many users in the network.

Traffic mix in the network refers to portions of streaming, browsing and messaging users in the system, denoted by V_S, V_B and V_M, respectively. Multiple traffic mixes are tested to see whether there is any impact on the optimal values for the SSS parameters of each service class. Given the total number of users in the network N_{tot} and the traffic mix, the total number of users N_i in the network within each service class is given by:

$$N_i = round(V_i N_{tot}), \qquad i \in \{S, B, M\} \tag{10.1}$$

where *round* means taking the closest integer value, S refers to streaming, B to browsing

Table 10.1 Input parameters for the model [5].

Symbol	Value	Explanation
R_S^{min}	20 kb/s	Min. acceptable throughput for streaming
R_R^{max}	20 kb/s	Max. throughput used by streaming TBF
R_B^{min}	Varies	Min. acceptable throughput for browsing
R_B^{max}	30 kb/s	Max. throughput used by browsing TBF
R_M^{min}	2 kb/s	Min. acceptable throughput for messaging
R_M^{max}	30 kb/s	Max. throughput used by browsing TBF
R_{tot}	10 kb/s	Throughput provided by one cell
N_{tot}	Varies	Total number of users (TBFs)
V_S	Varies	Portion of streaming users (TBFs)
V_B	Varies	Portion of browsing users (TBFs)
V_M	Varies	Portion of messaging users (TBFs)
SSS_S	1	Scheduling step size for streaming TBFs
SSS_B	Varies	Scheduling step size for browsing TBFs
SSS_M	12	Scheduling step size for messaging TBFs
p	1/300	Probability of a call in any single cell
s_{min}	0.90	Minimum network level satisfaction level

and M to messaging users. Since V_i represents portions of all users, it holds that:

$$V_S + V_B + V_M = 1 \qquad (10.2)$$

The model

The network consists of N_C identical cells, each having six time slots for GPRS use. As one slot is assumed to provide a constant throughput of 10 kb/s, each cell offers a throughput of 60 kb/s (without EDGE). The probability of a user connecting a particular cell is equal for all cells and is denoted as p. Since all users are connected to some cell, p is given by:

$$p = \frac{1}{N_C} \qquad (10.3)$$

For each cell the number of streaming users follows the binomial distribution:

$$P(n_i) = \binom{N_i}{n_i} p^{n_i} (1-p)^{N_i - n_i} \qquad (10.4)$$

The number of streaming users in a cell is independent of the number of browsing or messaging users. Thus, the probability of having n_S streaming, n_B browsing and n_M messaging users in a cell is given by:

$$P(n_S, n_B, n_M) = \binom{N_S}{n_S} p^{n_S} (1-p)^{N_S - n_S}$$

$$\times \binom{N_B}{n_B} p^{n_B} (1-p)^{N_B - n_B} \binom{N_M}{n_M} p^{n_M} (1-p)^{N_M - n_M} \qquad (10.5)$$

where N_S, N_B and N_M are the total number of streaming, browsing and messaging users in the network and p is the probability of a call being located in one particular cell as defined in (10.3).

Once the probability of each call combination for a cell is known, the expected value of happy users within each service class in the cell can be calculated. As the cells are identical, the expected number of satisfied users in the network is obtained simply by multiplying by the number of cells. Thus:

$$h_i = N_C \sum_{\{(n_S,n_B,n_M)|R_i \geq R_i^{\min}\}} P(n_S, n_B, n_M)\, n_i \qquad (10.6)$$

where h_i, $i \in \{S, B, M\}$ refers to the expected number of happy users within any service class, N_C is the number of cells in the network, P is as defined in (10.5), R_i denotes the throughput of service class i calculated by (10.1), $R_i^{\min}$ is the chosen minimum throughput for service class i that provides adequate QoE and n_i refers to either n_S, n_B or n_M.

The portion of satisfied users s_i within service class i is given by:

$$s_i = \frac{h_i}{N_i}, \qquad i \in \{S, B, M\} \qquad (10.7)$$

It is required that:

$$s_i \geq s_i^{\min} = 0.90 \qquad \text{for all } i \qquad (10.8)$$

that is, for all three service classes, streaming, browsing and messaging. It is assumed that any service having too low a user satisfaction rate is not widely accepted by users. Thus, the requirement for a sufficiently high satisfaction rate is independent for each service class.

To further clarify the model, all input parameters are listed in Table 10.1. With a given set of input values, the output of the model is either 'positive results' or 'negative results', depending on whether the condition specified in (10.8) is true or not, correspondingly. A positive result indicates that network level quality was within acceptable limits. If a negative result is obtained, N_{tot} is decreased and the model recalculated with the new input, until a positive result is received. The maximum value of N_{tot} that still gives a positive result is called the *maximum load* of the network with all other inputs fixed N_{max}. The maximum load can be found for all scheduling strategies – that is, all feasible values of SSS_B. (*Note*: based on work presented in [4] and in Section 5.1 of our own book, it is evident that SSS_S can always be fixed to the best priority, 1, and SSS_M to the worst priority, 12.) Since there are only 12 feasible values for SSS_B, an extensive search method is used to find the value giving the highest maximum load. SSS_B can be forced to have identical values in all the cells in the network. It can be further forced to have identical values over all traffic mixes. When such constraints are used, they are noted in connection with results. The value of SSS_B that maximises N_{max} is called *optimal SSS_B*.

10.2.1.2 Numerical results and discussion

The optimal SSS_B is searched for all points on a grid in the space spanned by V_S, V_B and V_M. The distance between grid points is 0.05 (i.e., in 5% steps). Note that due to the constraint given by (10.2), this space is two-dimensional only. The grid is referred to as

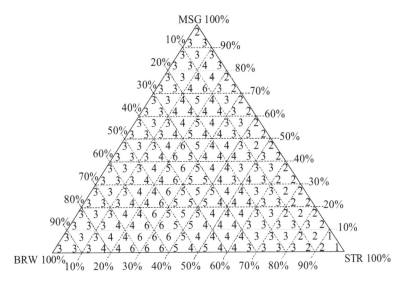

Figure 10.7 Optimal system-based SSS_B with R_B^{min} equal to 10 kb/s [5].
Reproduced by permission of © IEEE 2004.

'all traffic mixes' further in the text. A traffic mix with an $x\%$ margin refers to all traffic mixes where such points are removed, where:

$$V_i < \frac{x}{100} \quad \text{for any } i \in \{S, B, M\} \quad (10.9)$$

The grid in the space spanned by V_S, V_B and V_M can be represented in triangle form as shown in Figure 10.7. Each service type has its own vertex where 100% of the traffic belongs to that particular service type. Starting from the vertex, the portion of the traffic type in the traffic mix decreases along the line from the vertex to the midpoint of the opposing edge. Figure 10.7 shows the optimal SSS_B values for all traffic mixes with a 5% grid and R_B^{min} fixed at 10 kb/s. Additionally, in the results reported in Figure 10.7, SSS_B is supposed to have identical values in all the cells of the network.

From Figure 10.7 it can be seen that the highest values (worst priority) for SSS_B are in the area where streaming traffic is between 25% and 40% and there are more browsing users than messaging. A low SSS_B is optimal only when there is little streaming or browsing traffic present. This indicates that browsing traffic can have a relatively good priority when there is no real competition between streaming and browsing service types due to the small presence of either type. When there is a significant proportion of both traffic types, then the browsing priority has to yield in priority to make room for streaming.

N_{max} is obtained for each traffic mix. The N_{max} value depends on the constraints set on SSS_B. Four different priority strategies are investigated. In the *benchmark scenario*, with no differentiation, each service receives equal priority. We call *static differentiation strategy* the case where SSS_S and SSS_M are set according to Table 10.1, and SSS_B is optimised to maximise N_{max}, but is bound to a single value in all cells and all traffic mixes. It corresponds to real network situations where SSS values are used but are at the BSC

Table 10.2 Gain from three QoS differentiation strategies [5].

R_B^{min} (kb/s)	Margin (%)	SSS_B	Static (%)	Dynamic (%)	Adaptive (%)
6	0	6	163	166	178
6	5	6	137	138	145
6	10	8	112	113	114
6	'Evolution'	7	211	212	220
10	0	3	120	123	127
10	5	4	107	109	112
10	10	4	95	96	99
10	'Evolution'	4	162	165	172
12	0	3	98	101	108
12	5	3	88	89	96
12	10	3	80	81	87
12	'Evolution'	3	121	121	134
8–12	0	3	115	117	—
8–12	5	3	102	104	—
8–12	10	4	91	93	—
8–12	'Evolution'	3	151	154	—

level and cannot be set on a cell-by-cell basis. Furthermore, SSS values are kept fixed irrespective of the average traffic mix in the network. The *dynamic differentiation strategy* differs from the static one in that SSS_B is individually optimised for each traffic mix. Still, it is constrained to have an equal value in all cells of the network, but this value is changed as traffic mix changes in the network. Finally, in the *adaptive differentiation strategy*, SSS_B is optimised individually for each cell. In a real network it corresponds to the case when SSS values are automatically optimised at the cell level every time a new TBF is set up on the cell or an existing one leaves. Perfect optimisation is difficult to achieve since, in a cell where there is insufficient capacity to satisfy all users, it has to be decided whether, for example, streaming TBFs are given their required throughput and browsing users left below their satisfaction limit. This decision has to be made without knowing the overall network situation. In this study the decision criterion was such that the service with most TBFs in that particular cell is given the required throughput first. In the event of an equal number of TBFs per service, streaming is favoured over browsing and browsing over messaging.

N_{max} is specific for each traffic mix. Therefore, when evaluating how much difference the above four priority strategies make to N_{max}, we have to consider some particular traffic mix or rather a combination of several mixes. Table 10.2 presents the percentage gain in N_{max} using different differentiation strategies compared with the benchmark scenario. Gains are listed for different values of R_B^{min} and for four different traffic mix combinations. Traffic mixes are defined by margin, as explained by (10.9), except for a traffic mix combination called 'evolution', which is the set of four traffic mixes listed in Table 10.3. 'Evolution' represents a fictitious evolution path where the portion of streaming and browsing traffic types increase and the portion of messaging types decreases.

Table 10.3 Set of four traffic mixes, named 'Evolution' [5].

Service	Mix 1	Mix 2	Mix 3	Mix 4
Streaming	10%	15%	20%	25%
Browsing	40%	45%	50%	50%
Messaging	50%	40%	30%	25%

The results in Table 10.2 show that the load of the GPRS network can be increased significantly while maintaining an identical QoE when service differentiation is introduced. Gains from the static differentiation strategy vary from 80% to 211%, depending on R_B^{min} and on the traffic mixes over which the results are averaged. Gains from the dynamic and adaptive strategies compared with the benchmark case of no differentiation are very close to gains from the static differentiation strategy. This means that the benefit from these more complex strategies is very marginal.

In Table 10.2 a lower R_B^{min} always yields larger gains, since this allows browsing traffic to be treated at a lower priority (higher SSS), thus leaving more resources for the streaming and background services. Increasing the margin decreases gains from all differentiation strategies. This is due to the fact that traffic mixes that have a high proportion of messaging users can lead to extremely high gains. Increasing the margin excludes these extreme cases, thus decreasing the gain.

10.2.2 QoS optimisation in EGPRS radio access networks

R99 QoS can rightly be called 'enhanced quality of service' (EQoS) when compared with the QoS of R97/98. The additional parameters, which were introduced to the EQoS in addition to the deficit round robin scheduling algorithm described in Section 5.1, can be fine-tuned in such a way that the overall network performance increases significantly.

Figure 5.8 gives an example EQoS scenario gain over QoS and describes a simulation case with six services, SWIS, PoC, WAP, streaming, MMS and WWW browsing. Every service has its own satisfaction criteria specifying when an end-user is satisfied with QoE. Table 10.4 shows the satisfaction criteria of each service.

The x-axis of Figure 5.8 represents the overall network throughput during the simulations. The y-axis represents the percentage of satisfied users. The 90% line shows the decision point to compare the two differentiation strategies – that is, QoS and EQoS. There we can see the EQoS gain of approximately 12% to 13% of the network capacity in terms of overall network throughput. By mainly reducing the performance of both streaming and PoC services, but still satisfying over 90% of each service, EQoS was able to increase the number of satisfied WAP users, which was the limiting service in both cases. In absolute figures, for this case the 12% to 13% gain represents 1 Gb/s of absolute network throughput.

Every service had a different proportion of the offered network load, meaning that the number of users of each service in the simulated scenario varied. Table 10.5 reports the service distribution for the whole picture in the simulation.

Table 10.4   User satisfaction criteria for each service.

Service	Satisfaction criteria	Comment
SWIS	Rebuffering	Buffer size is 1 s. User constant data rate (speech + video) is 40 kb/s. User is satisfied when the buffer does not need to be refilled.
PoC	Rebuffering	Buffer size is 1 s. User data rate is 8 kb/s. User is satisfied when the buffer does not need to be refilled.
WAP	Average throughput over the whole session	User is satisfied when the average throughput over the whole session (potentially many pages) is over 10 kb/s.
STR (streaming)	Rebuffering	Buffer size is 8 s. Stream is encoded in 20 kb/s constant rate. User is satisfied when the buffer does not need to be refilled.
MMS	Message reaches destination withing 5 min	Message size is at least 3 kB and at most 200 kB.
WWW	Average throughput over the whole session	User is satisfied when the average throughput over the whole session is over 25 kb/s.

Performance results were attained using a dynamic system level simulator allowing simulations of different scenarios in GSM/GPRS/EGPRS networks. Using this tool every traffic unit (burst) transmission in a single time slot was simulated. Interference from the traffic of other terminals affecting the transmitted data was also taken into account during every simulation step. This was done for each terminal at every simulation step. Calculation of received power levels was done separately for both downlink and uplink directions. Mobile terminal positions in the scenario were updated at every time step, and the received powers and total interference accordingly calculated. Non-averaged carrier interference (CIR) with BER to BLER mapping was processed and radio resource management algorithms depending on the previous calculations were executed. More information on the tool can be found in [6].

The introduction of EQoS gives certain possibilities for network capacity improvement. R99 together with a more sophisticated scheduling system and admission control assures highest possible network utilisation.

Table 10.5   Traffic portions (mix) in the simulation.

Service	Network traffic proportion (%)
WAP	64.72
MMS	5.65
PoC	10.6
STR (streaming)	2.1
SWIS	2.26
WWW	14.67

10.3   QoS optimisation in UTRA networks

The selection of wireless services that WCDMA networks offer is diverse: speech, video, SWIS (or VS), PoC, streaming applications, web browsing, MMS and file transfer, to mention a few (for more information see Chapter 2). Services have different quality requirements in terms of throughput and delays. The quality of circuit-switched services, such as speech and video, or video sharing is easy to manage with guaranteed bit rates and connection retention, as long as the cell is not close to overload. Many features associated with packet scheduling make it difficult to sustain satisfactory quality levels of packet-switched communications offered on a best-effort (BE) basis. For those service applications, allocated resources are not retained for long periods and transmission bit rates are not guaranteed. In this case QoS management relies upon service prioritisation and discriminatory treatment as described in Section 5.2. Capacity allocation, maintenance and release is performed following the QoS class priority order, which enables time-critical packet-switched applications to be offered satisfactorily on a best-effort basis by giving them top priority. Moreover, packet scheduling related parameters such as *minimum* and *maximum* allowed bit rates can be defined on a per-priority-class basis.

10.3.1   QoS-sensitive parameters

As indicated above, sustaining satisfactory quality levels of PS communications is more challenging than those of guaranteed bit rate services. Parameters that have a large influence on QoS can thus be sought among those of the packet scheduler.

Choices that spring to mind for the algorithms presented in Section 5.2 include:

- *Minimum allowed bit rate.*
- *Maximum allowed bit rate.*
- *Inactivity timer* – the time interval for a connection to be inactive before changing its state from Cell_DCH to Cell_FACH.
- *Maximum queuing time* for a capacity request to stay in the queue, after which the request is rejected and removed from the queue.
- *Minimum DCH allocation time* before an allocated bit rate is downgraded in order to leave room for newer capacity requests of higher priority.

Setting the values of the above parameters in different ways with different traffic classes and traffic handling priorities enables QoS differentiation that facilitates targetting the optimal performance. Selection of the best choices of parameter values is not a straightforward task, however. Optimum values depend on the satisfaction criteria defined for services and on the traffic characteristics and load in the network.

10.3.2   QoS optimisation in WCDMA radio access networks

In this section, we demonstrate the benefits of QoS optimisation. A dynamic WCDMA system simulator is applied to explore the effects of changing parameter values on QoS

Table 10.6 Mapping of services onto traffic priorities [7].

Priority		Service	Bit rates (kb/s)
CS conversational		Speech	12.2
		Video	64
PS streaming		SWIS (or VS)	64
PS interactive			
	THP1	PoC	0, 8, 16, 32, 64
	THP2	Streaming applications	0, 8, 16, 32, 64, 128
	THP3	WAP and MMS	0, 8, 16, 32, 64, 128, 144, 256, 384
PS background, also referred to as THP4		Dialup	0, 8, 16, 32, 64, 128, 144, 256, 384

[7]. The parameters were analysed in two ways. First, the same value of each parameter was used with all services regardless of their traffic handling priority. Parameters were evaluated one-by-one with four different values. In the testing of one parameter, other parameter settings were kept fixed. Second, the parameters were differentiated so that services with different traffic handling priorities could have different values of the same parameter. Each parameter was divided into four subparameters specific to the traffic handling priority. Within each parameter, subparameters were tested with two values, one low and one high, and different combinations were evaluated.

10.3.2.1 Simulation scenario

The scenario traffic mix included services from all traffic classes with different possible QoS profiles – that is, conversational, streaming, interactive and background, the first of which was circuit-switched (CS) and the rest packet-switched connections. The services and traffic classes are shown in Table 10.6 in decreasing order of priority. The specific traffic characteristics of services were based on different distributions of call lengths, numbers of objects in sessions, object sizes, inactivity periods between the objects and shares of calls (see Table 10.7).

Once a session was completed, user satisfaction data were collected. Different services had different satisfaction criteria. First, a user was considered unsatisfied if the call was blocked. Blocking also covered those packet-switched services whose capacity requests were rejected by the PS. Second, users running streaming-type applications controlled by the PS were considered unsatisfied if the terminal streaming buffer ran out during the session. Assumed buffer sizes were 8 s for streaming applications and 1 s for SWIS and PoC. Finally, users of other PS-controlled services were unsatisfied if active session throughput (AST) did not meet specific levels, which were 32 kb/s for WAP, 8 kb/s for MMS and 64 kb/s for Dialup.

Table 10.7 Traffic models [7].

Service	Streaming rate (kb/s)	No. of objects	Object size (kB)	Inactivity period (s)	Share (%)
Speech Mean Min, max	12.2	1	Neg-exp 137 0, 400	—	20
Video Mean Min, max	64	1	Neg-exp 960 0, 2000	—	7
SWIS Mean Min, max	64	1	Neg-exp 80 32, 2400	—	10
PoC Mean Min, max	8	Geom 8 1, 30	Neg-exp 6 0.5, 40	Neg-exp 60 1, 1200	18
Str.app. Mean Min, max	64	1	Uniform 1680 160, 3200	—	12
WAP Mean Std Min, max	—	Geom 3 1, 50	Log.norm 12 16 10, 50	Neg-exp 20 1, 600	13
MMS Mean Min, max	—	1	Neg-exp 20 3, 200	—	5
Dialup Mean Std Min, max	—	Geom 1 1, 50	Log.norm 250 1,300 20, 20 000	Pareto* 2, 3600	15

* The Pareto shape and scale parameters were 1 and 2.

10.3.2.2 Results

In the first batch of simulations, parameters were not differentiated among the traffic handling priorities. During the testing of one parameter, other parameter values were held fixed at a reference value. QoS was measured with satisfaction ratios on a per-service basis. By definition, the best result was obtained when the satisfaction ratio of the poorest performing service was maximised. Tested values were as follows. Reference values are shown underscored and best values in boldface:

- Minimum allowed bit rate: **8**, 32, <u>64</u> and 128 kb/s.
- Maximum allowed bit rate: 64, **128**, 256 and <u>384</u> kb/s.
- Inactivity time: **0**, <u>5</u>, 10 and 20 s.
- Maximum queuing time: <u>4</u>, **10**, 20 and 30 s.
- Minimum DCH allocation time: 1, **3**, 5 and <u>15</u> s.

The best total satisfaction ratio was 95% and that of the lowest performing service was 79% (PoC). The result was obtained by setting the inactivity timer to 0 s. A natural follow-on test was to trial simultaneously the values of parameters that individually gave the highest satisfaction ratios. The joint performance was tested, which gave 89% total satisfaction and 74% satisfaction with the poorest performing service. Thus, the result was no better than the one obtained with a zero or disabled inactivity timer. The result indicated that parameters and their effects on QoS were not independent of each other, which highlights the importance of optimising parameter values simultaneously.

In the second batch of simulations, QoS differentiation was enabled by division of the services into different traffic classes and handling priorities, which specified the order in which to allocate and release capacity. Traffic handling priorities only applied to interactive and background traffic classes. Each THP-specific subparameter was tested with two different values, low and high, and all $2^4 = 16$ combinations were analysed. Tested values are reported in Table 10.8. During the testing of one parameter, the other parameter values were kept fixed at the reference value as shown above.

Table 10.9 shows the average user satisfaction ratio (Mean) and the user satisfaction ratio of the poorest performing service (Min) for the best value combination of each parameter. The best combination was defined as that maximising lowest satisfaction. Parameter differentiation showed improved performance over the undifferentiated case. The best total satisfaction ratio was 95% and that of the poorest performing service (PoC) was 87%. The parameter values that individually gave the highest satisfaction

Table 10.8 Tested values with enabled differentiation [7].

Parameter	Low value	High value
Minimum allowed bit rate	THP1: 8 kb/s THP2: 64 kb/s THP3: 32 kb/s THP4: 64 kb/s	THP1: 16 kb/s THP2: 128 kb/s THP3: 64 kb/s THP4: 128 kb/s
Maximum allowed bit rate	THP1: 8 kb/s THP2: 64 kb/s THP3: 64 kb/s THP4: 128 kb/s	THP1: 16 kb/s THP2: 128 kb/s THP3: 384 kb/s THP4: 384 kb/s
Inactivity timer	1 s	20 s
Maximum queuing time	1 s	20 s
Minimum allocation time	1 s	20 s

Table 10.9   Parameter values producing the best results [7].

Parameter	THP1	THP2	THP3	THP4	Mean	Min
Minimum allowed bit rate	8 kb/s	64 kb/s	32 kb/s	64 kb/s	92%	74%
Maximum allowed bit rate	16 kb/s	64 kb/s	384 kb/s	384 kb/s	93%	81%
Inactivity timer	1 s	1 s	1 s	20 s	95%	87%
Maximum queuing time	1 s	1 s	20 s	20 s	95%	82%
Minimum allocation time	1 s	1 s	20 s	20 s	91%	71%

ratios – that is, those shown in Table 10.9 – were tested simultaneously. Superior performance was obtained as the result. The total satisfaction ratio thus obtained was 98% and that of the poorest performing service (varied between WAP, PoC and streaming applications) was 97%. Obviously, many parameters needed to be fine-tuned to adequate levels in order to reach the best performance, in which value differentiation among the traffic handling priorities played a significant role.

Spectral efficiency gain is defined as the difference in spectral efficiencies between system load levels, in which the poorest performing service has a specified ratio of satisfied users. Spectral efficiency (SE) is defined as average cell throughput (average cell capacity) divided by exploited bandwidth (3.84 MHz) at a given QoE [8], [10]. Figure 10.8 shows the average cell capacity and user satisfaction percentages of the poorest performing service obtained with four different load levels and with the best undifferentiated or differentiated configurations. Setting the limit of the adequate satisfaction ratio to 90%, the spectral efficiency gain obtained by THP-based differentiation of PS

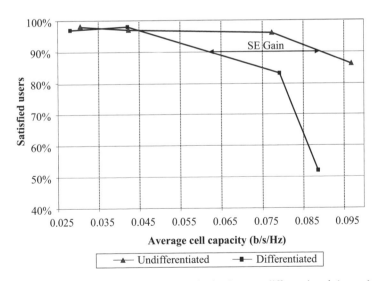

Figure 10.8   Lowest user service satisfaction with the best undifferentiated (upper) and differentiated (lower) configuration [7].

parameters was $(0.09 - 0.06)/0.06 = 50\%$. Improvement was not specific to the poorest performing service. Calculating the gain with the second-lowest service yielded a 55% improvement in spectral efficiency.

10.3.3 Genetic algorithms in QoS optimisation

In the previous section, parameters were tested independently one by one using just two different values for each priority-specific subparameter. The benefit was that the maximum number of tested combinations and simulation rounds was 16 with each parameter. However, parameters and their effects on QoS are not independent of each other. Thus, it would be reasonable to optimise all parameters simultaneously and use a wider range of parameter settings. The resulting value space is huge and beyond even an exhaustive search. The use of evolutionary search algorithms solves the dimensionality problem well. In this section the use of genetic algorithms is described, as is simultaneous optimisation of crucial QoS-sensitive parameters. The genetic algorithm is an optimisation method that makes use of a population of reproducing test solutions and works by enforcing conditions that improve the value of a fitness (or objective) function defined for the optimisation task. See [9] and [10] for more information on the topic.

10.3.3.1 Genetic algorithm implementation

The flow chart of the implemented genetic algorithm is shown in Figure 10.9. The initial population of the genetic algorithm consisted of 10 member vectors of 14 components. The components represented the values of 5 parameters, 3 of which were divided into 4 traffic-handling priorities (as shown in Table 10.10). The fitness of each member of the initial population was first computed. The fitness function was defined as the difference between the observed ratio of satisfied users of the poorest performing service and the target ratio, which was 90% for all services. Formally, fitness computation was based on the standard statistic of testing the similarity of two ratios [9], [10]:

$$Satisfaction(i) = \frac{\dfrac{SU(i)}{ST(i)} - TS(i)}{\sqrt{\dfrac{TS(i) \cdot (1 - TS(i))}{ST(i)}}} \qquad (10.10)$$

$$Fitness = - \min_{i=1,\dots,N} Satisfaction(i) \qquad (10.11)$$

where N is the number of services (8), $SU(i)$ is the observed number of satisfied users of service i, $TS(i) = 0.9$ is the target satisfaction ratio and $ST(i)$ denotes the total number of users of service type i. The lower the fitness value the better the network configuration. Fitness values below 0 meant that service performance was above the target. $Satisfaction(i)$ is a statistic that yields positive or negative values if the ratio of satisfied calls is higher or lower than the target set for the ratio.

Population members were given rank values between 1 and 10 according to the fitness value, so that the best member was ranked 10, the second best 9 and so on.

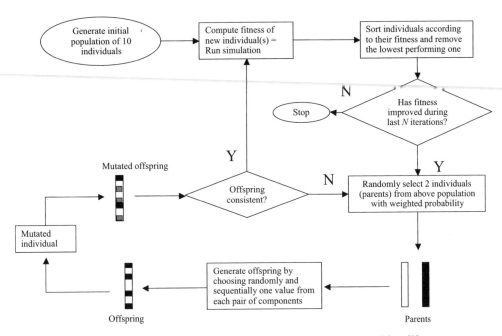

Figure 10.9 Flow chart of the implemented genetic algorithm [9].

Reproduced by permission of © IEEE 2005.

Table 10.10 Allowed parameter values in genetic optimisation [9].

Parameter	THP	Range	Service
Minimum allowed bit rate	T1	8, 16 kb/s	PoC
	T2	64, 128 kb/s	Streaming
	T3	32, 64, 128, 144, 256, 384 kb/s	WAP/MMS
	T4	64, 128, 144, 256, 384 kb/s	Dialup
Maximum allowed bit rate	T1	8, 16 kb/s	PoC
	T2	64, 128 kb/s	Streaming
	T3	32, 64, 128, 144, 256, 384 kb/s	WAP/MMS
	T4	64, 128, 144, 256, 384 kb/s	Dialup
Inactivity timer	T1	1, 2, 5, 10, 20, 30 s	PoC
	T2	1, 2, 5, 10, 20, 30 s	Streaming
	T3	1, 2, 5, 10, 20, 30 s	WAP/MMS
	T4	1, 2, 5, 10, 20, 30 s	Dialup
Minimum allocation time	Not differentiated	1, 2, 5, 10, 15, 20 s	All services
Maximum queuing time	Not differentiated	1, 2, 5, 10, 15, 20 s	All services

Reproduced by permission of © IEEE 2005.

The probability of a member being selected as a parent for generating offspring was defined as rank divided by the sum of all ranks $(1 + 2 + \cdots + 10 = 55)$. Two members were randomly selected from the population as parents of offspring. Offspring component values were chosen randomly from parent components. Moreover, each component was mutated with 10% probability into another value in the defined range. If offspring were already among the tested solutions or the generated minimum allowed bit rate was higher than the maximum allowed bit rate, then new parents were selected and offspring generated. If the fitness of offspring was better than that of the worst population member, then the population member was replaced by the offspring. The process was continued as long as there was no improvement in best fitness for 100 cycles, after which the member with the best fitness determined the optimum solution.

10.3.3.2 Simulation results

The simulation scenario was the same as in Section 10.3.2 and the traffic models were those shown in Table 10.7. However, the simulator described in Section 7.1.2 was used. Moreover, the parameters were optimised for the five different traffic mixes shown in Table 10.11.

Figure 10.10 shows the ratio of satisfied users of deployed services as a function of the iteration number for Mix 3, which clearly benefitted from optimisation. The figure shows that increasingly better parameter value combinations were found as the process advanced. Figure 10.11 shows the best fitness values obtained at each iteration step for all traffic mixes. The figure shows that good parameter values were found in a few iterations, but continuing the optimisation for 100 iterations could improve the fitness further. The traffic volume with Mixes 1 and 4 was rather small and satisfaction was above the target even with the initial parameter values. However, satisfaction could be improved to some degree. Mixes 2 and 5 produced high traffic in the network and the final satisfaction remained below the target, although it was much improved from the initial level. Table 10.12 shows the parameter values with the best fitness value as a result of optimisation. The results were combined in two reference cases, Ref. 1 and Ref. 2,

Table 10.11 Traffic mixes [9].

Service	Proportion of calls (%)				
	Mix 1	Mix 2	Mix 3	Mix 4	Mix 5
Speech	32	44	40	40	20
Video	0	3	0	2	7
SWIS	3	0	10	5	10
PoC	14	9	10	25	18
Streaming	2	10	10	4	12
WAP	38	17	10	10	13
MMS	8	5	10	7	5
Dialup	3	12	10	7	15

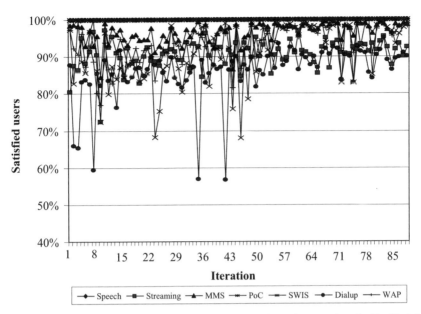

Figure 10.10 Percentage of satisfied users as a function of iteration number for Traffic Mix 3 [9].
Reproduced by permission of © IEEE 2005.

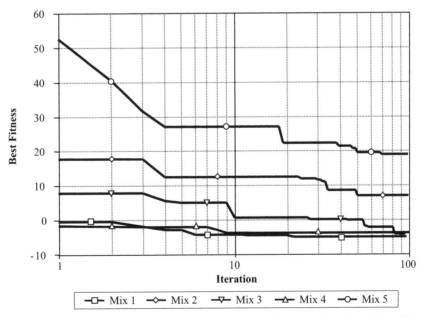

Figure 10.11 Best fitness values as a function of iteration number for all traffic mixes analysed [9].

Reproduced by permission of © IEEE 2005.

Table 10.12 Best parameter values obtained with the genetic algorithm [9].

Mix	Min. allowed BR (kb/s)				Max. allowed BR (kb/s)				Inactivity timer (s)				Min. alloc. time (s)	Max. queuing time (s)
	T1	T2	T3	T4	T1	T2	T3	T4	T1	T2	T3	T4		
1	8	128	32	64	16	128	64	144	1	1	2	5	5	10
2	8	64	32	64	16	64	384	128	1	5	30	1	20	5
3	16	64	32	64	16	128	144	144	2	1	5	1	20	20
4	8	64	64	128	16	128	256	144	1	2	5	5	2	10
5	8	64	32	128	16	128	144	256	2	1	1	20	15	5
Ref. 2	8	64	32	64	16	128	144	144	1	1	5	5	15	10
Ref. 1	64	64	64	64	64	128	144	144	5	5	5	5	15	10

shown in the table. They collected the most common, best values obtained with Mixes 1 to 5 such that Ref. 1 was undifferentiated – that is, the parameter values were the same for all THPs – and Ref. 2 was differentiated. Figure 10.12 shows the lowest user satisfaction with four different load levels of Mix 4. The consensus parameters of Ref. 2 produced a 4% higher spectral efficiency than those of Ref. 1, which corroborated the benefit of

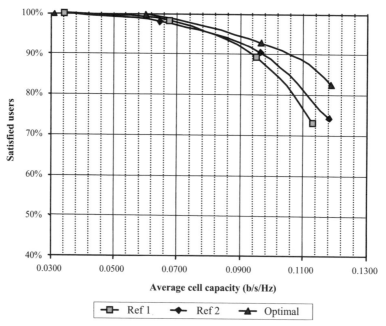

Figure 10.12 Lowest user service satisfaction with undifferentiated reference case (Ref. 1) and differentiated reference case (Ref. 2), and the best parameter values from the genetic algorithm with Mix 4 [9].

differentiating parameter values among the THPs. The best parameter values of Mix 4 from the genetic algorithm produced 8% and 12% higher spectral efficiencies than Ref. 1 and Ref. 2, respectively. The trends were similar with other mixes as well. The result indicated that it is beneficial to adopt different network configurations with different traffic conditions.

10.3.3.3   Practical considerations

Genetic algorithms have been shown to be robust for online parameter adaptation of various dynamic systems [11]. The use of optimising WCDMA network parameters in a simulator environment has just been described. Figure 10.10 showed that some, although infrequent, parameter value combinations during optimisation yielded poor user satisfaction with the Dialup service, for instance. In online optimisation of a running WCDMA network such performance drops are naturally disadvantageous. Network performance thus needs to be monitored actively during the whole process. In practice, performance data for fitness function calculations need to be collected for a sufficiently long period – for instance, one or even a few days after each parameter change – in order to get a statistically significant sample. However, a substantial decrease in performance can be detected with a much smaller sample than that required for differentiating between high satisfaction ratios. In the occurrence of problem performance, the iterations can be stopped and new parameter values can be generated for evaluation. It is beneficial to analyse a drop in performance simply to find the reasons that created it and then utilise the findings to modify the rules on how new offspring can be recombined from population members before continuing with the next cycle. Such manual adaptations rule out a fully automated process, but the benefits can easily outweigh the effort needed. Another way to maintain the stability of an optimised network is to only allow small changes in offspring generation. However, convergence is likely to slow down as the trade-off.

10.3.4   Simple fuzzy optimisation

Adoption of genetic algorithms may face difficulties in a network planning organisation due to the random character of optimisation. Experts want to understand the justification for parameter adjustments and to see good reasons for a specific action to yield better performance. Rule-based methods may thus be preferred over stochastic ones. In this section we demonstrate the design of a simple service optimisation method that applies fuzzy rules that can be used to incorporate simple knowledge in parameter control. Basically, the expert defines the specific rules of action as a response to specific performance measurements. Fuzzy logic is a means of combining several performance measurements intuitively to give a warning that parameter adjustment is needed.

10.3.4.1   Delay control parameter

Five parameters were introduced in Section 10.3.1 to control the QoS of packet-switched services that have different THP values. (In this chapter, background is also referred to as

THP4.) Defining rules to control them independently is a demanding task. Instead, the task is simplified by controlling one THP-specific parameter, referred to here as the delay control parameter (DCP), whose values range from 0 to 1. Each value of the DCP is mapped onto a specific combination of the *minimum allowed bit rate, maximum allowed bit rate, minimum allocation time, maximum queuing time* and *inactivity timer*. Moreover, mapping of the DCP onto parameters is linear in nature, and the lower the DCP value the higher the parameter values, allowing better or sustained resource allocation. The parameters share the property that the higher their values the shorter the expected packet transfer delay for a specific THP. Thus, small values of the DCP are expected to produce shorter packet transfer delays than large values.

We define the mapping from the DCP onto a specific THP-specific parameter in such a way that value 0 produces the maximum and value 1 produces the minimum defined value of the parameter (and the value decreases linearly between the maximum and the minimum). The remaining task is to define the appropriate minimums and maximums for each parameter. Mapping can be formulated as:

$$Parameter(DCP\_THPi) = (max - min) * (1 - DCP\_THPi) + min \qquad (10.12)$$

in which DCP_THPi is the delay control parameter of $THPi$, max is the maximum value of the range and min is the minimum value of the parameter range. Table 10.8 shows some plausible ranges for parameters. For instance, if $DCP\_THP3 = 0.45$, the above formula gives the following values for the parameters in Table 10.8:

$$MinAllowedBitrate(0.45) = (64-32) * (1-0.45) + 32 = 50 \text{ kb/s (rounded to 64 kb/s)},$$

$$MaxAllowedBitrate(0.45) = (384-64) * (1-0.45) + 64 = 240 \text{ kb/s (rounded to 256 kb/s)},$$

$$InactivityTimer(0.45) = (20-1) * (1-0.45) + 1 = 11 \text{ s},$$

$$MaxQueuingTime(0.45) = 11 \text{ s},$$

$$MinAllocationTime(0.45) = 11 \text{ s}.$$

With $DCP\_THPi = 0$ or 1, the formula gives the maximum or the minimum value of the parameter, respectively. The values need to be rounded to the nearest allowed value. Thus, the minimum and maximum allowed bit rates are rounded to 64 kb/s and 256 kb/s in the above example (intermediate bit rates may be taken from Table 10.10).

10.3.4.2 Discrete control rule

Call satisfaction is quantified for optimisation by *Satisfaction(i)* of (10.10). A plain ratio of satisfied calls (or sessions) to all calls would be fine if the number of calls was sufficient for a statistically accurate estimate of the ratio. However, it is better not to make such an assumption. The satisfaction measure takes into account the number of calls. If the number of calls is sufficient, *Satisfaction(i)* is normally distributed. The next question is whether the ratio's deviation from the target is significantly high or just a random occurrence. Assuming that *Satisfaction(i)* is normally distributed, its values with magnitude >2 mean that there is less than a 2% probability of detecting the satisfaction ratio erroneously as poor or excessive. Thus, magnitudes of *Satisfaction*(i) >2 can be considered significant, for instance. However, to assume the normal distribution of *Satisfaction*(i) validly, there must be an adequate number of samples to calculate it. We adopt the typical requirement for the number of samples used in statistical hypothesis testing, which gives the condition:

$$ST(i) \geq \frac{5}{1 - TS(i)} \tag{10.13}$$

where $ST(i)$ is the number of sampled calls (or sessions) of bearer service type i and $TS(i)$ is the expected ratio of satisfied calls among all calls that have *THPi*.

A simple discrete rule to control the DCP can be formulated as follows. If the satisfaction of a THP is below target, improvement in the situation can be achieved by decreasing the DCP value of the THP. Such an improvement is more likely attained if the DCP values of THPs with overly high satisfaction are increased at the same time. The reasoning yields three conditions and two actions for service optimisation:

FOR $i = 1$ To #THP {

IF $ST(i) \geqslant \dfrac{5}{1 - TS(i)}$ THEN {

    IF *Satisfaction*(i) < -2 THEN decrease *DCP_THPi* by a step $\qquad$ (10.14)

    ELSE IF *Satisfaction*(i) > 2 THEN increase *DCP_THPi* by a step

}}

10.3.4.3 Rule 'fuzzyfication'

The conditions of the discrete control rule above are either true or false, which allows control actions only when the particular conditions are fully met. Using fuzzy truth-values in the conditions the rule can be made more flexible, which may help responding to problem situations more quickly.

In the discrete control rule the truth-value for a sufficient number of calls is 0 when the number of calls is fewer than $5/(1 - TS(i))$ and the truth-value is 1 when call numbers are higher than the limit. We make the truth-value smoother by giving fractional truth-values for the sufficiency of call numbers lower than $5/(1 - TS(i))$. With decreasing call numbers, truth-values decrease linearly to 0. We define truth-value 0 as being obtained with call numbers of 10 or lower, which are statistically too low for confident analysis with almost any value from the observed ratio. We define the truth-value for a sufficient number of calls formally as:

$$GoodSample(N) = \begin{cases} 0 & \text{if } N < 10, \\ \dfrac{N - 10}{A - 10} & \text{if } 10 \leq N < A, \\ 1 & \text{if } N \geq A, \end{cases} \tag{10.15}$$

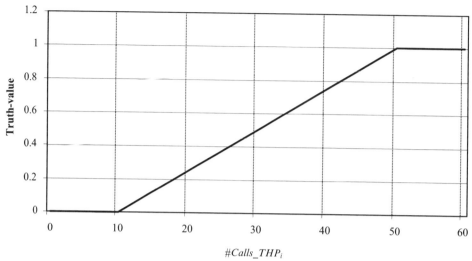

$\#Calls\_THP_i$

Figure 10.13 Truth-value for the sufficiency of a sample count that has a 90% target satisfaction ratio.

where N is the number of samples (also denoted as $\#Calls\_THPi$), $ST(i)$ was defined in (10.13), and $A = 5/(1 - TS(i))$. Figure 10.13 shows truth-values that have $TS(i) = 0.9$.

In the discrete control rule, the truth-value of poor satisfaction is 1 when $Satisfaction(i)$ is lower than -2. The rule is made smoother by allowing linearly decreasing truth-values when $Satisfaction(i)$ goes from -2 to 0:

$$Poor(S) = \begin{cases} 1 & \text{if } S \leq -2, \\ \dfrac{-S}{2} & \text{if } -2 < S \leq 0, \\ 0 & \text{if } S > 0, \end{cases} \qquad (10.16)$$

where $S = Satisfaction(i)$. Moreover, the truth-value of excessive satisfaction is 1 when $Satisfaction(i)$ is higher than 2. The rule is made smoother by allowing linearly increasing truth-values when $Satisfaction(i)$ goes from 0 to 2:

$$Excessive(S) = \begin{cases} 0 & \text{if } S \leq 0, \\ \dfrac{S}{2} & \text{if } 0 < S \leq 2, \\ 1 & \text{if } S > 2. \end{cases} \qquad (10.17)$$

The truth-values are illustrated in Figure 10.14. Now, fuzzy conditions can be combined to fuzzy rules:

$$\text{IF } GoodSample(N) \text{ AND } Poor(S) \text{ THEN decrease } DCP\_THP(i) \qquad (10.18)$$

and

$$\text{IF } GoodSample(N) \text{ AND } Excessive(S) \text{ THEN increase } DCP\_THP(i) \qquad (10.19)$$

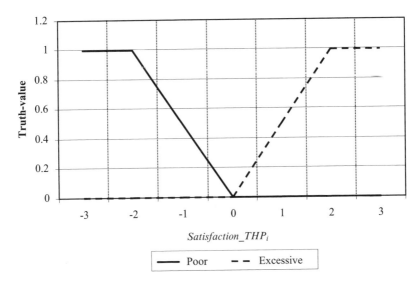

Figure 10.14 Truth-values of THP satisfaction, which are increasingly poor with values lower than 0 and increasingly excessive with values higher than 0 until truth-value 1 is reached.

In practice, the AND operation among fuzzy truth-values can be obtained using the minimum operator.

We still need to translate the fuzzy truth-values to get a meaningful output for $DCP\_THP(i)$. With truth-value 1, $DCP\_THP(i)$ can be changed by steps of 0.2, say, with each step decreasing linearly with the truth-value:

$$DownStep = -0.2\,MIN(GoodSample(N), Poor(S)) \qquad (10.20)$$

and

$$UpStep = +0.2\,MIN(GoodSample(N), Excessive(S)) \qquad (10.21)$$

Equations (10.20) and (10.21) can be summed to obtain the final step value:

$$Step = UpStep + DownStep \qquad (10.22)$$

and

$$DCP\_THP(i) = DCP\_THP(i) + Step \qquad (10.23)$$

The discrete rule in (10.14) yields steps of magnitude 0.2 only, whereas the fuzzy rules in (10.23) can also yield any magnitude lower than 0.2. Compared with the discrete control rule, fuzzy control can respond to small deviations from the target by making small control steps, allowing faster solutions to satisfaction problems.

10.3.4.4 Discussion

The limitations of the simple rules described are obvious. All parameters are expected to increase or decrease the packet transfer delay in a similar manner when changed in small, similar steps. In practice, the effects of parameters on packet transfer delay, not to mention QoS, are complex and not necessarily obvious even to an expert. Designing rules that are sufficiently general for all needed scenarios thus faces many challenges. Although stochastic optimisation with methods like genetic algorithms may seem random and undesirable as such, designing a rule-based method may also require a great deal of trial and error.

10.4   QoS optimisation in core and backbone networks

This section gives an insight into the problem of finding a means of core network (CN) and backbone network (BN) performance improvement. Section 10.4.1 analyses several parameters and configuration options for traffic management in CN elements, such as priority queuing and IP packet marking. Some recommendations for improving the performance of 'GPRS attach', 'PDP context activation' and 'Routing Area update' procedures are also given. In Section 10.4.2, we present some ideas on how the traffic path between core network elements could be optimised. Section 10.4.3 describes a performance monitoring and configuration process that can be integrated in the end-to-end QoS optimisation procedure presented in Section 10.5.

10.4.1   Parameter optimisation

The following sections show how the performance of traffic management and signalling procedures may be improved by tuning the parameters available in CN elements – that is, the SGSN and GGSN.

10.4.1.1   Traffic management in 2G SGSN

Queuing mechanisms in the CN are the main factors to be considered for QoS optimisation. If weighted fair queuing is implemented the operator will have the opportunity to control selection of the relative weights assigned to different priority queues. Prioritisation of traffic is beneficial in a congestion situation, where the lack of resources, either in the radio part or in the core network, could cause performance degradation. Control of weighting factors depends on the specific vendor implementation; nevertheless, the purpose of this section is to highlight the benefits optimised weights can provide to QoS.

Let us consider an example. Figure 10.15 shows a simplified GPRS scenario with non-guaranteed bit rate (NGB) traffic. The main assumptions are:

- Mobile M1 belongs to multislot Class 10 (Rx/Tx slots = 4/2) [12].
  ○ QoS attributes associated with M1: interactive TC, THP 1, ARP 1.
- Mobile M2 belongs to multislot Class 4 (Rx/Tx slots = 3/1).
  ○ QoS attributes associated with M2: interactive TC, THP 3, ARP 3.
- Two users are in the same cell and there are no resource limitations at U_m.
- Congestion is at the Gb interface, where only a permanent virtual circuit (PVC) with a committed information rate (CIR) equal to 48 kb/s is defined. In practice, due to protocol overheads the available bandwidth for traffic is reduced by 10%.
- Tests consist of parallel FTP downloads of a reference file, performed by both mobiles.
- During tests, throughput at the TCP level is measured.

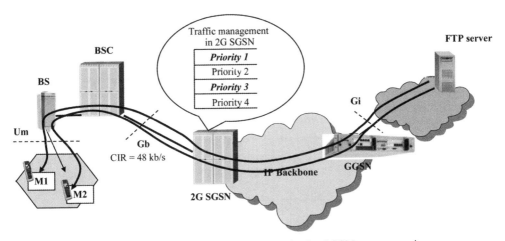

Figure 10.15 GPRS traffic management in the SGSN: test scenario.

The NGB priority queues supported in the 2G SGSN, corresponding to different TC and THP attribute combinations, are:

- Priority 1: interactive TC, THP 1.
- Priority 2: interactive TC, THP 2.
- Priority 3: interactive TC, THP 3.
- Priority 4: background TC.

Table 10.13 reports possible sets of weights assigned to the above queues. Weights are relative to the total available bandwidth at the Gb interface. If some priority classes are not present in the traffic mix the amount of bandwidth reserved for present priority classes is scaled, accordingly.

Following the above assumptions, in the SGSN the traffic generated by two users will be handled through the queues with Priorities 1 and 3. As for the protocol overhead, the bandwidth available for NGB traffic at the Gb interface is about 42 kb/s. In the simple test scenario, expected bandwidth subdivision between the two users can be derived from the parameter settings of the weighted fair queuing algorithm as follows:

- Weight Set 1: the available bandwidth is equally shared between the two priority queues. Expected experienced throughput is about 21 kb/s for both users. This is confirmed by the measurement results shown in Figure 10.16.

Table 10.13 Weights assigned to priority queues.

Priority	Weight Set 1 (%)	Weight Set 2 (%)	Weight Set 3 (%)
1	25	50	70
2	25	30	15
3	25	15	10
4	25	5	5

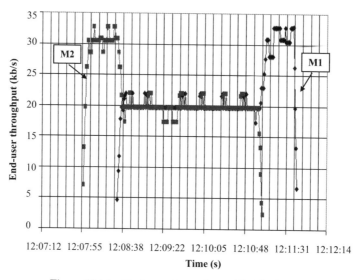

Figure 10.16   End-user throughput: Weight Set 1.

- Weight Set 2: since only two users are present in this scenario the weights need to be rescaled. The new values are about 77% for the first user and 23% for the second user. Expected end-user throughput is about 32 kb/s and 10 kb/s, respectively. This is in line with the performance results depicted in Figure 10.17.
- Weight Set 3: the scaled weights are about 87% for the first user and 13% for the second user. End-user throughput is about 36 kb/s and 6 kb/s, respectively, as expected and confirmed by the measurement data illustrated in Figure 10.18.

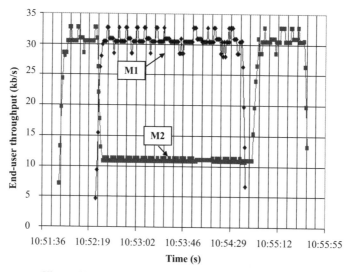

Figure 10.17   End-user throughput: Weight Set 2.

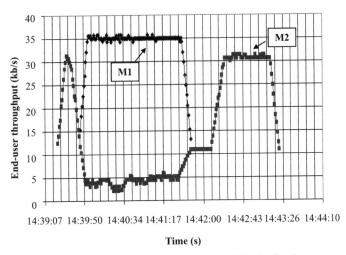

Figure 10.18   End-user throughput: Weight Set 3.

These tests highlight the beneficial influence of the traffic management parameters on QoS. The example describes a simplified scenario, with a very simple traffic mix. In real networks, optimisation activity has the aim at finding weights that better suit the traffic mix present in the network. Optimal weights depend on the association between subsets of PDP context attributes, better services and particular priority values.

10.4.1.2   DSCP marking in SGSN and GGSN

Core network elements mark IP packets according to the GRPS/UMTS traffic classes since IP networks, such as the GPRS backbone, are unaware of bearer service attributes, and use the differentiated services concept [13]. Differentiated services use the DSCP in the IP header to identify the priority of the packet. Routers use this information to provide consistent QoS in the IP backbone. Details on how this mechanism works on different core network elements can be found in Chapter 6. Core network elements usually perform a default mapping between QoS profiles and different DSCP values. A possible implementation solution for 3G SGSN is reported in Table 10.14. Several combinations are possible for the operator to use; however, the marking of IP packets must be consistent over the entire end-to-end path.

10.4.1.3   Traffic management in GGSN

Mechanisms for traffic metering and policing in the GGSN were described in Chapter 6. These functions are based on a subset of attributes of the QoS profile associated with the requested APN. Parameters are stored in the HLR. Optimal parameter settings are essential for satisfactory QoS provisioning.

Table 10.14   Mapping between 3GPP traffic class and DSCP values.

	Classifier		Action	
Traffic class	Traffic handling priority (THP)	Allocation retention priority (ARP)	Per-hop behaviour (PHB)	DSCP (HEX)
Conversational	—	ARP1	EF	B8
Conversational	—	ARP2	EF	B8
Conversational	—	ARP3	EF	B8
Streaming	—	ARP1	AF41	88
Streaming	—	ARP2	AF42	90
Streaming	—	ARP3	AF43	98
Interactive	THP1	ARP1	AF31	68
Interactive	THP1	ARP2	AF32	70
Interactive	THP1	ARP3	AF33	78
Interactive	THP2	ARP1	AF21	48
Interactive	THP2	ARP2	AF22	50
Interactive	THP2	ARP3	AF23	58
Interactive	THP3	ARP1	AF11	28
Interactive	THP3	ARP2	AF12	18
Interactive	THP3	ARP3	AF13	38
Background	—	ARP1	BE	00
Background	—	ARP2	BE	00
Background	—	ARP3	BE	00

For instance, consider the following GPRS test scenario (conclusions are also valid in the case of UMTS):

- A single user in a cell performs an FTP download of a reference file with sufficient radio resources allocated for maximum throughput.
- The QoS profile associated with the requested APN has, among others, the following parameters: interactive TC, THP 1, ARP 1, maximum bit rate downlink 16 kb/s.

In this case, the limiting factor is the maximum bit rate in the downlink direction, which limits the reachable throughput to 16 kb/s. Tests results are depicted in Figure 10.19. The TCP throughput measured at the client shows clearly the effects of the traffic management functionalities of the GGSN. These mechanisms are implemented for each PDP context, so the correct association between APN and QoS profile parameters is fundamental, since it could make ineffective all prioritisation mechanisms in the end-to-end chain.

As described in Chapter 6, the GGSN is the edge element for GPRS and UMTS packet data services. Depending on the vendor's implementation choices, the vendor can be aware of the different services requested by the users, since it can perform inspection and analysis of the applications carried by IP packets. In this scenario the key element is the access point name (APN), which defines the entry to external packet data networks (both Internet and private networks). (The APN is a label which according to DNS naming

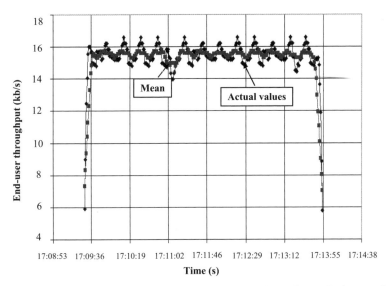

Figure 10.19 Effects of GGSN policing/shaping on end-user throughput: the instant throughput is sampled at intervals of 1 s, mean throughput is calculated over a sliding window of 15 s, excluding the initial transient.

conventions reveals the physical access point to the external packet data network [14].) The APN defines two logical pipes: one between the user equipment (UE) and the GGSN, and another from the GGSN to data networks. In the GPRS access network, the first pipe uses a GTP tunnel, while the other pipe might use some other tunnel (such as GRE or L2TP) or plain IP. The APN binds these two pipes together and creates a connection between the UE and the destination point in the data networks. When the *intelligent edge* concept is supported in the GGSN, APN models and their configuration affect the mapping of QoS attributes onto data pipes at the Gn interface.

We can distinguish between the standard APN model and the single APN model. The standard access model is the one that is usually used when the service awareness concept is not supported: in the GGSN the APN binds the Gn and Gi data flows with no possibility of enforcing a certain QoS level on traffic related to a particular service. This is handled by the HLR QoS profile and by the capability of the UE to allocate its own specific attributes. From this point of view, if this model is used no optimisation activity needs to be carried out in the GGSN related to QoS. In this case, if the operator wants to differentiate the QoS between IP flows, separate APNs are required. In intelligent edge mode, only one APN is required for each PDP session. Differentiation between IP flows is performed at the GGSN edge, where the routing of packets at the Gi interface and the provisioning of QoS at the Gn interface is based on L4/L7 parameters. During the 'PDP context activation' procedure, the SGSN defines the maximum QoS (see Chapter 3). Services cannot exceed these values. When the GGSN receives the 'PDP context activation request' message from the SGSN, it determines the default parameters set, which defines the default QoS for traffic. Based on the L4 analysis, the GGSN can upgrade the QoS according to its L4/L7 parameter configuration. Whenever there is no traffic requiring a special QoS the GGSN downgrades the QoS back to the default level.

In short, in the standard access point name model:

- Services are differentiated by means of different APNs and distinct QoS profiles defined in the HLR. These profiles can be seen as upper bounds.
- If the UE is not QoS-aware, QoS differentiation will be based only on the different HLR profiles.
- If the UE is QoS-aware, QoS differentiation will also be based on the requested attributes.
- In the GGSN there is no QoS-specific optimisation to be done, unless there is DSCP marking, as already discussed in Section 10.4.1.2.

In the single access point name model supporting the service awareness concept:

- Only one access point name can be used.
- The profile in HLR related to the single APN can be seen as an upper bound.
- Based on L4/L7 classification, packet routing at the Gi interface can be associated with packets related to different services, and different QoS parameter sets can be imposed on traffic matching the rules.
- L4 flow specifications have an important role in QoS optimisation in the CN. Rules should be optimised according to QoS differentiation between services.

From previous considerations it is obvious that, if these mechanisms are supported, the configuration of the GGSN has to be consistent with the QoS provisioned in other parts of the network.

10.4.1.4   Signalling procedures

This section analyses different signalling procedures, such as 'GPRS attach', 'PDP context activation' and 'routing area update', which can trigger security functions such as authentication, ciphering and IMEI checking procedures [15]. Depending on the vendor's implementation of the SGSN, security functions can be controlled by repetition rate parameters. The optimisation process aims at reducing the signalling latency introduced by core network elements.

The 'GPRS attach' and 'PDP context activation' procedures are illustrated in Figures 10.20 and 10.21. Figure 10.22 depicts the security functions during the 'routing area update' procedure. Optimal settings of parameters such as *authentication repetition rate* and *IMEI checking repetition rate* could improve, on average, 'GPRS attach' and 'PDP context activation' delays. Security function repetition rates may be set as follows.

- Authentication and ciphering procedures:
  - o 'GPRS attach' – the procedure cannot be avoided since it prevents illegal SIM cards from registering with the network;
  - o mobile-originated 'PDP context activation' – this procedure can be avoided. Authentication has to be done when the subscriber enters the network. If ciphering is in use, the authentication procedure is necessary only to change the cipher key.

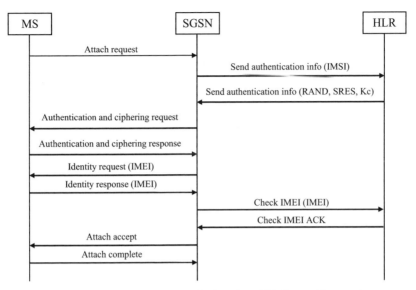

Figure 10.20 Security functions during the 'GPRS attach' procedure.

Usage of authentication increases the network load between the SGSN-VLR and the HLR;
  o normal 'routing area update' – in this case the procedure can be avoided. Like PDP context activation, authentication has to be done when the subscriber enters the network. The authentication and ciphering procedure increases the time necessary for routing area update, thus also increasing breaks in data, which is much more sensitive if LLC transfer mode is unacknowledged.
• IMEI checking procedure:
  o 'GPRS attach' – if IMEI checking is activated in the SGSN, the procedure has to be performed every time to prevent the use of illegal mobiles;

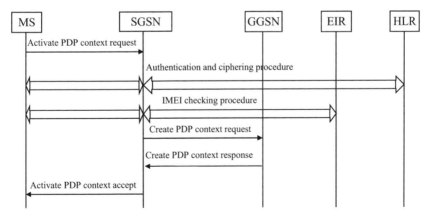

Figure 10.21 Security functions during the 'PDP context activation' procedure.

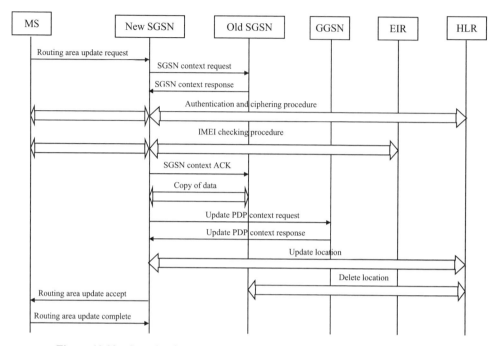

Figure 10.22   Security functions during the 'routing area update' procedure.

○ mobile-originated 'PDP context activation' – this procedure could be avoided, since it causes a lot of unnecessary load;
○ normal 'routing area update' – the procedure could be avoided.

10.4.2   Routing configuration

Figure 10.23 shows the messages exchanged in the core network during the 'PDP context activation' procedure. During the procedure, GGSN selection mechanisms may increment the delay introduced by the core network. Once the APN is selected, the SGSN must resolve this name to derive the IP address of the corresponding GGSN. This can be done in two different ways: by sending a DNS query to the DNS server, or simply by looking up the APN from its cache memory, as long as SGSN implementation supports this option and the information in the cache is still considered valid. In the former case, the network architecture should be taken into account when configuring the DNS server. If the SGSN is collocated with one or more GGSNs, it is important to configure the DNS server so that the IP addresses of the collocated GGSNs are the first on the list. This implementation benefits both signalling and user plane traffic, which is not affected by geographical link delays. The latter solution not only reduces the number of DNS queries, but also makes it possible to select the nearest GGSN by means of subnets. Use of the cache should be carefully evaluated when DNS queries are also performed using a round robin algorithm. The combination of DNS queries and a resolver cache

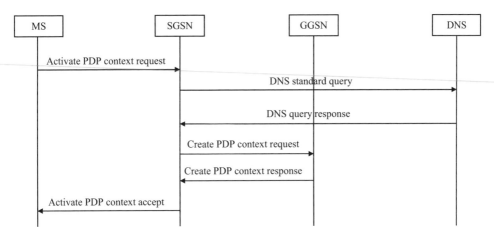

Figure 10.23 'PDP context activation' procedure.

with round robin entries returned by the DNS could produce a negative effect, spreading the traffic all over the network.

Figure 10.24 describes two possible implementations of the GGSN selection. In the first case (dashed line), during PDP context activation the GGSN of Core Site 1 is selected, keeping the traffic within the same core site. The external IP network is accessed directly from Core Site 1 and usage of geographical links is avoided. Using the second option (continuous line), the GGSN of Core Site 2 is selected in the PDP context activation phase and, therefore, user traffic is routed on the IP backbone over geographical links, which most probably will deteriorate QoS.

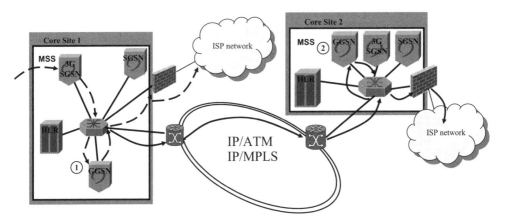

Figure 10.24 GGSN selection: (——) optimal choice in which traffic is not routed over the IP backbone; (– – –) suboptimal choice in which traffic is routed over the IP backbone.

10.4.3   GPRS core network and GPRS backbone troubleshooting

The core network and the IP backbone can be the cause of QoS degradation: delays can be introduced or IP packets can be dropped by network elements. For this reason there is the need to have a defined process in order to locate the source of the problem.

The process includes all the activities that are usually performed during specific tests carried out in a controlled environment – that is, using test mobiles with end-to-end applications, such as FTP, web browsing, WAP or streaming. Key ingredients for monitoring and analysing CN behaviour are traces of protocols collected at different interfaces. A complete measurement setup is illustrated in Figure 10.25, where two different GGSN core sites are considered. A protocol analyser should be placed at the Gb and Iu interfaces (points 1 and 2), and an IP sniffer should be inserted at the Gn and Gi interfaces of all SGSN and GGSN sites (points 3–8). Another sniffer should be placed at Gi close to the application servers (points 9 and 10).

In practice, not all the above measurement points are needed: the number of probes depends on the network architecture and on the level of detail needed for data analysis. If analysis results highlight unexpected behaviour in the GPRS core network, new measurement points can be added allowing better resolution in different sections of the backbone. Figure 10.26 presents an example optimisation process. It describes the case when there is a reliable transport layer, such as TCP. Nevertheless, it can be adapted to any other application, even if UDP transport is used.

If the SGSN and GGSN are located in the same core site, there is no need to trace user traffic at both points 3 and 4 (or 5). In this case only one measurement point is sufficient, because it is only the site switch that assures connectivity between the SGSN and GGSN. Measurement points 7 and 9 at the GGSN1 site, as well as 8 and 10 at the GGSN2 site,

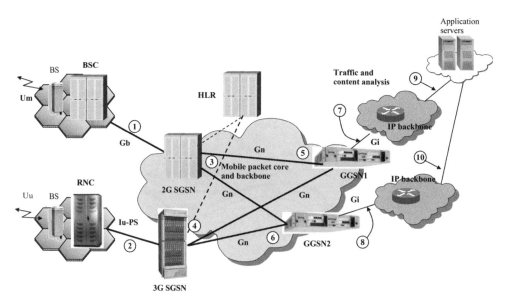

Figure 10.25   Core network performance monitoring for QoS troubleshooting and optimisation: measurement points at different interfaces.

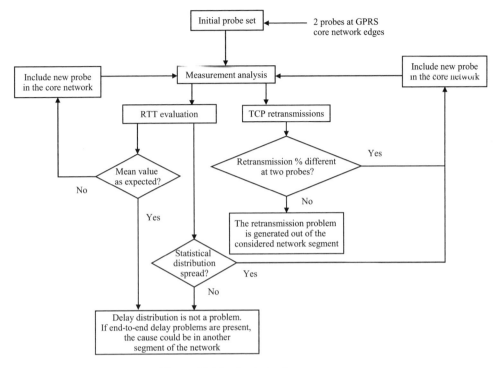

Figure 10.26 Optimisation process.

are needed if the traffic and data analysis solution requires the two network elements to be separated, otherwise measurement points 7 and 8 need only be considered. Traces taken outside the GPRS backbone can help in evaluating, for example, the contribution to round trip times introduced by the GPRS core network and backbone itself. Different measurement points at both edges of the GPRS core network are needed if the connection between the SGSN and GGSN (both 2G and 3G) is geographical – that is, uses IP over ATM or IP over MPLS technologies.

Starting with traces taken in the packet core and backbone, the delays introduced by core network elements and potential causes of packet losses can be assessed. Data analysis, based on dedicated tests, can be improved with performance indicators extracted from network elements, as described in Chapter 9. Assessment of delays and packet losses can be performed only if the traces taken at different interfaces or reference points, as suggested in Figure 10.25, are available. Synchronisation of protocol analysers and packet sniffers can be a difficult task to carry out, but usually it is not mandatory since delays can be evaluated at every single section as time differences between measurements made at different interfaces. The analysis procedure depends on the application used during tests. If the service under evaluation uses a reliable transport protocol, such as TCP, the round trip time (RTT), the percentage of retransmitted packets and throughput can be easily measured at different levels of the protocol stack. For example, TCP allows estimation of the round trip time as the time difference between the timestamp of a TCP segment and that of its acknowledgment [16]. Several methods for RTT computation can be implemented in servers and clients [17], [18]. Based on the above definition it is possible to calculate the TCP RTT at each measurement point. Estimation of this indicator can be made both in the downlink and uplink directions – that is:

- RTT downlink: measurement point ↔ client.
- RTT uplink: measurement point ↔ server.

Furthermore, the number of retransmitted TCP packets can be derived at each measurement point. Moreover, the correlation between two or more measurements made at different reference points can be based on:

- Delta round trip time: the difference between RTT evaluated at two different reference points – for example, points 3 and 5 (or 3 and 6) in Figure 10.25.
- Packet loss: the number of TCP retransmitted packets at two different reference points. Comparing the values at the reference points, we get an indication of the amount of packet loss between the two measurement points.

If the application runs over an unreliable transport (e.g., UDP), round trip time cannot be estimated with the information available at the transport layer. Packet loss, throughput and inter-packet arrival time, which give useful information on the effect of the backbone on packet delays, can be assessed at the application layer.

Implementation of the QoS functions in the backbone network and the related technologies were explained in Chapter 6. For troubleshooting and optimisation purposes, the following considerations can be useful to get information about missing performances:

- QoS implemented in backbone network elements should be in line with 3GPP attributes and consistent with the mapping adopted at edge network elements, such as SGSNs and GGSNs. For this reason it is important to check the QoS settings in switches and routers: it might happen that the default behaviour in the IP network elements does not correspond with the behaviour expected in the end-to-end path and configured in GPRS core network elements (i.e., the SGSN and GGSN).
- Regarding site connectivity, Ethernet port configuration should be checked. It would be preferable not to set the speed port in auto-negotiation mode between routers and between the SGSN/GGSN and switches. This could cause problems when the selected speed is lower than the maximum allowed. It would be preferable to force the port speed up to the maximum allowed value.
- Site connectivity is based on the concept of a virtual LAN (VLAN), which makes it possible to keep different kinds of traffic (e.g., Gn, Gi, O&M, etc.) separated at the IP level. Correct implementation of site connectivity results in easier setup of test measurement points.
- Regarding site interconnectivity, it is important to check routing at the Gn interface.

Another issue which has to be considered is the dimension of IP packets. The maximum transmission unit (MTU) represents the maximum dimension of an IP packet. In the end-to-end path, this parameter could be automatically tuned by means of the MTU discovery procedure presented in [19]. In such a case the maximum dimension of IP packets will be the lowest over the whole chain. This means that the slowest network element or network portion will affect overall system performance. Several tests have shown that having a large number of small IP packets is not beneficial for throughput. In the case of a very small MTU, it could be useful to check all network elements over the whole path, in order to find the possible bottleneck. The maximum dimension of IP packets is an issue common to all IP networks. Nevertheless, in the GPRS core network another problem arises at the Gn interface due to the presence of GTP. The additional protocol layers used for tunnelling user traffic introduce extra headers, which reduce the number of user data payloads transported in one IP packet. If the IP packets at the Gi interface have the maximum dimension allowed at the physical layer (1500 bytes for Ethernet), they will be fragmented because they exceed the maximum allowed dimension at the Gn interface. Of course, this will cause additional load in the GGSN that performs IP packet fragmentation in the downlink direction, and in the SGSN that reassembles the fragments. Therefore, for optimal performance at the Gn interface, IP packet fragmentation should be avoided; in most cases, payload length optimisation is an issue related to servers and not directly related to the core network.

If controlled tests cannot be performed as described in Figure 10.25, some information on how to proceed with optimisation can be obtained from the network element counters described in Chapter 9. More precisely, in a 2G SGSN it is possible to determine the network service virtual channel (NS-VC) starting with the counters related to NS-VC passed data. Depending on the measurement period, the corresponding KPI can give information about the bytes per second transferred to the BSC through that particular NS-VC. If NS-VC load gets close to the CIR, this should be increased. If the CIR cannot be increased, it is necessary to create a new NS-VC to the BSC. In both cases the result is an increment in the available bandwidth at the Gb interface. Useful information can also be obtained from counters related to NS-VC discarded data and BSSGP-discarded data due to CIR overflow, if available.

The same approach is also valid in the 3G SGSN for the Iu interface, using counters related to passed and discarded data.

10.5   Service application performance improvement

The approach to optimisation is actually evolving towards a concept of bearer performance improvements in which underlying network layers interact with several applications. These could simply be circuit-switched applications, such as speech, circuit-switched data calls or WAP over circuit-switched bearers, or applications offered through the packet-switched domain of the network, such as WAP, web browsing and MMS, as well as PoC, VoIP, VS, gaming, streaming and business connectivity services. Each of these applications works with its own protocol stack using different protocol settings that may depend on the QoS profile and, thus, may communicate with underlying layers in many different ways.

This section presents some crucial aspects of 'application-driven' optimisation. Besides this, it reports some examples of how a network or parameter configuration that yields optimal performances while offering a particular service application – for example,

Table 10.15 Example of end-user measurements for some applications.

Application	Key performance indicators
WAP	Session setup duration (s) Delay in homepage retrieval (s) Delay in first page retrieval (s)
WEB browsing	Delay in test page download (s) HTTP download session setup duration (s)
MMS	Successfully submitted/terminated messages (%) Delivery time to: handset, application, legacy handset (s)
PoC	Start-to-talk delay (s) Voice through delay (s) Subjective voice quality (MOS)
Streaming	Session setup duration (s) Initial buffering time (s) Number and length of rebuffering (s)

browsing – may be suboptimal, or non-optimal, when a different service application – for example, WAP – is deployed.

Similar to other optimisation processes, also in our case the first step is to define a list of measures and target values for each application. An example of performance indicators is reported in Table 10.15, and related target values may be found in Chapters 2 and 9.

The next step is to define the optimisation process in order to reach the targets. For this purpose, the following aspects need to be taken into account:

- The impact on QoE of the QoS profile provisioned in the HLR.
- Parameter settings performing well for one application could be suboptimal, or non-optimal, for other services
- The traffic characteristics of diverse applications that may perform differently at the physical layer.
- Flow control between radio and core network elements.
- Utilisation of a performance enhancing proxy (PEP).

10.5.1 Impact of parameter settings

For each of the services deployed, the whole protocol stack may be divided between the transport network layer and application layer. The structure of the first depends on the technology adopted, while the second depends on the application. (See Chapters 3–6 for more information.) The behaviour of such protocols depends on the QoS profile provisioned for that particular application and QoS functions available in the network elements. The old R97 attribute 'reliability class' basically defined acknowledged/ unacknowledged mode for the RLC, LLC and GTP (in R97 for Reliability Class 1).

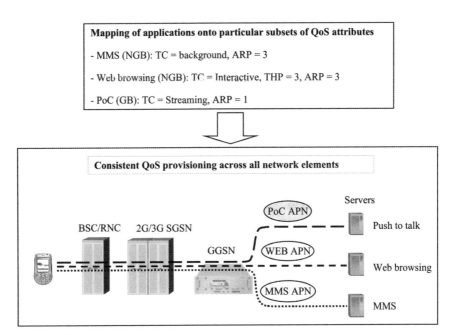

Figure 10.27 Mapping of applications onto particular subsets of QoS attributes and implementation of QoS in network elements.

This attribute has now been mapped onto the R99 attributes of 'SDU error ratio', 'residual BER' and 'delivery of erroneous SDUs' (see Chapter 3). The QoS profile is provisioned in the HLR for every APN and can be set differently for different APNs. For example, the QoS profile for MMS – that is, a non-real time application in which the user is unaware of the delay in message delivery – normally presents a background class attribute, while, in order to have good speech quality in PoC, it is wise to adopt at least an interactive class profile for such a user. In real time services the planner could be tempted to provision the user with the best possible profile, but he/she has to take into consideration that all the network elements must be able to support it. For example, there could be some handsets that still do not support the highest traffic classes, like streaming or conversational class, or some network elements may limit the QoS profile during the 'PDP context activation/modification' procedure, based on their capability and on the level of traffic in the network. This entire scenario helps us to understand that the optimisation process should be primarily based on correct and consistent QoS provisioning of every accessible service to users. This means that every service should be mapped onto the right QoS attribute combination, and accordingly treated across all network elements, at the same time as diverse applications share the same network resources.

The concept of QoS-based application optimisation is illustrated in Figure 10.27. QoS provisioning is not, however, the only way to achieve such a goal.

Another possibility to improve service performance is to tune network element parameters to adapt their values to application requirements, as stated above.

Let us consider PoC as an example service application. For PoC the application layer consists of AMR frames. Eight of them build one AMR packet 110 octets in size.

Considering the headers introduced by lower layers, the required bit rate at the physical layer is at least 8 kb/s. (*Note*: This also depends on implementation of application layer protocols that may be vendor-dependent.) This means that an 8-kb/s radio access bearer (if using 3G coverage) should be enough to carry such user data. In this particular case we are getting very near to the capacity of the RAB, and, most probably, usage of a 16-kb/s RAB would provide better quality in terms of mobile objective score (MOS). This choice could be performed by setting the *maximum bit rate for uplink* and *maximum bit rate for downlink* in the QoS profile of the HLR to the desired value.

The minimal capacity needed in 2G radio access networks (GPRS or EDGE) could be, for example, a single time slot with CS-1 (9.05-kb/s capacity) fully dedicated to the packet-switched domain in the case of GPRS, or a single time slot with MCS-1 (8.8 kb/s) in the case of EDGE. (All these assumptions relate to the load generated by one user when transmitting a talk burst, while the inputs needed for complete dimensioning would also include PS territory size and number of users/time slots, as explained in Chapters 5 and 7.) In Figure 10.27 the usage of streaming class is suggested for PoC. When using such a class, the PoC user has dedicated time slots in the PS territory that may have higher priority against interactive and background users. This allows the system to maintain good performances (i.e., a transfer delay lower than 250 ms with R99 QoS attributes) for streaming class users.

If we dedicate one time slot for PoC to the streaming class in an EDGE network, it is worth using the Erlang-B formula to understand how many users can be served. If the allowed blocking probability is 2%, then the traffic in Erlangs that could be managed is:

$$T = ErlB(p, ch) = 0.019 Erl \qquad (10.24)$$

where $ErlB$ is the Erlang-B formula, p is blocking probability and ch is the number of time slots (1 in this case). Assuming that every user develops 5 mErl of traffic in the busy hour, then up to three users will be served by this resource. Adding a new time slot can increase the number of satisfactorily served users. If two time slots are used, then the traffic managed with the same blocking probability will be higher, dealing with up to 44 users. In the event streaming class is still unsupported by some network elements in the chain or by the UE, then interactive class should be used instead, admitting some delay in data transfer. In this case, if we allow a certain level of queuing time – say, 1 s – and we expect a talk burst duration of 5 s, traffic with the same blocking probability will be:

$$T = ErlC(p, ch, q, d) = 0.025 Erl \qquad (10.25)$$

where $ErlC$ is the Erlang-C formula, q is the queuing time allowed before getting the token and d is time duration.

If we consider other applications, such as web browsing or FTP, where the most important requirement is to reach the highest possible throughput, the most suitable choice is to select, for example, a 384-kb/s RAB in case of UMTS and, in case of GPRS or EDGE, to set the size of the packet-switched domain in order to respect UE multislot class.

So, with the above examples it is clear how the QoS profile of a certain application may influence the required capacity, and also affect quality in terms of transfer delay and jitter. Thus, either the PS territory domain or the RAB attributes should be set depending on the traffic characteristic handled, traffic class used and service application carried.

Table 10.16   QoS profile for push to talk 3G/2G.

PoC over WCDMA, interactive, RAB 16/16 kb/s

Allocation/Retention priority   1

Profile handling

Class	Traffic class	Interactive
Order	Delivery order	N
DelErr	Delivery of erroneous SDU	Y
SduMax	Maximum SDU size	1500
DwnMax	Maximum bit rate for downlink	16
UpMax	Maximum bit rate for uplink	16
Ber	Residual BER	$10\exp(-5)$
SduErr	SDU error ratio	$10\exp(-4)$
Prior	Traffic handling priority	1

PoC over (E)GPRS, interactive

Allocation/Retention priority   1

Profile handling

Class	Traffic class	Interactive
Order	Delivery order	N
DelErr	Delivery of erroneous SDU	Y
SduMax	Maximum SDU size	1500
DwnMax	Maximum bit rate for downlink	16
UpMax	Maximum bit rate for uplink	16
Ber	Residual BER	$10\exp(-5)$
SduErr	SDU error ratio	$10\exp(-6)$
Prior	Traffic handling priority	1

Table 10.16 reports an example QoS profile for PoC over WCDMA and (E)GPRS networks. Table 10.17 shows an example QoS profile for video streaming.

10.5.2   Impact of traffic characteristics

Another topic to be considered is how different applications interact with the physical layer. There are services that present very bursty behaviour (like WAP, web browsing and push to talk) and services that present very smooth behaviour (like FTP and streaming).

The first behave in such way that data are sent in many little bursts, as we can see from Figure 10.28. In 3G networks, sometimes the bytes sent are so few that the RRC state goes from Cell_PCH to Cell_FACH without passing through Cell_DCH state. The state of the UE, when new data reach it, affects delay. This means that the parameters that manage the transitions between these states could adversely affect end-user perception or bursty applications, where we have a lot of radio bearer setup and release. The same concept applies to GPRS and EDGE. In this case, the connection between RLC entities located in the MS and BSC is the TBF. If the TBF is immediately released after a burst, then a new establishment is needed when new data come: such an establishment takes

Table 10.17   QoS profile for video streaming.

Streaming service application

Allocation/Retention priority		1
Profile handling		
Class	Traffic class	Streaming
Order	Delivery order	N
DelErr	Delivery of erroneous SDU	Y
SduMax	Maximum SDU size	1500
DwnMax	Maximum bit rate for downlink	256
UpMax	Maximum bit rate for uplink	32
Ber	Residual BER	$10\exp(-6)$
SduErr	SDU error ratio	$10\exp(-5)$
Prior	Traffic handling priority	1

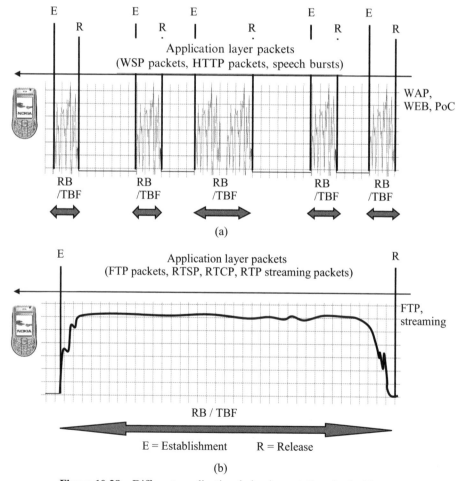

(a)

(b)

E = Establishment   R = Release

Figure 10.28   Different application behaviour at the physical layer.

time, especially if the UE falls in the CCCH after the previous data transfer. So, it is better to maintain the UE on the dedicated channel (PDTCH/PACCH) for applications like WAP, PoC and web browsing, by lengthening the period of time the UE stays on the dedicated channel before releasing resources. Of course, the drawback is occupation of the resource when waiting for new data to come.

For such services as FTP, behaviour at the physical layer is different. Here, a radio bearer or a TBF is established and maintained hopefully until the end of data transfer. For this reason, speed in connection establishment is not so important, and it is other parameters (e.g., TCP-related parameters) that affect end-user perception.

Finally, we have applications like streaming in which the connection hopefully lasts until the end of the download, but with more inter-working between the client and server. Depending on the implementation, the client may frequently send reports to the server in order to execute its own form of flow control. Such data must be carried by uplink connections at the physical layer that must be established if the previous connection has already been released, as indicated in Figure 10.29.

In the example, establishment always happens by sending a 'Packet DL ACK/NACK' message to the BSC and a connection is set up for every report. In order to reduce the signalling required for connection setup, we should, for example, try to delay release of the previous one and wait for new data with the old connection still up and running.

10.5.3 Impact of flow control

Finally, application optimisation should be reached by taking care of the right parameter settings throughout all network elements. Below is an example of how throughput during

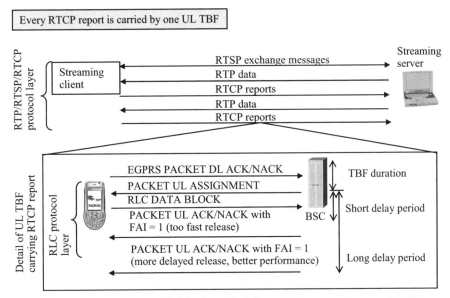

Figure 10.29   Client reports carried in the uplink by a TBF: streaming over EDGE case.

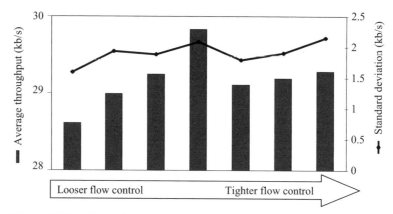

Figure 10.30 Throughput optimisation when tuning flow control action.

a web browsing session can be affected by the flow control mechanism that the BSC performs toward the SGSN in the case of a GPRS bearer.

The performance results illustrated in Figure 10.30 were attained by fine-tuning the flow control parameters in the BSC and downloading a test webpage ten times for every parameter configuration, so every bar in the chart corresponds to an average of ten downloads. The test was executed on a test BSC with no load at different interfaces in the network. The results show how fine-tuning the flow control algorithm may generate variations in end-user perception. Indeed, the throughput indicated is the application layer bit rate. Here, as in other cases, there is not a common rule on how to achieve best possible performance. Optimal values (of flow control in this case, but may be of any other parameter) depend on many factors: from the application that is running to the related QoS profile down to protocol layer settings. Most probably, different parameter settings are needed to optimise the flow control behaviour for applications that run over UDP instead of TCP. That is why the best network setting usually comes from a trade-off between different services, or by choices driven by network providers.

10.5.4 Impact of performance enhancing proxies

As already pointed out in the above sections, many services are based on TCP. This protocol is an acknowledgement-based transport layer protocol, commonly used in any Internet protocol suite. Because it was not originally designed for wireless networks with small available bandwidths and relatively long delays its behaviour is not optimal. A common way to overcome specific application (e.g., HTTP) and transport layer (TCP) limitations over a 3G network is to apply performance-enhancing proxies (PEPs). (See Chapter 3 for more information.) A PEP acts on Layer 4 and above of the OSI stack, between the HTTP and TCP stack. If used, the PEP is connected between the GGSN and application server (e.g., exchange mail server, web server) and breaks up an HTTP and TCP peer-to-peer connection (see Figure 10.31) acting as a server towards 3G clients and as a client towards the application server.

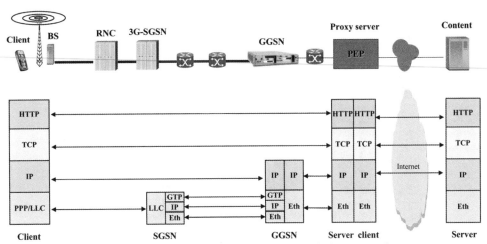

Figure 10.31 Implementation of PEP in wireless network.

Since TCP/IP traffic is not best suited to wireless connections, a PEP can speed up data traffic by improving TCP/IP behaviour by imposing optimal TCP parameters that affect end-user perception. Also, a PEP can remove from webpages any content that increases download times – such as HTML comments, banners and GIF animations – and compress the data packet by means of some compression mechanism to reduce the amount of data to be transmitted. A PEP can also join multiple small objects to make bigger ones, thus improving radio link utilisation. Additionally, image resolutions can be reduced before sending them across the wireless link. In this way the transfer of web content from the Internet/intranet to mobile clients can be achieved in a format that is optimised for wireless transmissions.

Figure 10.32 shows the download of different pages with different object sizes. Tests were executed using a 3G downlink bearer of 128 kb/s. These tests show that for small page/file size, a PEP does not provide any relevant improvement in end-user perception of throughput. As the page/file size grows, so do the benefits from a PEP. For large pages/ files, a PEP often provides roughly twice as much throughput at the application layer as normal 3G connections without a PEP. Therefore, a PEP really helps in enhancing user experience by compressing content, HTTP pipelining, content manipulation and, option- ally, protocol conversion. However, if the PEP is not properly configured and integrated in the overall network, it may become a bottleneck in the end-to-end path.

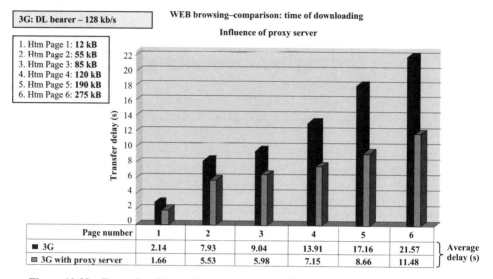

Figure 10.32 Example of throughput optimisation by the use of PEP in 3G network.

References

[1] 3GPP, R6, TS 32.101, Telecommunication Management; Principles and High Level Requirements, v. 6.1.0.

[2] TMF, GB910, Telecom Operations Map, Approved Version 2.1, March 2000.

[3] J. Laiho, A. Wacker and T. Novosad (eds), *Radio Network Planning and Optimisation for UMTS*, John Wiley & Sons, 2nd Edition, 2006, 630 pp.

[4] A. Kuurne, D. Fernández and R. Sánchez, Service based prioritization in (E)GPRS radio interface, *IEEE Vehicular Technology Conference, Fall, 26–29 September 2004*, pp. 2625–2629, Vol. 4.

[5] A. M. J. Kuurne and A. P. Miettinen, Weighted round robin scheduling strategies in (E)GPRS radio interface, *IEEE Vehicular Technology Conference, Fall, 26–29 September 2004*, pp. 3155–3159, Vol. 5.

[6] T. Halonen, J. Romero and J. Melero (eds), *GSM, GPRS and EDGE Performance*, John Wiley & Sons, 2nd Edition, 2003, 615 pp.

[7] K. Valkealahti and D. Soldani, QoS sensitivity to selected packet scheduling parameters in UTRAN, *Proc. IEEE Vehicular Technology Conference, Spring 2005, May 31–June 1, Stockholm, Sweden*.

[8] ETSI, TR 101112 (UMTS 30.03), Selection Procedures for the Choice of Radio Transmission Technologies of the UMTS, v. 3.2.0.

[9] D. Soldani and K. Valkealahti, Genetic approach to QoS optimisation for WCDMA mobile networks, *Proc. IEEE Vehicular Technology Conference. Spring 2005, May 31–June 1, Stockholm, Sweden*.

[10] D. Soldani, QoS management in UMTS terrestrial radio access FDD networks, dissertation for the degree of Doctor of Philosophy, Helsinki University of Technology, October 2005, 235 pp. See *http://lib.tkk.fi/Diss/2005/isbn9512278340/*

[11] D. E. Goldberg, *Genetic Algorithms in Search, Optimisation, and Machine Learning*, Addition-Wesley, 1989.

[12] 3GPP, R98, TS 05.02, Multiplexing and Multiple Access on the Radio Path, v. 7.7.0.

[13] IETF, RFC 3086, Definition of Differentiated Services per Domain Behaviours and Rules for Their Specification, April 2001.

[14] 3GPP, R5, TS 23.060, General Packet Radio Service (GPRS); Service Description; Stage 2, v. 5.10.0.

[15] 3GPP, R5, TS 24.008, Mobile Radio Interface Layer 3 Specification; Core Network Protocols; Stage 3, v. 5.13.0.

[16] IETF, RFC 792, Transmission Control Protocol, September 1981.

[17] IETF, RFC 2018, TCP Selective Acknowledgment Options, October 1996.

[18] IETF, RFC 2883, An Extension to the Selective Acknowledgment (SACK) Option for TCP, July 2000.

[19] IETF, RFC 1192, Path MTU Discovery, November 1990.

Index